Farm Production in England
1700–1914

Farm Production in England 1700–1914

M. E. Turner

J. V. Beckett

B. Afton

OXFORD
UNIVERSITY PRESS

OXFORD
UNIVERSITY PRESS

Great Clarendon Street, Oxford OX2 6DP

Oxford University Press is a department of the University of Oxford.
It furthers the University's objective of excellence in research, scholarship,
and education by publishing worldwide in
Oxford New York

Athens Auckland Bangkok Bogotá Buenos Aires Cape Town Chennai
Dar es Salaam Delhi Florence Hong Kong Istanbul Karachi
Kolkata Kuala Lumpur Madrid Melbourne Mexico City Mumbai
Nairobi Paris São Paulo Shanghai Singapore Taipei Tokyo Toronto Warsaw
and associated companies in Berlin Ibadan

Oxford is a registered trade mark of Oxford University Press
in the UK and in certain other countries

Published in the United States
by Oxford University Press Inc., New York

© Michael Turner, John Beckett, Bethanie Afton 2001

The moral rights of the author have been asserted
Database right Oxford University Press (maker)

First published 2001

All rights reserved. No part of this publication may be reproduced,
stored in a retrieval system, or transmitted, in any form or by any means,
without the prior permission in writing of Oxford University Press,
or as expressly permitted by law, or under terms agreed with the appropriate
reprographics rights organizations. Enquiries concerning reproduction
outside the scope of the above should be sent to the Rights Department,
Oxford University Press, at the address above

You must not circulate this book in any other binding or cover
and you must impose this same condition on any acquirer

British Library Cataloguing in Publication Data

Data Available
Library of Congress Cataloging in Publication Data

Turner, Michael Edward.
Farm production in England 1700–1914 / M.E. Turner, J.V. Beckett, B. Afton.
p. cm.
Includes bibliographical references (p.).
1. Agriculture--England--History. I. Beckett, J.V. II. Afton,
B. (Bethanie), 1948-III.
Title.
S455 .T77 2001   630'.942'09033--dc21   00-065227

ISBN 0-19-820804-9

1 3 5 7 9 10 8 6 4 2

Typeset by Newgen Imaging Systems (P) Ltd., Chennai, India
Printed in Great Britain on acid-free paper by
T. J. International Ltd, Padstow, Cornwall

Preface

As agricultural historians we have long been fascinated by the problem of the 'agricultural revolution'. Generations of historians have been unable satisfactorily to date, or indeed to document, this elusive event, although no one denies that it took place. Nearly a decade ago we came to the conclusion that it would remain a mystery unless new material could be found which might help us to understand the course of agricultural change in the eighteenth and nineteenth centuries. Did such material exist? The only source of which we were aware that had not been thoroughly tested was farm records. In 1993–4 we investigated the quantity and quality of these records, and ran a sensitivity analysis of the contents of a substantial sample of them. For financial support for this survey we are grateful to the University of Hull. Some of our conclusions were published in the *Agricultural History Review* in 1996. As a result of this work we became convinced that a fuller and more substantive study of the archives would enable us to say something positive about long-run change in farm production and output. The main body of our research took place between 1995 and 1998, with the aid of a substantial grant from the Leverhulme Trust, to which we should like to express our sincere thanks on this occasion.

We should like to thank the staff of the many repositories in which we have worked during the course of our research, particularly to Michael Bott and his staff at the University of Reading Library, which houses a major collection of farm records. Our thanks are due also to the librarian of the Perkins Collection of agricultural literature at the University of Southampton. We should also like to thank the staff of the Rural History Centre at Reading University.

We have benefited greatly from discussions with colleagues working in related areas. Professor E. J. T. Collins of the University of Reading has given us considerable support, including suggesting various sources with which he became acquainted during his own early work on farm records in the 1960s. Our thanks are due also to Professor Patrick O'Brien, some of whose ideas sparked off this research in the first place, and who has been consistently helpful in giving advice and encouragement. Some of our findings were floated at a symposium in Sheffield in 1995, and we are particularly grateful to contributors on that occasion, especially Dr Peter Dewey, Dr John Hare, Professor Alan Howkins, Professor G. E. Mingay, Dr A. D. M. Phillips, Professor Brian Short, Professor F. M. L. Thompson, and Dr Charles Watkins. More mature reflections were presented at conferences in London, Madrid, Shrewsbury, and Washington. We are grateful to participants on those occasions, particularly to Professor Mark Overton at Shrewsbury and, at Washington, Professors David Mitch, Joyce Burnette, and Gregary Clark. Others we should like to thank for help and advice at various stages of our work are: Professor Michael Anderson, Mrs Janice Avery, Dr Paul Brassley, Dr Rob Bryer, Dr Stephen Toms, Dr Tom Williamson, Dr Angus Winchester, and Dr Julian Wiseman.

M. E. T.
J. V. B.
B. A.

Contents

List of Figures — viii
List of Tables — x
Abbreviations — xii

Introduction — 1

1. Agricultural Production, Output, and Productivity, 1700–1914 — 9
2. The Farmers and their Records — 27
3. Farming Practice and Techniques — 66
4. The Wheat Question — 116
5. Barley and Oats — 150
6. Livestock — 173
7. Farm Production and the Agricultural Revolution — 210

Appendix 1. Farm Records, 1700–1914 — 231
Appendix 2. Measurement and Weighting Problems in Farm Records — 245

Bibliography — 248

Index — 269

List of Figures

2.1	Location of farms, 1700–1914	64
2.2	Chronological distribution of archive collections, 1700–1914	65
3.1	Traditional crops grown on English farms, 1700–1914	68
3.2	Temporary grasses and forage legumes grown on English farms, 1700–1914	71
3.3	Root crops grown on English farms, 1700–1914	73
3.4a	Seeds of recently introduced crops purchased or sold on English farms, 1700–1914	77
3.4b	Seeds for traditional crops purchased or sold on English farms, 1700–1914	78
3.5	Soil conditioners used on English farms, 1700–1914	82
3.6	Traditional manures and fertilizers used on English farms, 1700–1914	84
3.7	Nitrogen-fixing crops used on English farms, 1700–1914	86
3.8	New manures and artificial fertilizers used on English farms, 1700–1914	87
3.9a	Traditional arable crops used for livestock feeds, 1700–1914	102
3.9b	Non-traditional arable crops used for livestock feeds, 1700–1914	102
3.10	Purchased livestock feeds used on English farms, 1700–1914	103
4.1	Boxplot of English wheat yields	131
4.2	Wheat yields on good-quality soils	146
4.3	Wheat yields on medium-quality soils	146
4.4	Wheat yields on mixed good/medium-quality soils	147
5.1	Boxplot of English barley yields	152
5.2	Boxplot of English oats yields	157
5.3	Index of grain yields	165
5.4	Index of pulses yields	166
5.5	Seeding rates of the grain crops	170

6.1*a*	Carcass weights for ewes	188
6.1*b*	Annual average carcass weights for ewes	188
6.2*a*	Carcass weights for lambs	189
6.2*b*	Annual average carcass weights for lambs	189
6.3*a*	Carcass weights for wethers	190
6.3*b*	Annual average carcass weights for wethers	190
6.4*a*	Carcass weights for 'sheep'	191
6.4*b*	Annual average carcass weights for 'sheep'	191
6.5*a*	Carcass weights for calves	197
6.5*b*	Annual average carcass weights for calves	197
6.6*a*	Carcass weights for cows and heifers	198
6.6*b*	Annual average carcass weights for cows and heifers	198
6.7	Annual average carcass weights for cows and heifers: regional examples	199
6.8*a*	Carcass weights for adult cattle	200
6.8*b*	Annual average carcass weights for adult cattle	200
6.9*a*	Carcass weights for 'bacon' pigs	203
6.9*b*	Annual average carcass weights for 'bacon' pigs	203
6.10*a*	Carcass weights for hogs	204
6.10*b*	Annual average carcass weights for hogs	204
6.11*a*	Carcass weights for porkers	205
6.11*b*	Annual average carcass weights for porkers	205
6.12*a*	Carcass weights for 'pigs'	206
6.12*b*	Annual average carcass weights for 'pigs'	206

List of Tables

1.1	Net agricultural output in England and Wales, 1700–1900	21
2.1	Geographical distribution of archives used, 1700–1914	62
3.1	Principal arable crops growing on English farms, 1700–1914	67
3.2	Rotations in four fields at Sutton Court Farm, East Sutton, Kent, 1727–1746	74
3.3	Rotations at Deanshanger, Northamptonshire, 1843–1850	76
3.4	Examples of crop varieties noted in the farm records	79
3.5	Examples of livestock breeds noted in the farm records	98
3.6	Inventory of farm stock taken Lady Day 1765, Coton Hall Farm, Bridgnorth, Shropshire	109
3.7	Expenditure and income, Coton Hall Farm, Bridgnorth, Shropshire, Lady Day 1765 to Lady Day 1768	110
3.8	Products sold from John West's holdings at Dunsholme and Washingborough, Lincolnshire	112
3.9	Average annual income and expenditure from Upton Farm, Sompting, Sussex	114
4.1	English wheat yields, c.1675–1914 (bushels per acre)	117
4.2	English wheat yields based on probate inventories, 1673–1749 (bushels per acre)	123
4.3	Estimates of wheat yields drawn from Hampshire inventories (bushels per acre and index numbers)	124
4.4	Wheat yields—summary statistics	129
4.5	Demonstrating the influence of one farm on the trend of wheat yields (bushels per acre and percentage differences)	132
4.6	Comparison of new wheat estimates with Board of Agriculture estimates of 1790–1809 (bushels per acre)	135
4.7	A comparison of the wheat new wheat estimates with P. G. Craigie (bushels per acre)	136
4.8	A comparison of the new wheat estimates with those from the mid- and late nineteenth century	137
4.9	Estimated arable acreage and fertilizer inputs	140
4.10	English land utilization classification	142

5.1	Barley yields—summary statistics	153
5.2	Oats yields—summary statistics	158
5.3	United Kingdom grain yields, $c.$1850–$c.$1914	161
5.4	Crop yields—summary statistics	163
5.5	Seeding rates, crop yields, and yield rates	167
5.6	Seed and yield rates at Hurstpierpoint, East Sussex	169
6.1	The composition of cattle carcasses at Coton Hall Farm, Bridgnorth, Shropshire, 1744–1769	178
6.2	Eighteenth-century estimates of sheep carcass weights	182
6.3	Average weight of the different kinds of animals, $c.$1700–$c.$1914 (lbs)	186
7.1	Locating the 'agricultural revolution' in England	218
7.2	Labour productivity	227
A2.1	Variations in measurements	246

Abbreviations

AgHR	*Agricultural History Review*
AHEW	*Agrarian History of England and Wales*, general editor H. P. R. Finberg, later Joan Thirsk, 8 vols. (Cambridge: Cambridge University Press, 1967–2000)
AO	Archives Office
BPP	British Parliamentary Papers
Campbell and Overton (1991)	B. M. S. Campbell and M. Overton (eds.), *Land, Labour and Livestock: Historical Studies in European Agricultural Productivity* (Manchester: Manchester University Press, 1991)
EcHR	*Economic History Review*
JEH	*Journal of Economic History*
JRASE	*Journal of the Royal Agricultural Society of England*
JRSS	*Journal of the Royal Statistical Society*
PRO	Public Record Office, Kew
RC	Royal Commission
RHC	Rural History Centre, University of Reading
RO	Record Office
RUL	Reading University Library
TBA	M. E. Turner, J. V. Beckett, and B. Afton, *Agricultural Rent in England, 1690–1914* (Cambridge: Cambridge University Press, 1997)

Introduction

One of the key questions in agricultural history is concerned with a straightforward and simple fact. Before the eighteenth century, as often as not, population tended to outrun the means of agricultural supply. Contemporary commentators were fairly sure about this inability of the home agricultural industry to feed the nation. Malthus and Ricardo in the early nineteenth century feared that catastrophe was at hand, because the limits of agricultural production were close to being reached in their own generation. Rising prices and rent, which they took to be signs of lean times ahead, put pressure on wages and profits, redistributed income to landowners, and threatened the onset of a 'stationary state'. We now know that their thinking was unnecessarily gloomy: population continued to grow, the urban proportion increased, the industrial revolution took place, and at no point after 1815 was food supply a critical issue except during the World Wars of the twentieth century. Up to $c.1770$ there had developed a net surplus of basic food output. This was subsequently reversed, but by the mid-nineteenth century many more mouths were being fed than in earlier generations without resort to major imports of foodstuffs.

Breaking the Malthusian trap was a vital point in the development of modern society and it was held by many commentators to have been possible because of an 'agricultural revolution' which ran in parallel with the industrial revolution to turn Britain into the first industrial nation. The accumulation of evidence as to population, urbanization, industrial growth, foreign trade, and other factors, all pointed inexorably in the same direction: the Malthusian cycle had been broken, and Britain had been launched on a path of industrialization. *What* had been achieved was clear, and subsequent studies of population,[1] foreign trade, patents, and other related issues[2] collectively built up a picture of accelerating industrial production. *How* it had been achieved was, and still is, rather more of a mystery, notably because so little is known about farm output during the eighteenth and nineteenth centuries. While we may be able to date the introduction of a new plough, the adoption of a new rotation, the enclosure of a particular estate, or the substitution of root crops for fallow, these are essentially qualitative measures of change. All too often they have been treated as proxies for output, necessary substitutes in the absence of reliable quantitative data.

The shortfall in statistical data is well known. Eighteenth-century commentators recognized the need to measure farm output. None was more enthusiastic than Arthur Young who, on his numerous tours, collected details of the farms he visited and recycled them in the *Annals of Agriculture* and other publications. Reporters for the Board of Agriculture in the 1790s and beyond were encouraged to collect any data that they could extract from farmers, and the numerous volumes of the first and second edition *General Views* published between 1794 and 1817 contain a wealth of individual observations. Like Young and other 'collectors' the reporters gathered what they could without any systematic or particularly regular methodology, and the same was true of many of the Prize Essays published in the *Journal of the Royal Agricultural Society of England* from its inception in 1840. Unfortunately this desire by agricultural experts to collect data on farm output was not shared by the government at Westminster. Occasional national surveys were conducted such as the evidence gathered during the corn crises of the 1790s,[3] but it was only in 1832 that Parliament took the first serious steps towards estimating the produce of the soil. This came about when the Board of Trade sought Treasury approval to ascertain the wealth, commerce, and industry of the United Kingdom.[4] Other information was collected at the

---

[1] E. A. Wrigley and R. S. Schofield, *The Population History of England, 1541–1871* (London: Edward Arnold, 1981).

[2] On patents as an indication of the pace of industrial change see R. J. Sullivan, 'The Revolution of Ideas: Widespread Patenting and Invention during the English Industrial Revolution', *JEH* 50 (1990), 349–62; T. Griffiths, P. A. Hunt, and P. K. O'Brien, 'Inventive Activity in the British Textile Industry, 1700–1800', *JEH* 52 (1992), 881–906; C. Macleod, 'Strategies for Innovation: The Diffusion of New Technology in Nineteenth-Century British Industry', *EcHR* 45 (1992), 285–307.

[3] M. E. Turner, 'Agricultural Productivity in England in the Eighteenth Century: Evidence from Crop Yields', *EcHR* 35 (1982), 489–510; id., 'Counting Sheep: Waking up to New Estimates of Livestock Numbers in England, c.1800', *AgHR* 46 (1998), 142–61.

[4] Ministry of Agriculture, Fisheries and Food (MAFF), *A Century of Agricultural Statistics: Great Britain 1866–1966* (London: HMSO, 1968), 2.

time of tithe commutation in the 1840s,[5] and for income tax returns beginning in 1842.[6]

If Arthur Young had collected farm output data because he wanted to champion the cause of improved farming, the government began to recognize by the 1840s that with the population growing—as it knew from the national censuses beginning in 1801—food production was a national issue. The surveys of the 1790s had been in the context of unusual conditions, particularly a combination of poor harvests and the interruption of supplies from the Continent in wartime. By the 1840s it was increasingly clear that even normal conditions needed to be regulated. In 1844 Mr T. Milner Gibson moved a petition in the House of Commons to collect agricultural statistics. It was, he argued, pointless to have a regulated census of people if there was no means of knowing whether the country could support them. Milner Gibson spoke as a free trader who understood that an appreciation of the agricultural output of the kingdom relative to the population was critical to an understanding of precisely how dependent on imports the kingdom had become. Gladstone, as President of the Board of Trade, responded. He accepted the arguments regarding the efficacy of collecting the basic statistics. However, once the Corn Laws were repealed in 1846 the argument seemed less pressing, and the collection of statistics did not proceed, except in Ireland where an annual census began in 1847.[7]

It was twenty years, and after a great deal of debate, before Milner Gibson's proposals came anywhere near fulfilment. He unsuccessfully introduced a bill in Parliament in 1847, which was lost after criticism of his proposed procedures and their attendant costs. In the 1850s James Caird took up the baton, complaining in 1851 that 'there are statistical returns on almost every other subject connected with the business or welfare of the country, but that which may be well regarded as the most important of all—the

---

[5] R. J. P. Kain, *An Atlas and Index of the Tithe Files of Mid-Nineteenth-Century England and Wales* (Cambridge: Cambridge University Press, 1986).

[6] Sir Josiah Stamp, *British Incomes and Property* (London: P. S. King & Sons, 1927). In addition the earlier property tax and income tax of the French wars offer a number of insights into agricultural development as well as some valuable raw data, for which see J. R. McCulloch, *A Statistical Account of the British Empire*, i (London: The Society for the Diffusion of Useful Knowledge, 1837), 631; P. K. O'Brien, 'British Incomes and Property in the Early Nineteenth Century', *EcHR* 12 (1959), 255–67.

[7] *Hansard's Parliamentary Debates*, 3rd ser. 75 (1844), 92–103; *Census of Ireland for the Year 1841*, BPP, XXIV (1843); Saorstat Eireann, *Agricultural Statistics 1847–1926: Reports and Tables* (Dublin: Department of Industry and Commerce, 1930); L. Napolitan, 'The Centenary of the Agricultural Census', *JRASE* 127 (1966), 82.

annual supply of food—is still left to conjecture'.[8] To a degree there was a pilot project for a wider annual survey with the sample collection of data in 1854 in nine English and two Welsh counties. This produced details of crop acreage and animal numbers, but the returns were regarded as unsatisfactory on the grounds that ten of the eleven counties involved provided insufficient data to justify the government proceeding with what it had hoped would be a national scheme.[9] Caird continued to press the issue in Parliament through the 1850s and early 1860s,[10] but eventually the systematic collection of agricultural data was introduced less from a spirit of general enquiry and more by the incidence of the cattle plague (rindepest) of 1865.[11] Statistics were collected for the first time in 1866, but the government still failed to create the machinery to find out and tabulate, let alone analyse, the data. Official *estimates* of crop yields were not made until as late as 1884, before which date we must rely on the independent calculations made (but only for wheat) by J. B. Lawes and J. H. Gilbert.[12]

---

[8] James Caird, *English Agriculture in 1850–51* (2nd edn. London; Longman, 1852; repr. London: Frank Cass, 1968), 520.

[9] The 1854 survey was preceded by a trial exercise on Norfolk and Hampshire in England, and Haddingham, Roxburgh, and Sutherland in Scotland. The history of this 1854 data-gathering experiment is given in a number of critical essays by J. P. Dodd, but see in particular his 'The Agricultural Statistics for 1854: An Assessment of their Value', *AgHR* 35 (1987), 159–70. See also S. Wade Martins, *A Great Estate at Work: The Holkham Estate and its Inhabitants in the Nineteenth Century* (Cambridge: Cambridge University Press, 1980), 260–2, for a detailed breakdown of the 1854 survey for Norfolk at the level of the Poor Law Unions. The 1854 survey can be found in *Reports by Poor Law Inspectors on Agricultural Statistics (England), 1854*, BPP, LIII (1854–5), 495. See also *AHEW* vi. 1042–4, for more accessible full details at the county level, and a summary in *AHEW* vii. 1768–9. See also G. E. Fussell, 'The Collection of Agricultural Statistics in Great Britain: Its Origin and Evolution', *Agricultural History*, 18 (1944), 164, for notes on Scotland.

[10] Caird was an early critic, see particularly James Caird, 'On the Agricultural Statistics of the United Kingdom', *JRSS* 31 (1868), 127–45. See also the series of articles by R. H. Rew presented to the Committee Appointed to Inquire into the Statistics Available as a Basis for Estimating the Production and Consumption of Meat and Milk in the United Kingdom. These were presented as 'Production and Consumption of Meat and Milk: Second Report', *JRSS* 67 (1904), 368–84; 'Production and Consumption of Meat and Milk: Third Report', ibid. 385–412; 'Observations on the Production and Consumption of Meat and Dairy Products', ibid. 413–27. This series of articles was preceded by his 'An Inquiry into the Statistics of the Production and Consumption of Milk and Milk Products in Great Britain', *JRSS* 55 (1892), 244–86. See also his 'The Nation's Food Supply', *JRSS* 76 (1912), 100–1. See also P. G. Craigie, 'Statistics of Agricultural Production', *JRSS* 46 (1883), 1–58; id., 'On the Production and Consumption of Meat in the United Kingdom', *Report of the British Association for the Advancement of Science* (London, 1884), Section F, 841–7; J. B. Lawes and J. H. Gilbert, 'Home Produce, Imports, Consumption, and Price of Wheat, over Forty Harvest-Years, 1852–53 to 1891–92', *JRASE* 3rd ser. 4 (1893), 77–133; R. C. Turnbull, 'The Household Food Supply of the United Kingdom', *Transactions of the Highland and Agricultural Society*, 15 (1903), 197–211.

[11] MAFF, *A Century*, 3.

[12] Lawes and Gilbert, 'Home Produce'. See also M. J. R. Healy and E. L. Jones, 'Wheat Yields in England, 1815–59', *JRSS* 125 (1962), 574–9, for relatively modern estimates of mid-19th-century wheat yields based on contemporary sources.

Underlying these debates was a fundamental objection by the farming community to telling the government about farm output. In general, whether it was during the inquiries of the 1790s, or the debates of the 1850s, the fear was that the government would use the information for tax purposes. During a parliamentary debate in 1864 on a motion moved by Caird, one speaker expressed the view that tenant farmers known to him had come round to the view that statistics should be collected, thus presumably reflecting what had formerly been a natural state of objection and defiance.[13] When, in August 1866, C. S. Read enquired in the Commons about the progress of the collection of statistics, he was told that 'an impression seemed to prevail among some farmers that the returns were connected with some system of taxation, and though they were entirely misled on that part, the impression naturally prevented them from making returns'.[14]

A combination of government inertia and farmer resistance persisted until the mid-1860s.[15] From then onwards we can begin to piece together the picture of national agricultural output,[16] but before the 1860s there is a void. Series and indices have been compiled for prices and wages, and for rents, but farm output and productivity remain the great unknowns in English farming. While a method has been devised of extracting plausible figures from probate inventories, these dwindle to insignificant numbers from about 1730, and thereafter historians have relied almost entirely on Arthur Young and a number of other contemporary collectors for information on output. Given the reluctance of the farming community to broadcast its results, even these figures must be treated with caution. Was Young told the truth? Farmers had every incentive to keep a tight rein on their output figures, both to frustrate the government over taxation and also to make sure their landlords lacked the evidence on which to propose a rent increase. Yet, surely, farmers' own records would include calculations and estimates designed to show them the truth of their situation?[17] Surprisingly, the question seems never to have been asked or, if asked, not answered.

[13] *Hansard*, 3rd ser. 175 (1864), 1362–85.   [14] Ibid. 184 (1866), 2039.
[15] Collection problems in 1866 suggest that 1867 is the first reliable year. In general see J. T. Coppock, 'The Statistical Assessment of British Agriculture', *AgHR* 4 (1956), 17–20; J. T. Coppock and R. H. Best, *The Changing Use of Land in Britain* (London: Faber & Faber, 1962), chapter 1; Fussell, 'The Collection of Agricultural Statistics', 161–7; MAFF, *A Century*, 1–4; Napolitan, 'The Centenary', 81–96; E. Thomas, 'The June Returns One Hundred Years Old', *Agriculture*, 73 (1966), 245–9; J. A. Venn, *The Foundations of Agricultural Economics* (Cambridge: Cambridge University Press, 1933), chapter 20, 424–40.
[16] MAFF, *A Century, passim*; *AHEW* vii. 224–320.
[17] M. E. Turner, 'Weighing the Fat Pig: Agricultural History and the National Income Accounts', Inaugural Lecture, University of Hull (25 Jan. 1993), 10–13, 16, 26–7.

Of course asking the question depended on the availability of archives. The survival of farm records was known in general terms from the work of E. J. T. Collins and E. L. Jones at Reading University. Between 1964 and 1973 they actively sought to establish a collection of these documents. This involved first targeting those families who, because they were older-established members of the local farming community, were likely to have records, and then persuading them by telephone and personal visits to deposit the records at Reading. Failing that, efforts were made to copy the documents for use by historians. The resulting collection contains records relating to over 750 farms and is the largest of its type in the United Kingdom. Collins and Jones published the results of their work in the mid-1960s,[18] and a catalogue of the Reading holdings was produced in 1973.[19] A further result of the activities at Reading was to alert County Record Offices to the potential value to historians of farm records. This encouraged a more active collecting policy, as is clear from the relevant lists in the National Register of Archives maintained by the Historical Manuscripts Commission, and a somewhat dated but still useful calendar of farm records in public and private hands compiled by the Rural History Centre at Reading.[20]

Although the records exist, they have not been systematically used, so to test their quality we undertook a sensitivity analysis of a sample of the archives at Reading University. To offset the bias among the Reading archive, which includes a disproportionate number of farms in the south Midlands, and southern and eastern England, records were also sampled in repositories elsewhere in the country. Altogether we were able to identify more than 160 collections with material that could potentially provide an insight into farm production in England 1700–1914. This gave us the green light to proceed, and once we began to examine the material many more useful records came to light (Table 2.1).[21] As a result, it is our contention that the data exist to take a fresh look at an old problem, farm production and output in the eighteenth and nineteenth centuries.

---

[18] E. J. T. Collins, 'Historical Farm Records', *Archives*, 7 (1966), 143–9; E. L. Jones and E. J. T. Collins, 'The Collection and Analysis of Farm Record Books', *Journal of the Society of Archivists*, 3 (1965), 86–9.

[19] University of Reading Library, *Historical Farm Records: A Summary Guide to Manuscripts and Other Material in the University Library and Collected by the Institute of Agricultural History and the Museum of English Rural Life* (Reading: University of Reading, 1973).

[20] M. E. Turner, J. V. Beckett, and B. Afton, 'Taking Stock: Farmers, Farm Records, and Agricultural Output in England, 1700–1850', *AgHR* 44 (1996), 21–34.

[21] Ibid.

We begin in Chapter 1 by reviewing the literature, both to establish what is known of output in this period and to point to some of the gaps in our knowledge. In Chapter 2 we look at the farmers and their records. They kept diaries, memoranda books, and a great many accounts in one shape or form. Individual farmers could be running a substantial enterprise, deploying considerable sums of capital, and employing large numbers of labourers. They needed to keep track of their inputs and outputs, and even if they chose to do so in a manner which today might be regarded as both crude and uninformative they presumably had their reasons for collecting the data they assembled, and this was surely to make an assessment of their financial and farming position. Nor can the haphazard nature of some of these records detract from the significance they held for their compilers.

Having established that there is a database from which to work, we look in Chapters 3 to 6 at what the farm records reveal. In Chapter 3 we examine the processes and practices of farming in this period, and how these are revealed in the records, including the extent to which farmers innovated in the way they worked the land. Essentially this is a qualitative study since we cannot quantify innovation in a meaningful way. By contrast, in Chapters 4 to 6 we have extracted from the records data on grain yields (particularly wheat, barley, and oats) and animal carcass weights. In these chapters we locate, for the first time, the reality behind the claims of farmers and the landed interest more generally about output and production. From their own records we can trace the course of yields and weights, and the picture we paint differs significantly from the received version of events, a version primarily derived from proxies such as rents and prices rather than the actual output of the English farm.

When he undertook his survey of the sources in the 1960s, Collins suggested that farm records would provide evidence relating to a whole series of subjects:

The levels of farm output, receipts, expenditure, profits and investment; the influence of price movements on individual farming systems; the ways in which farmers raised capital and in more prosperous times how they invested it; crop and milk yields; lambing and calving rates; the size of the wool clip and livestock weights; innovations in crop variety, animal breeds, farming techniques and implements, rotations and land utilisation; land reclamation and enclosure; and marketing methods.[22]

[22] Collins, 'Historical Farm Records', 145.

We started with this list which, perhaps not surprisingly, turned out to be over-ambitious, and we have not been able to answer all the questions we had in mind at the outset of our work. We hoped originally to employ a fully articulated national income accounting approach to the agricultural history of this period, but it soon became clear that this was unlikely to emerge from the farm records. We set out to measure, not to estimate, guess, or infer, English agricultural output, but we have ended up emphasizing those essential but not all-embracing aspects of farm output that constitute, at the aggregate level, English agricultural output. Our emphasis switched from national agricultural income accounting to farm production, and within that adjustment two essentials of production were identified in output terms, crop yields and animal carcass weights. We may not have been able to achieve all that we had hoped, but what we have found not only makes a significant contribution to our understanding of farm output in the eighteenth and nineteenth centuries, but also points definitively in the direction of a process of change which was nothing short of an agricultural revolution in the first half of the nineteenth century.

CHAPTER 1

Agricultural Production, Output, and Productivity, 1700–1914

To an earlier generation of historians the agricultural revolution was the time when the old common-field systems of farming were swept away in the enclosure movement, the yeomen of England disappeared from the countryside, and great landowners let their land in large farms at rack rents to tenant farmers who worked it by employing waged labour. In the words of R. E. Prothero, Lord Ernle, whose *English Farming Past and Present* appeared for the first time in 1912 and has subsequently become the classic account of agricultural change, 'it was the large landlords who took the lead in the agricultural revolution of the eighteenth century, and the larger farmers who were the first to adopt improvements. Both classes found that land was the most profitable investment for their capital.'[1] The capitalist system, in other words, destroyed the old rural community in favour of a new monster concerned with efficiency and output, and hence incomes. Just as the industrial revolution had brought factories and the division of labour, so the agrarian revolution had brought large units of production (in terms of estates and farms) and a division of responsibility between the owner (fixed capital), the farmer (circulating capital), and the labourer.

---

[1] R. E. Prothero (Lord Ernle), *English Farming Past and Present* (6th edn. London: Heinemann, 1961), 161.

The Agrarian Revolution

Ernle reflected a widely held view that an agricultural revolution had occurred, and that it needed only to be *described* not necessarily to be *measured*. He and other late nineteenth- and early twentieth-century writers simply researched its main institutions and changing structures: the role of enclosure, the disappearance of the open fields, the growth of large farms, changes in tenure by which capitalists and capitalist structures emerged triumphant. The emphasis throughout was on *description*: in Ernle's words:

the chief characteristics in the farming progress of the period... may be summed up in the adoption of improved methods of cultivation, the introduction of new crops, the reduction of stock-breeding to a science, the provision of increased facilities of communication and of transport, and the enterprise and outlay of capitalist landlords and tenant-farmers.[2]

There was no need to measure the impact of these changes because it was plain for all to see: the population had been fed, and this was evidence enough.

Ernle's explanation of agricultural change was more or less accepted for half a century, but by the 1960s such broad-brush explanations no longer seemed so convincing. Just as industrial historians started to look for far distant causes of the industrial revolution, so agricultural historians began to question the assumption that the two revolutions had gone together in unison over the period *c.*1750–*c.*1850. A. H. John in 1960, and E. L. Jones in 1965, stressed the significance of changes prior to 1750.[3] In Jones's words: 'Between the middle of the seventeenth century and the middle of the eighteenth century, English agriculture underwent a transformation in its techniques out of all proportion to the rather limited widening of its markets.'[4] J. D. Chambers and G. E. Mingay, in a textbook published in 1966, retained the traditional dating but devoted nearly one-quarter of their book to the period prior to 1760 on the grounds that this was when a great deal of

---

[2] Prothero, *English Farming*, 149.
[3] A. H. John, 'The Course of Agricultural Change, 1660–1760', in L. S. Pressnell (ed.), *Studies in the Industrial Revolution* (London: Athlone Press, 1960), 125–55; id., 'Agricultural Productivity and Economic Growth in England, 1700–1760', *JEH* 25 (1965), 19–34; id., 'Aspects of Economic Growth in the First Half of the Eighteenth Century', *Economica*, 28 (1961), 176–90; E. L. Jones, 'Agriculture and Economic Growth in England 1660–1750: Agricultural Change', *JEH* 25 (1965), 1.
[4] Jones, 'Agriculture and Economic Growth', 1.

change had occurred.[5] Even more radical was the approach of Eric Kerridge who argued in 1967 that the agricultural revolution took place between 1560 and 1767, with its main achievements all occurring before 1720 and mostly before 1673.[6] At almost the same time, F. M. L. Thompson published an article in which he argued for a second agricultural revolution beginning in the wake of the Napoleonic wars.[7] Within a decade the old certainties had collapsed.

To resolve such discrepancies it made sense to move from the essentially qualitative discussions contained in the 1960s contributions, to a more rigorous quantitative assessment of change across time. As was well known, the difficulty here lay in the evidence. Until the 1860s the government had made no systematic attempt to collect data on farm output, and it was the 1880s before a serious attempt was made to measure productivity. Occasional censuses such as the 1801 crop returns, and some pilot surveys in the 1850s, offered historians a cross-sectional view of agricultural practice, but nothing more.[8] Individual references to farms and regions could be found in the work of contemporary commentators, notably Arthur Young, from the 1770s onwards, but beyond this it was assumed that there was nothing of value to be said. Nowhere was this more clearly stated than in the volumes of the *Agrarian History of England and Wales*. Volume iv, covering the period 1500–1640, appeared in 1967, too early for the contributors to make a reasoned contribution to the newly developing debate, but volumes v (1640–1750) and vi (1750–1850) were not published until the 1980s. As such, historians expected them to offer a definitive judgement about the changing fortunes of agriculture, and they certainly offered an enormous digest of the subject-matter of agrarian history, together with a range of relevant statistical data. J. A. Chartres, in a chapter on the marketing of agricultural produce in the period 1640–1750 for volume v, made tentative estimates of the output of cereals for the years 1695 and 1750, and of livestock for 1695. Yet most of the information came from an existing database of contemporary estimates, which were few in number. These were coupled

---

[5] J. D. Chambers and G. E. Mingay, *The Agricultural Revolution, 1750–1880* (London: Batsford, 1966).
[6] E. Kerridge, *The Agricultural Revolution* (London: Allen & Unwin, 1967).
[7] F. M. L. Thompson, 'The Second Agricultural Revolution, 1815–80', *EcHR* 21 (1968), 62–77.
[8] M. E. Turner, 'Arable in England and Wales: Estimates from the 1801 Crop Returns', *Journal of Historical Geography*, 7 (1981), 291–302; id., 'Counting Sheep: Waking up to New Estimates of Livestock Numbers in England, c.1800', *AgHR* 46 (1998), 142–61; J. P. Dodd, 'The Agricultural Statistics for 1854: An Assessment of their Value', *AgHR* 35 (1987), 159–70.

with plausible interpretation regarding seeding rates and other features of farming life. In total, the discussion of output takes up less than five pages in a volume running to 952 pages. The contributors to volume v apparently accepted the general editor's view, expressed elsewhere, that 'the historian has to accept the impossibility of measuring agricultural change in the past in a totally satisfactory way. The documents... never offer a complete picture, and the final verdict on the scale of change has to be a matter of personal judgement'.[9] The words 'productivity' and 'output' do not appear in the index to volume v, although some output data are given in an appendix.[10]

Nor did the situation change with the publication in 1989 of volume vi of the *Agrarian History*, covering the period 1750–1850. B. A. Holderness's chapter on prices, productivity, and output ran to over 100 pages and included a valiant attempt to extend the output figures to 1800 and 1850. The chapter is, however, stronger on prices than on output, and there is little directly on productivity. In part, this reflected the nature of the material from which Holderness worked. A. H. John, who compiled the statistical appendix, mainly recycled material collected by contemporary writers including Arthur Young, Sir James Caird, and J. R. McCulloch.[11] No attempt was made to use 'farm accounts and the like' despite their acknowledged significance.[12]

The approach to measuring farm production and agricultural output adopted in volumes v and vi of the *Agrarian History* was unsatisfactory for various reasons, not the least of which was the significant disparities between the findings of Chartres and Holderness where they overlap at *c.*1750. Despite their heavyweight billing, and the wealth of quantitative data they contained, the *Agrarian History* volumes failed to address what many historians saw as the key questions arising from the debates of the 1960s: the timing and nature of the agricultural revolution or revolutions. Joan Thirsk, editing volume v, wrote of the 1640–1750 period that 'agriculture was led along new paths which opened out into the agricultural revolution'; but

---

[9] J. Thirsk, *England's Agricultural Regions and Agrarian History, 1500–1750* (Basingstoke: Macmillan, 1987), 61.

[10] *AHEW* v (2). 442–8, 880–5.

[11] J. R. McCulloch, *A Statistical Account of the British Empire*, i (London: The Society for the Diffusion of Useful Knowledge, 1837), 528–9. McCulloch is recognized as a recycler from Young, the Revd Henry Beeke, John Middleton, and others, rather than as a collector of new material himself. But in this function he was probably the first commentator seriously to collate the material of others and thereby address issues of agricultural output.

[12] *AHEW* vi. 84–189, 1038–58. Holderness's calculations of output are on pp. 145 and 155.

G. E. Mingay, who edited volume vi covering the period 1750–1850, far from picking up the baton laid down by Thirsk, argued that changes beyond 1750 merely prepared the way for even more significant developments in the nineteenth century: 'in many ways the hundred years that ended in 1850 may be seen as a base, or rather a preparation, a limited but essential preparation, for the greater changes yet to come.' Of the achievements of the hundred years after 1750 'it could hardly be said that they amounted to an agricultural revolution'.[13] Without a firm statistical base such obfuscation was perhaps inevitable, but a whole generation of historians seemed unwilling to come to terms with concepts of productivity in an *explicit* fashion, accepting *implicitly* the traditional agenda. Mingay, writing in 1989, and after completing his editorial work for volume vi, approached the issue in lofty terms: 'it may be, as has been conjectured, that the overall rise in output was of the order of some 40 per cent, i.e. rather less than the growth of the market, but this is merely a reasoned guess.' His scepticism is transparent: the agricultural revolution, whatever it was and whenever it occurred, was certainly not something that could be accurately measured.[14]

Since the 1980s the approach adopted in the *Agrarian History* volumes has come to seem increasingly untenable. Simply arguing that the population was fed, and that therefore there must have been an agricultural revolution which raised output, but then denying that this could be satisfactorily measured and moving swiftly on to examine the qualitative and institutional elements of the revolution, has come to look like weak methodology. Furthermore, the reticence of the various *Agrarian History* authors to measure farm production on a systematic basis, in an attempt to produce reliable output figures across time, has not been shared by others. Scholars have pressed into service Gregory King's questionable guesstimates to try to produce data for the late seventeenth century.[15] Others have plundered the

---

[13] *AHEW* v (2), xxvii; *AHEW* vi. 953, 971.

[14] G. E. Mingay, 'Agricultural Productivity and Agricultural Society in Eighteenth-Century England', in G. Grantham and C. S. Leonard (eds.), *Agrarian Organization in the Century of Industrialization* (Research in Economic History, Supplement no. 5, 1989), 31–2. Mingay originally read this paper to a conference in 1984 but no changes were made to the text when it appeared in 1989.

[15] The list is long and includes specialist research and textbooks alike. The following is indicative of the dependency on King. G. S. Holmes, 'Gregory King and the Social Structure of Pre-industrial England', *Transactions of the Royal Historical Society*, 27 (1977), 41–68; P. H. Lindert and J. G. Williamson, 'Revising Britain's Social Tables, 1688–1913', *Explorations in Economic History*, 19 (1982), 385–408; N. F. R. Crafts, *British Economic Growth during the Industrial Revolution* (Oxford: Oxford University Press, 1985), esp. 7–17; P. Mathias, *The First Industrial Nation: An Economic History of Britain 1700–1914* (London: Methuen, 1969), 23–31; D. C. Coleman, *The Economy of England*

1801 crop returns for output figures during the French wars.[16] Yet others have developed techniques which relied less, or sometimes not at all, on measurements of agricultural production and output, and more on alternative indicators of the performance of agriculture. These included extracting plausible figures from sixteenth- and seventeenth-century probate inventories, or employing the record of prices in relation to other parameters such as population to explore likely patterns of agricultural (essentially food) supply and demand.[17]

The Timing of the Agricultural Revolution

The contested ground is mainly over the chronological turning point during the eighteenth and nineteenth centuries. Was it a history of even, or of disjointed, growth? There are a number of competing series to choose from. One of them demonstrates faster growth in the nineteenth than in the eighteenth century,[18] whereas another shows much greater growth in the early eighteenth century and again in the early nineteenth century, but relative stagnation in between,[19] and a third shows a more even growth, though

1450–1750 (Oxford: Oxford University Press, 1977), 6; P. Deane and W. A. Cole, *British Economic Growth 1688–1959* (2nd edn. Cambridge: Cambridge University Press, 1969), 2. This dependency can relate to King's estimates of land use distribution and other indicators of output, if not actually King's supposed measurements or estimates. In this category there are a number of references in *AHEW*, vi by Holderness in one of which (pp. 138–9) he refers to M. K. Bennett's use of both Gregory King and Charles Smith, in M. K. Bennett, 'British Wheat Yield Per Acre for Seven Centuries', *Economic History*, 3 (1937), 12–29. Fussell also refers to King's estimates fairly casually in G. E. Fussell, 'Population and Wheat Production in the Eighteenth Century', *History Teachers' Miscellany*, 7 (1929), in two parts, 65–8, 84–8. Chartres, in *AHEW* v (2). 442–5, also uses the output statistics from King.

[16] Turner, 'Arable in England and Wales'.

[17] M. Overton, 'Estimating Yields from Probate Inventories: An Example from East Anglia, 1585–1735', *JEH* 39 (1979), 363–78; R. C. Allen, 'Inferring Yields from Probate Inventories', *JEH* 48 (1988), 117–25; P. Glennie, 'Measuring Crop Yields in Early Modern England', in Campbell and Overton (1991), 255–83; M. Overton, 'The Determinants of Crop Yields in Early Modern England', ibid. 284–322; M. Overton, 'Re-estimating Crop Yields from Probate Inventories', *JEH* 50 (1990), 931–5. See also G. Clark, 'Yields Per Acre in English Agriculture 1250–1860', *EcHR* 44 (1991), 445–60, esp. 447–8.

[18] R. C. Allen, 'Agriculture during the Industrial Revolution', in R. Floud and D. McCloskey (eds.), *The Economic History of Britain since 1700*, i: *1700–1860* (2nd edn. Cambridge: Cambridge University Press, 1994), 100–3. Allen used data from Deane and Cole, *British Economic Growth*, but because it is based on population growth and assumptions about relatively constant per capita consumption, against a backdrop of a relatively modest proportionate reliance on imports, it was bound to show rising output roughly in line with population growth.

[19] Allen, 'Agriculture during the Industrial Revolution', reporting Crafts, *British Economic Growth*, and R. V. Jackson, 'Growth and Deceleration in English Agriculture, 1660–1790', *EcHR* 38 (1985), based on demand for food with assumptions regarding income and price elasticities.

picking up after 1750 and then more strongly after 1800.[20] However, a consensus has begun to emerge which supports the argument that the agricultural revolution was all over by the third quarter of the eighteenth century. Output, it is argued, grew quickly over the period 1650–1740, with an impressive performance by the agricultural sector down to 1760. But it is what happened after that which is interesting. According to Jackson this phase of strong growth was followed 'by fifty years of much slower growth, perhaps even of near stagnation'.[21] Crafts calculated a rate of growth of 0.6 per cent per annum 1700–60 but only 0.13 per cent per annum 1760–80.[22] Other evidence seemed to lend support to these figures. Although wheat yields in Norfolk and Suffolk, according to Overton, increased steadily to 1801, there was in general a slowdown in the rate of growth in output per acre over the period 1750–1830.[23]

Two further approaches added support to this view of agriculture as innovative and highly productive over the century or so prior to 1750. The first was Allen's concept of two agricultural revolutions—the yeoman's in the seventeenth century and the landlord's in the eighteenth.[24] In the first of these revolutions productivity strides were made from the Middle Ages to the eighteenth century, and particularly in the seventeenth century, by yeomen farmers mainly working in the open fields. Had these farmers continued in post, as it were, their 'greater proprietary interest' would have meant that they had 'a greater incentive to increase production than tenancy at will gave the capitalist farmer'. Consequently 'agricultural output might have grown faster in the eighteenth century if the yeomen had retained ownership of their land'.[25] But of course they did not: they were driven out by the enclosure movement. In Allen's view this was a landlord-driven agricultural revolution which was regressive in its impact. As an exercise in redistributing income to the large landowners it acted as a disincentive to independent

---

[20] Allen, 'Agriculture during the Industrial Revolution', based on his own derivations from the estimates of others, which in turn were based on the kinds of estimates we report below. Allen, it should be noted, believes that a more significant agricultural revolution occurred before 1700.

[21] Jackson, 'Growth and Deceleration', 333.   [22] Crafts, *British Economic Growth*.

[23] M. Overton, *Agricultural Revolution in England: The Transformation of the Agrarian Economy 1500–1850* (Cambridge: Cambridge University Press, 1996). An earlier suggestion on the same lines was in M. E. Turner, 'Agricultural Productivity in England in the Eighteenth Century: Evidence from Crop Yields', *EcHR* 35 (1982), 489–510, esp. 506.

[24] R. C. Allen, *Enclosure and the Yeoman: The Agricultural Development of the South Midlands 1450–1850* (Oxford: Oxford University Press, 1992); id., 'The Two English Agricultural Revolutions', in Campbell and Overton (1991), 236–54.

[25] Allen, *Enclosure*, 310–11.

yeomen to invest time, money, and effort. They were reduced to tenant status. Therefore enclosure had the effect of slowing down agricultural growth and leaving most people less well off than had it never occurred.[26] In these debates the question of the base starting point had not properly been addressed, and to this extent, growth rates on their own can be misleading.

A second approach has also laid considerable stress on the importance of change before $c.1770$.[27] Clark believes that the agricultural revolution was a phenomenon of the period prior to 1770, and that 'the agricultural revolution thus pre-dates the industrial revolution'.[28] His estimates suggest that factor productivity in agriculture in England and Wales in 1700 and 1770 was running at 92 and 90 per cent of the level which was attained by 1850. While this allows for some growth over the century or so of the industrial revolution, it was only modest, hardly of an order of magnitude to constitute a revolution in its own right.[29] It follows that the real agricultural revolution occurred either before 1770 or even 1700, or after 1850.

This is all well and good, but however carefully compiled the statistics, quantitative historians are still faced with the same problems that confronted their qualitative predecessors when it comes to explaining why the agricultural revolution was apparently all over and done with before the great rise in population towards the end of the eighteenth century put pressure on resources as never before. The fact that disaster did not follow from the slowing down of the sector in the second half of the century is explained by those who support this timing in terms of an increase in the cultivated acreage, a rapid rise in the output of wheat, and a shift by the 1770s from the grain surpluses which were offset by exports, to grain deficits which were covered by imports. Jones has suggested that by 1800 about 90 per cent of the population was fed by domestic agricultural production.[30] Chambers had once argued that the worst years of dearth during the French wars left the country short of about nine weeks' consumption of basic grain.[31] This was the worst case, though Mokyr elevated it to the general situation when he

---

[26] Allen, *Enclosure*, 21.
[27] G. Clark, 'Agriculture and the Industrial Revolution, 1700–1850', in J. Mokyr (ed.), *The British Industrial Revolution: An Economic Perspective* (Oxford: Westview Press, 1993), 249.
[28] Clark, 'Yields Per Acre in English Agriculture', 459.
[29] Clark, 'Agriculture and the Industrial Revolution', 246–8, 255.
[30] E. L. Jones, 'Agriculture, 1700–80', in R. Floud and D. McCloskey (eds.), *The Economic History of Britain since 1700*, i: *1700–1860* (1st edn. Cambridge: Cambridge University Press, 1981), 68.
[31] Chambers and Mingay, *The Agricultural Revolution*, 115–16; M. Olsen, *The Economics of the Wartime Shortage* (Durham, NC: University of North Carolina Press, 1963), 65.

argued that about one-sixth of all grain consumption was imported by the early nineteenth century, and then about 22 per cent by 1841.[32] Other indicators, including rising prices and rents, are interpreted as additional evidence that output was not keeping pace with demand. The concerns of the classical economists were not, it would seem, unfounded.

So how was disaster avoided? Rising prices, it is argued, and an increase in output to counter problems caused by poor harvests in the 1790s and wartime conditions, stimulated agriculture, and growth rates improved to exceed those of the period pre-1760.[33] Growth picked up from 0.75 per cent per annum over the period 1780–1801, to 1.18 per cent per annum 1801–31.[34] Since by the 1840s grain prices were little higher than in the 1770s, output would seem to have risen in such a way as to keep prices relatively low compared to total population, given the fact that prior to 1850 there is no statistically significant evidence of changing consumer preference.

Is this almost too neat for comfort? The idea, now in the textbooks, that 'growth slowed just as population growth began to accelerate'[35] cries out for some supportive evidence, as does the argument that over the period 1760–1830 all that seems to have taken place was a steady spread and application of new techniques. Such arguments simply do not add up. The population of Britain increased from 8.5 million in 1770 to 21 million by 1851, but in 1851 only about one-fifth of British food was imported: it looks therefore as if there was a 147 per cent increase in the mouths to feed but only a 20 per cent shortfall in the supply of food. At constant levels of per capita consumption, about 16 million of the 21 million in 1851 were fed by home supplies (i.e. the large residual of four-fifths were in fact fed), which is about twice as many actual mouths as were fed in 1770.

Since it is inconceivable that the cultivated land area doubled, or that the labour force in agriculture doubled, or that the capital input doubled, or that some combination of all three led to a doubling of the factor inputs, depending how much less than a doubling of inputs occurred we can point inferentially to an increase in agricultural productivity. Certainly the land input did not double, although the adjustment in the bare fallow component, whether related or not to enclosure, effectively increased the land area; the agricultural

---

[32] J. Mokyr (ed.), *The Economics of the Industrial Revolution* (London: Allen & Unwin, 1985), 147–8.
[33] Jackson, 'Growth and Deceleration'.   [34] Crafts, *British Economic Growth*.
[35] M. J. Daunton, *Progress and Poverty: An Economic and Social History of Britain 1700–1870* (Oxford: Oxford University Press, 1995), 35.

labour force, far from doubling, seems to have stayed much the same or risen slightly to its peak size as recorded in the 1861 census; and it seems unlikely that the capital input doubled either, despite the capital costs of enclosure and drainage. Unless we could show, which we cannot, a substitute of grain imports for home-grown non-grain foodstuffs, there is no suitable explanation: increases in output, even when supplemented by imports, were simply not sufficient—on present information—to cover the gap. There may have been a switch from other crops into food, as Clark has suggested, but we do not have enough hard data to demonstrate such a change.[36] However, if we relax the assumption about constant consumption then we must argue not simply for a decline in living standards, but for a dramatic and inconceivable decline in those living standards.[37]

In reality the post-1750 period was of key importance for English agriculture. Some of the necessary statistics to show that this was the case are available in volume VI of the *Agrarian History*. Even allowing for the problems we have previously identified of the sources on which this volume is based, from the conjectures and sometimes guesses regarding the statistical base of agriculture it looks as though between 1750 and the 1840s the acreage under wheat in England and Wales may have doubled, at the very least, largely before 1812 and at the expense of rye and barley (i.e. not amounting to a net doubling of the grain area). We cannot be precise but the output of wheat alone probably increased by between 50 and 75 per cent in the course of the eighteenth century. Overall between 1750 and 1850 wheat output increased by about 225 per cent, barley by 68 per cent, and oats by 65 per cent. The weight of meat went up from about 6 million hundredweight in 1750 to about 12 million in 1850, and the weight of wool also doubled over the period. Between 1750 and 1850 the general output of English agriculture rather more than doubled.[38] In addition, improvements in labour productivity meant that each agricultural worker's output was capable of feeding 2.7 non-agricultural workers in 1841 compared to 1 in 1760.[39]

---

[36] Clark, 'Agriculture and the Industrial Revolution'.

[37] The most recent reworking of the standard of living debate, even in its most pessimistic moments, does not support such an interpretation: C. H. Feinstein, 'Pessimism Perpetuated: Real Wages and the Standard of Living in Britain during the Industrial Revolution', *JEH* 58 (1998), 625–58.

[38] *AHEW* vi. 128–74. Caution is needed in relation to animal estimates given the doubts over sheep numbers, c.1800, expressed by Turner, 'Counting Sheep'.

[39] P. K. O'Brien, 'Agriculture and the Home Market for English Industry 1660–1820', *English Historical Review*, 100 (1985), 773–800.

Other evidence supports this view. Campbell and Overton have argued of Norfolk that the 1750–1850 period was of key importance for output and productivity. Norfolk farmers were among the earliest to cultivate clover in England and much has been made of the crop's revolutionary effects, but between 1660 and 1739 no more than 15 per cent of Norfolk's large farms grew clover and it accounted for only one-eighth of the legume acreage. By contrast, by the 1830s it comprised one-quarter of the sown acreage and around 90 per cent of all the legumes cultivated. A similar pattern was found with turnips; around 1710 turnips comprised about 7 per cent of the sown acreage in Norfolk, but the figure was nearer to 25 per cent by the 1830s.[40] Most recently Wade Martins and Williamson, accepting that there is little sound hard statistical data for the period 1650–1870, have looked in detail at changes in Norfolk agriculture using qualitative data. They have concluded that changes in both arable and pastoral farming practice played a key role in improving farming in Norfolk, and brought rapidly rising yields in the period after c.1750. They cannot produce the level of quantification beloved of cliometricians, but they do make a sustained and clear case for historians to look carefully at what was happening on the ground.[41] Thus, although they question Campbell and Overton's claims about stocking densities,[42] Wade Martins and Williamson accept that because animals in East Anglia were increasingly stall fed, this allowed for better husbanding of animal manure and urine, the value of which had been dissipated through leaching on open ground.[43] Although this looks like a return to the older tradition of agricultural history it may also offer a sounder basis for future research than simply recycling questionable quantitative data.

The development of cliometrics has led to one advantage above others; it allows new ways of handling the limited data currently available. A recent summary by Allen seemed to show that real agricultural output over the period from 1700 to 1851 grew by a factor of between 3.37 and 3.56.[44]

---

[40] B. M. S. Campbell and M. Overton, 'A New Perspective on Medieval and Early Modern Agriculture: Six Centuries of Norfolk Farming, c.1250–c.1850', *Past and Present*, 141 (1993), 53–61.

[41] S. Wade Martins and T. Williamson, *Roots of Change: Farming and the Landscape in East Anglia, c. 1700–1870* (Exeter: British Agricultural History Society, 1999).

[42] Campbell and Overton, 'A New Perspective'; M. E. Turner, 'English Open Fields and Enclosures: Retardation or Productivity Improvements', *JEH* 46 (1986), 687. In fairness, Wade Martins and Williamson also question Campbell and Overton's dating of the impact of turnips and clover in East Anglia.

[43] Campbell and Overton, 'A New Perspective', 74, 83, 88. Wade Martins and Williamson, *Roots of Change*, 171, 173.

[44] Allen, 'Agriculture during the Industrial Revolution', 100–3.

An alternative measure, but essentially derived from the same material, is Overton's summary of the literature in which he found a rather less impressive overall growth from 1700 to 1850. According to Overton, output grew by a factor of more than 2.5 but less than 3.[45] Although these two summaries are ostensibly based on the same literature, in neither case is the articulation of their methods of estimation clear enough for precise reproduction. In particular there is not enough clarity in the derivation of real growth from nominal data. When it comes to estimating production and output over the long run the current database is neither rich enough nor secure enough before the late nineteenth century, and there is a fundamental problem in determining what proportion of total production constituted output.

The first of these problems seems insoluble, but the second may be less of a problem if we determine that an appraisal of the performance of the national farm necessarily means an appraisal of total production net of those internally produced inputs which are used in the remaining value added processes. In the main these are the fodder crops and the retained seeds. Table 1.1 gathers together some of these quasi-production quasi-output estimates. Everything has been reduced to a monetary form. The grains are the cereal grains of wheat, barley, oats, and rye, and the animals, as near as we can determine, are cattle, sheep, pigs, and poultry/fowl, and their products (we cannot guarantee this in all cases but we assume this includes wool and milk). The estimate in 1700 represents Gregory King's England. It combines Chartres's separate inventory value of net livestock, and the production volume of deadstock—in fact grains—to which has been applied an appropriate set of prices.[46] Essentially we have taken the eleven-year average price for the four grains centred on the years specified, though sometimes in the case of rye—for which there is not always a price quoted—we have taken a price which is compatible with the historic relationship between wheat and rye prices.[47] For c.1750 we quote the grains-only estimate from Charles Smith, and also separate estimates from Chartres and Holderness. By 1800 we only have the Holderness estimate, though we also quote Arthur Young's

---

[45] M. Overton, 'Re-establishing the English Agricultural Revolution', *AgHR* 44 (1996), 6. The key table 1 is also reproduced as table 3.11 in his *Agricultural Revolution*, 86.

[46] The data refer to the 1690s and is from Chartres in *AHEW* v (2). 445, in which the animals are simply as he lists them, but for compatibility with later estimates we exclude rabbits, conies, and horses.

[47] These prices are a combination of Bowden's prices in *AHEW* v (2). 828–49, and those in B. R. Mitchell and P. Deane (eds.), *Abstract of British Historical Statistics* (Cambridge: Cambridge University Press, 1962), 471–3.

Table 1.1 Net agricultural output in England and Wales, 1700–1900

(a)

Date	Source	\multicolumn{3}{c}{In nominal price terms (£m.)}	\multicolumn{3}{c}{In constant price terms (£m.)}
		Grain	Animals	Total	Grain	Animals	Total
1700	Chartres	13.9	18.5	32.4	29.4	39.2	68.6
1750	Smith	15.3			33.0		
1750	Chartres	14.0			30.2		
1750	Holderness	13.1			28.3		
1770	Young	29.4	22.0	51.4	48.8	36.5	85.4
1800	Holderness	44.0			44.7		
1850	Holderness	51.9			77.1		
1870	Bellerby	44.5	48.1	92.6	60.2	65.1	125.3
1880	Bellerby	29.0	53.7	82.7	44.1	81.6	125.7
1890	Bellerby	24.3	50.8	75.1	47.1	98.4	145.5
1900	Bellerby	22.2	57.0	79.2	44.5	114.2	158.7
1910	Bellerby	20.8	63.9	84.7	36.7	112.9	149.6
1700	Allen	9.0	9.9	18.9	19.1	21.0	40.0
1750	Allen	11.6	15.7	27.3	25.1	33.9	59.0
1800	Allen	36.4	50.2	86.6	37.0	51.0	88.0
1850	Allen	37.7	53.2	90.9	56.0	79.0	135.1

(b)

	Annual rates of growth (% per annum)		
	Grain	Animals	Total
Chartres 1700 to Smith 1750	0.231		
Chartres 1700 to Chartres 1750	0.054		
Chartres 1700 to Holderness 1750	−0.076		
Chartres 1700 to Young 1770	0.727	−0.102	0.313

Table 1.1 *Continued*

	Annual rates of growth (% per annum)		
	Grain	Animals	Total
Holderness 1750 to Holderness 1800	0.918		
Holderness 1800 to Holderness 1850	1.096		
Holderness 1850 to Bellerby 1870			0.384
Holderness 1850 to Bellerby 1900	−1.093		
Bellerby 1870 to Bellerby 1910	−1.230	1.386	0.444
Allen 1700 to Allen 1750	0.548	0.965	0.777
Allen 1750 to Allen 1800	0.782	0.820	0.804
Allen 1800 to Allen 1850	0.833	0.880	0.860
Allen 1850 to Bellerby 1900	−0.460	0.739	0.323

Note See text and the associated footnotes to understand how the output has been reduced to a common monetary form, for an indication of the product prices employed, and for the method of constructing the price deflator.

grain estimate for *c.*1770, from which we have deducted Young's own estimate of the seed employed, and we also quote his animal production estimate. Finally, we conclude with both grain and animal estimates for England and Wales based on a disaggregation of United Kingdom estimates by J. R. Bellerby. In all cases we have deflated the nominal outputs with a combination of O'Brien's and Rousseaux's agricultural price indices.[48] Thus what we produce in the final three columns of the table are estimates of the real volume growth of grain output, and partially the equivalent volume growth rates of animal and total output.

Our research has enabled us to confirm the general size of the Holderness output figure for 1800. We have calculated output on the basis of the aggregate cereal output for the English counties which in turn is derived from estimates of the cereal acreages and yields which were reported in the 1801 crop returns and associated government inquiries of the 1790s.[49] We have applied the same prices to these estimates as we applied to the Holderness

[48] See usage of the same procedure TBA 206–8.
[49] Turner, 'Arable in England and Wales'; id., 'Agricultural Productivity in England'.

estimates. The value of our gross cereal output is £46.2 million against the £44 million we calculated from Holderness. His figure was based on gross output net of seed, while ours is gross output before a deduction for seed. This may have equalled as much as one-eighth or one-ninth of gross output. Therefore within fairly tight margins the estimates are roughly similar.

We have included in Table 1.1 our recomposition of Allen's own estimates of real output. That is, Allen has provided estimates of the real volume output of English and Welsh agriculture at constant 1815 prices. Unfortunately he has not specified the price deflator he employed. We have taken the price index specified above and recomposed what we think Allen's nominal outputs were in the years specified.[50] This is not a trouble-free procedure. The Allen-based recomposed estimates of nominal output are out of line with the Chartres estimates, for 1700 at least. This is not surprising since Allen relied on both Chartres and Holderness to derive his estimates.[51] The mismatch in the grain estimates is substantial, but a further worry is the difference in the animal estimate. A likely explanation for these differences lies in the price deflator employed, but without knowing what Allen used we have proceeded using, consistently, a combination of plausible deflators.

What is the outcome of this exercise? The figures suggest practically no growth in real grain output in the early eighteenth century, and then an implied and relatively massive growth in the third quarter of the century, which continued, if Holderness's estimates are reliable, into the first half of the nineteenth century. The livestock estimates are too few to say much about, although they show modest real growth in both centuries. The demise of grain and the relative rise of animal production seems well established by the end of the nineteenth century. Using Allen's manipulation of the data in conjunction with the more secure late nineteenth-century data suggests a much more steady growth in real output. Certainly this was the case until 1850, and not only was it steady, but at approaching 1 per cent per annum it looks quite impressive. It then culminates in the familiar relative collapse of agriculture in the late nineteenth century.

Whether or not we call this an agricultural revolution is a matter of taste. In Overton's view 'the increase in output ... was sufficient to break the "Malthusian trap" and allow population to expand beyond the pre-industrial

[50] We have done this before with Allen's estimates and also using the same price index deflator we employ here, for which see TBA 206–8.
[51] Allen's estimate includes grain items which we have omitted (peas and beans), but also includes oats, not simply net of seed but also of an unspecified estimate of horse feed.

ceiling'.[52] This he sees as the key issue, and the one most notably overlooked by the cliometricians: 'Defining the changes in the agricultural sector of the economy which are held to be significant in this way sets the empirical agenda: the measurement of output and productivity.'[53] In Overton's view it is possible to give some statistical backing to the idea of an agricultural revolution in the period c.1750–c.1850. However, with the exception of Norfolk he has had to rely on material from other sources, much of it speculative and open to considerable doubt. Yet by looking at the available material *in the long term* he and Campbell have concluded that despite the acknowledged shortcomings of the data, 'the evidence is consistent in showing that the achievements of medieval farmers were not significantly bettered until the eighteenth century in England, but that thereafter progress in output and both land and labour productivity was at an unprecedented level'.[54]

In this way, Overton has sought to fill the void left by the *Agrarian History* volumes, and to rescue the agricultural revolution from the hands of the cliometricians by stating clearly and on the basis of the best available evidence that there must have been something significant happening in the agricultural sector between 1750 and 1850, and that this was nothing less than an agricultural revolution. Yet the available hard data remain weak, and many of those who have entered the debate are uncomfortably aware of the gaps, particularly the lack of reliable statistics. The lack of clarity is perhaps best illustrated from the words used by Martin Daunton, in an undergraduate textbook on the British economy between 1700 and 1850 published in 1995. Under the heading of 'Agricultural Production', Daunton wrote that he had been struck by 'the dearth of reliable data... to measure agricultural output and the productivity of land and labour'. 'Historians have', he added, 'necessarily concentrated on the production of cereals which provided the basis of the agricultural economy', and thus 'it is crucial to calculate the output of grain'; either by estimating the total output by calculating yields per acre and the area under cultivation, or by using 'national

---

[52] The general argument is outlined in Overton, 'Re-establishing the English Agricultural Revolution', and developed in greater detail in chapter 3 of his *Agricultural Revolution*.

[53] Overton, 'Re-establishing the English Agricultural Revolution', 5. As we indicate above this is touched upon by Clark, 'Agriculture and the Industrial Revolution'. Note Clark's comment, 229, that the belief in an agricultural revolution running parallel to the industrial revolution is a belief 'produced by a few simple but seemingly ironclad arguments'.

[54] M. Overton and B. M. S. Campbell, 'Statistics of Production and Productivity in English Agriculture, 1086–1871', in B. J. P. van Bavel and E. Thoen (eds.), *Land Productivity and Agro-systems in the North Sea Area, Middle Ages–Twentieth Century: Elements for Comparison* (Turnhout: Brepols, 1999), 189–208.

aggregates and using a simple demand and supply equation to move from statistics of population and prices to the missing variable of agricultural output'. As a result he concluded that it was 'difficult to be certain about longer-term trends in the course of the eighteenth century'. But since 'there is no accurate estimate of the cultivated acreage over the eighteenth and early nineteenth centuries', he was forced to pursue the same methodology as other historians and economists, and resorted to supply and demand considerations, proceeding 'from the known level of prices and demand to the unknown level of agricultural output'.[55] There is, in other words, a significant gap still to be filled.

A New Approach?

If the so-called June returns had been invented in 1566 rather than 1866 agricultural historians would have had available a databank from which almost all the questions they might like to ask could be answered. In particular, the statistical record would exist from which to assess output and productivity in great detail, and thereby to release historians to estimate the relative significance of, for example, enclosure or changes in farm sizes. In the absence of such a blissful situation, they have worked the other way around, arguing that, since the population was fed, output must have increased, and that this increase has perforce to be explained in terms of the structure and the technology of agricultural practice.

This is an unsatisfactory way to proceed, and it has arisen largely because of the available data or, more precisely, lack of data. The result has been a belief that key questions in agricultural history cannot be satisfactorily answered. Patrick O'Brien concluded in 1977 that estimates of national capital formation were 'unlikely to include a satisfactory index for agriculture before the nineteenth century'. In his view, easily available and published material would simply never answer the questions which needed to be tackled. On a more positive note he predicted that the shortfall in evidence

---

[55] Daunton, *Progress and Poverty*, 29–31. Daunton's comment about the acreage is not strictly correct, and a consensus is fast growing about the level of grain yields in the late 18th century: Turner, 'Arable in England and Wales'; id., 'Agricultural Productivity in England'; R. C. Allen and C. Ó Gráda, 'On the Road again with Arthur Young: English, Irish and French Agriculture during the Industrial Revolution', *JEH* 48 (1988), 93–116, esp. 102–3; Overton, *Agricultural Revolution*, 79.

on output and productivity might be an omission rather than an impossibility: it could be done, he claimed, by 'further and more laborious research'.[56]

More than two decades on, and despite the publication of the *Agrarian History* volumes, Mark Overton attempted to pull together what he calls the 'discontinuous assortment' of statistics collected by the pre-1866 statisticians, but he concluded that on the basis of currently available material any figures for production and productivity must be 'speculative'. His own estimates, he added, would be 'superseded as historians rise to the challenge of measuring English agricultural performance'.[57]

The issue we have to address is the existence of a 'black hole' in terms of measured or measurable productivity indicators, particularly for the period between probate inventory data, which runs out towards the end of the seventeenth century, and the collection of annual farm returns from the 1860s. In cosmological terms a black hole is full of matter which cannot escape because of gravitational forces. What we have to try to show is that the analogous black hole in agricultural history is full of matter, and that although releasing it poses many problems the attempt is worth while. In the chapters which follow we first seek to release the material, then to assess its contents, and finally to suggest how it impacts on our understanding of the English agricultural and industrial revolutions.

[56] P. K. O'Brien, 'Agriculture and the Industrial Revolution', *EcHR* 30 (1977), 168, 170.
[57] Overton and Campbell, 'Statistics of Production'.

CHAPTER 2

The Farmers and their Records

The modern farmer is a businessman, running an interest which is as complex as an industrial or manufacturing company. His inputs and outputs are carefully measured, his animals go to market at precisely the right time, his milk quota is fixed by European Community regulations, his land is farmed (or set aside) according to rules and regulations which are monitored by satellites orbiting far above the earth—and he must convince his accountant that he is solvent. Farmers in the past were also businessmen, but they operated within very different parameters: without quotas, without satellites, without accountants. And in this environment they also operated without the modern obsession with paperwork. So how did they know what they had produced, over and above a crude cash calculation of year-on-year profit and loss, figures that they presumably worked out on the basis of whether or not they could pay their rent and still have a decent standard of living? What did farmers understand about the output of their land and about rising or falling levels of production? These are issues to which we can find appropriate answers only by consulting their own archives, the accounts, memoranda books, cropping books, diaries, and other records, which they generated in ever-increasing numbers over time.

But first, who were the farmers, the men and women who worked the land and did—or just as likely did not—collect written evidence of their activities? Before we can tackle key issues relating to agricultural production and output, innovation and farming practice, we have to find out something about these most central characters in the farming community. Perhaps surprisingly, the English farmer turns out to have been relatively neglected by historians. While considerable ink has been spilt on studies of the landowning community, and on the agricultural labour force, there is no substantial modern study of the farmer. It is perhaps not surprising, therefore, that the farmers' records have also been neglected. In this chapter we look initially at the farmers, and then at the records they have generated.

The Farmer

In English rural society, the farmer occupied a critical position between the larger landowner and the labourer. A broad definition would suggest that a farmer was a man or woman who worked the land, sometimes in conjunction with others (family or hired labour), taking entrepreneurial risks by investing in the full panoply of arable cultivation and harvesting, and keeping livestock, in order to produce food and raw materials for sale and consumption. The excess of income over expenditure or outgoings, loosely defined as rent, labour, and production costs, and including taxes and tithes, represented a form of profit. The farmer may have either owned all or part of the farm, or leased it from an owner in return for a rent or dues.

Contemporary social commentators including Gregory King (1688), Joseph Massie (1760), and Patrick Colquhoun (1804) appear to have had little difficulty defining and even counting farmers, as well as locating them firmly within the social structure of English rural society. Agricultural experts from Nathaniel Kent through Arthur Young, William Marshall, and the Board of Agriculture writers, as well as the authors of numerous articles in the *Journal of the Royal Agricultural Society of England*, considered the role and function of the farmer to be clear and unambiguous.[1] 'Farmers' were an occupational category in the census enumerators' books: the 1851 census records 208,119 'farmers and graziers' in England, with another 312,934 male and female relatives, including wives, working with them.[2]

---

[1] G. S. Holmes, 'Gregory King and the Social Structure of Pre-industrial England', *Transactions of the Royal Historical Society*, 27 (1977), 41–68; P. Mathias, 'The Social Structure in the Eighteenth Century: A Calculation by Joseph Massie', *EcHR* 10 (1957), 30–45; F. M. L. Thompson, 'The Social Distribution of Landed Property in England since the Sixteenth Century', *EcHR* 19 (1966), 505–17; P. H. Lindert and J. G. Williamson, 'Revising Britain's Social Tables, 1688–1913', *Explorations in Economic History*, 19 (1982), 385–408; N. F. R. Crafts, *British Economic Growth during the Industrial Revolution* (Oxford: Oxford University Press, 1985), esp. 11–14; TBA 60.

[2] Figures for England from *Census of Great Britain, 1851: Population Tables*. LXXXIII (1852–3).

Despite the obvious importance of the farmer in English rural society, the literature is remarkably thin. Many years ago Fussell wrote various volumes and articles about farming, and increasingly had a great deal to say about farmers, but essentially, he did so by recycling a rather meagre and superficial collection of knowledge slowly accumulated both by himself and by others before him.[3] More recently, Mingay and Holderness have written briefly on Victorian farmers[4] but, perhaps surprisingly, volumes v and vi of the *Agrarian History of England and Wales* virtually ignored the seventeenth- and eighteenth-century farmer. In volume v (1640–1750), there is no specific section devoted to farmers and the terms 'farmer' or 'farmers' does not appear in the index. As a social category 'farmers' are discussed in about six pages in volume vi (1750–1850) and then in the context only of social structure and farm sizes. From the index it is apparent that no section of the book is devoted specifically to farmers.[5] By contrast, nearly 100 pages are devoted to landlords and landownership, and two chapters covering nearly 200 pages to labour.[6] Only in volume vii (1850–1914) is a specific chapter devoted to the farmer.[7] There is no single volume on farmers to match Alan Armstrong's book on farm labour,[8] or the substantial literature that has been compiled apropos of the farm servant and the commoner.[9]

---

[3] Museum of English Rural Life, *G. E. Fussell: A Bibliography of his Writings on Agricultural History* (Reading: University of Reading, 1967).

[4] G. E. Mingay, *Rural Life in Victorian England* (London: Heinemann, 1976), chapter 3; id. (ed.), *The Victorian Countryside*, 2 vols. (London: Routledge & Kegan Paul, 1981).

[5] *AHEW* v (1) and (2); *AHEW* vi.

[6] In case it seems that we are singling out these two books for specific criticism, it is worth adding that farmers as a distinct category warrant only four entries from two separate articles in the index to the first twenty-five years of the *Agricultural History Review*, fewer than, for example, references to rabbits, and many fewer than to sheep, water meadows, prices, population, peasants, and so on, though the term peasant with its greater social and political overtones surely hides a lot of farmers. In Raine Morgan's *Dissertations on British Agrarian History* (Reading: University of Reading, 1981), which lists theses awarded higher degrees in British and foreign universities between 1876 and 1978, only six of the 1,502 entries specifically pick up the farmer. Two of these relate to Ireland, while of the other four only one was completed since the Second World War (entries 269, 362, 657, 684, 703, 1,256). No additions are to be found in the first two supplements to this list (*AgHR* 30 (1982), 150–5 and 37 (1989), 89–97), although there are two items in the third supplement (*AgHR* 42 (1994), 168–85). The Aug. 2000 Institute of Historical Research Web Site has four references to theses completed or in progress containing the word 'farmers', of which only two refer to places in England, and while it has twenty-one references to theses containing the word 'farming' only eleven refer to places or periods in England, though there are a few others covering Wales and Scotland.

[7] *AHEW* vii. 759–809. The chapter is by Mingay.

[8] A. Armstrong, *Farmworkers in England and Wales: A Social and Economic History, 1770–1980* (London: Batsford, 1988).

[9] A. Kussmaul, *Servants in Husbandry in Early Modern England* (Cambridge: Cambridge University Press, 1981); J. M. Neeson, *Commoners: Common Right, Enclosure and Social Change in England, 1700–1820* (Cambridge: Cambridge University Press, 1993).

Implicitly farmers appear time and time again in the study of agriculture, but they have been given little specific attention either as a social or as an economic group. They appear by default in the use of terms such as owner-occupier, peasant, and yeoman, but these are imprecise categories, and some of them were not used by contemporaries or, in the case of 'peasant', were used in a different context from farmer.[10] Of course 'farmer' is, in many respects, an all-embracing term for the many different people who farmed the land: large farmers and small farmers, those who paid a rack rent to a landlord and the great variety of those who did not—copyhold, customaryhold, and leasehold tenants and so forth. The complexities of tenure are slowly being made to yield up their secrets,[11] and we are well aware that 'farmer' came to be used as a shorthand term for individuals working the land only in the seventeenth and eighteenth centuries. The reason is clear from the etymology of the word. A farm was a fixed annual payment as a rent, originating in the thirteenth century. As a verb 'to farm', it gave rise to 'to rent', which became obsolete in the fifteenth century, and then gave way to 'to let or lease out' in the sixteenth century. In the meantime 'farmer' came to mean the collector of revenues, or the bailiff or steward (fourteenth century), before he or she became a cultivator of a farm (sixteenth century).[12] Thomas Wilson, in discussing the commonalty in 1600, distinguished the greater yeomanry, the 'yeomen of meaner ability which are called freeholders', in distinction to 'copyholders and cottagers'. He did not use the term 'farmer'.[13]

'Farmer' came into more general usage in the language of agriculture only when the range of occupations had widened to a point where a generic term was needed to distinguish those who worked the land from those with occupations that did not involve working the land. Wilson may not have used it in 1600, but Gregory King clearly understood its meaning by 1688. He did

---

[10] J. V. Beckett, 'The Peasant in England: A Case of Terminological Confusion?', *AgHR* 32 (1981), 113-23.

[11] M. E. Turner and J. V. Beckett, 'The Lingering Survival of Ancient Tenures in English Agriculture in the Nineteenth Century', in F. Galassi, K. Kauffman, and J. Liebowitz (eds.), *Land, Labour and Tenure: The Institutional Arrangements of Conflict and Cooperation in Comparative Perspective* (Madrid: Fundación Fomento de la Historia Económica, 1998), 97-114; J. V. Beckett and M. E. Turner, *Freehold from Copyhold and Leasehold: Tenurial Transition in England between the Sixteenth and Nineteenth Centuries* (forthcoming, Leiden, 2001); M. E. Turner, 'Corporate Strategy or Individual Priority? Land Management, Income and Tenure on Oxbridge Agricultural Land in the Mid-Nineteenth Century', *Business History*, 42 (2000), 1-26.

[12] C. T. Onions (ed.), *The Oxford Dictionary of English Etymology* (Oxford: Oxford University Press, 1966), 345.

[13] L. Stone (ed.), *Social Change and Revolution in England, 1540-1640* (London: Longman, 1966), 116, 130.

not use yeoman and nor did he refer to copyholders or customary tenants, or to servile tenancies. Instead he distinguished simply between freeholders, non-gentry owning their land but farming it, and farmers, which he was presumably using as a generic for all those farming the land but not enjoying an ownership stake.[14] Other evidence also points to the term farmer as a cultivator of the soil coming into common use at about this date. The label or description 'farmer' in probate inventories occurs no earlier than 1686 in Worcestershire, and 1709 in Norfolk and Suffolk.[15] It came into common parlance even later in Cumbria, only in the 1770s and 1780s.[16] Massie in the 1760s and Colquhoun at the turn of the nineteenth century used the term farmer, although both also distinguished husbandmen while leaving vague the status of copyholders and customary tenants. We may assume that these were within the generic 'farmer', now being used as a catch-all.[17]

Farmers may have been neglected in the literature, but they were clearly recognized and distinguished by contemporary commentators. The term 'farmer' embraced by the eighteenth and nineteenth centuries a great many different kinds of 'rural entrepreneur'. At the top of the farming ladder were the 386 farmers (0.2 per cent of the total) who in 1851 were working 1,000 or more acres. Such men (and occasionally women) lived in the style of country gentry, and were the real elite, mixing easily with their landlords, hunting and shooting in their company, and acting in a style which belied their tenant status. They were often as close to being farming dynasties as the landowners to whom they paid rent. Grading down through the ranks, 5 per cent of English farmers held tenancies of 300–1,000 acres, and 39 per cent held between 50 and 300 acres. On the bottom rungs of the ladder were the 54 per cent of farmers who leased 50 acres or less, the great body of copyholders, owner-occupiers, and smallholders, many of whom enjoyed incomes so modest as barely to provide a minimum subsistence.[18] Nor did these ratios change greatly. As late as 1908 slightly over two-thirds of holdings in England were of 50 or fewer acres, and only 4 per cent were greater than 300 acres.[19]

---

[14] Holmes, 'Gregory King'.
[15] We are grateful to Professor Mark Overton for this information.
[16] L. Ashcroft (ed.), *Vital Statistics* (Kendal: Cumberland and Westmorland Antiquarian and Archaeological Society, 1992). We are grateful to Dr Angus Winchester for help on this point.
[17] Mathias, 'The Social Structure'.
[18] Figures for England from *Census of Great Britain, 1851: Population Tables*. LXXXIII (1852–3).
[19] Figures for England from *The Agricultural Output of Great Britain, 1908*, BPP, Cd. 6277 (1912–13), 4. The figures are not exactly comparable because the 1851 material is from the census

While the larger farmers enjoyed opulence and social position, at the foot of the farming ladder were farm labourers desperately clinging to the lowest rung in the hopes of turning themselves (and their families) into small farmers. Some engaged in part-time farming while also selling their labour, and they were often to be found scraping around for sufficient savings to raise the investment capital needed if they were ever to convince a landlord to take them on as a tenant. Few climbed the ladder from working intensively a handful of acres with just their own labour and that of their families, to achieve the status of a gentleman farmer with a large holding, a substantial farmhouse, and an army of labourers. Even the bottom rungs were falling off by the end of the nineteenth century. As the Duke of Bedford commented: 'the agricultural labourer has, in most places, his own special grievance. He complains, and perhaps justly, that for him there is no rung on the social ladder on which he can place his foot.'[20] How many who did make the first rung or two, only to slide off in hard times, will never be known.[21]

Almost certainly the majority of farmers, particularly those with the greatest holdings, were the eldest sons of farmers, and consequently they inherited both a tenancy and the accompanying stock or capital. Landlords seldom evicted good tenants, and they were particularly loath to part with established families who worked the land well and paid their rent promptly. Consequently farmers' sons often inherited the tenancy in a manner not dissimilar to the way the landlord acquired ownership of the property, and in both cases the training was minimal. It consisted primarily of the son learning the business from his father, perhaps by spending time on all the tasks required on a farm. A. G. Street was the son of a Wiltshire tenant farmer who left school at 16 in 1907 and returned home to work on his father's farm: 'my father made me do every job on the farm at some time or another in order that I might, from personal knowledge, be able to estimate whether a man was working well or ill at any particular job.... One assimilates knowledge unknowingly.'[22] To broaden the son's experience, his father might send him as a pupil or apprentice to work on the farm of someone recognized locally as a distinguished practitioner of the most progressive methods.[23]

---

and counts farms, while the 1908 data were compiled from the *Annual Agricultural Returns* and count holdings.

[20] Duke of Bedford, *A Great Agricultural Estate* (London: John Murray, 1897), 10.
[21] *AHEW* vii. 766.   [22] Quoted ibid. 623.
[23] Mingay, *Victorian Countryside*, i. 223; *AHEW* vii. 767.

As such this was practical training for practical men, but the general distrust of formal academic training within the farming community came over time to be seen as a problem. Farmers simply did not share in the progress which was being made by the 1850s and beyond in science.[24] Nor did they learn accounting and business practices, so that few of them had the experience and knowledge to assess in a business context which crop or kind of livestock or fertilizer was the most profitable for them to use. Rather, they tended to take a casual, almost haphazard approach to decision making, primarily determined by their capacity to meet their bills, pay their rents, and keep their bank manager sweet.[25] This lack of book learning came by the end of the nineteenth century to be recognized by the farming community as a deficiency. Of 121 Nottinghamshire parish authorities consulted in a survey carried out in 1890, 71 believed that technical instruction applied to agriculture should be given priority when educational expenditure was under consideration. As a result, the first full-time course on agriculture in the county began two years later and led eventually to the foundation of the Midland Agricultural College.[26] Christopher Turnor, who owned 24,000 acres in Lincolnshire, complained in 1911 that farmers' sons needed a sound education if they were to make the most of their holdings,[27] and the Rothamsted director, Daniel Hall, noted that 'what the ordinary farmer needs above all things is better education; and by this we mean not so much additional knowledge of a technical sort, but the more flexible habit of mind that comes with reading, the susceptibility to ideas that is acquired from acquaintance with a different atmosphere than the one in which he ordinarily lives.'[28]

The Farmers' Records

The traditional methods which were so entrenched in farming inevitably helped to determine both the quantity and quality of the records generated. As a loose generalization farms, and farm records, divide according to the

---

[24] *AHEW* vii. 598–624, and esp. 620.   [25] Ibid. 803.
[26] B. H. Tolley, 'M. J .R. Dunstan and the First Department of Agriculture at University College, Nottingham, 1890–1900', *Transactions of the Thoroton Society*, 87 (1983), 71–9.
[27] C. Turnor, *Land Problems and National Welfare* (London: The Bodley Head, 1911), 70–1.
[28] A. D. Hall, *A Pilgrimage of British Farming, 1910–1912* (London: John Murray, 1913), 440.

type of holding into two broad categories—those kept on farms under the direct control of a larger estate, and those kept by tenant farmers and owner-occupiers. The distinction is important because it may influence the way the records are used by historians. On estates, farms could have been under direct management. Home farms were often run directly by the estate in order to supply it with farm produce. This was generally both for human and animal consumption—cheese, butter, milk, meat, grain for bread or beer, for people, and oats, grazing, hay, and other items for animals. Other farms were under estate management because no tenant was available to rent the holding—they were 'in hand'. Farms were also taken back temporarily into direct management in order to improve them. This may have arisen for any number of reasons, including the possibility that tenants had exhausted the soil, or the estate was making improvements, such as erecting new farm buildings or inserting field drainage, in order to raise the rents and/or attract tenants. Many landowners required managers, bailiffs, or stewards to keep records, and such records often form part of estate archives.

Generally, and this expression when used of any set of farm records is used loosely, estate farm records were well kept, neat, orderly, and logical. The landowner who required records tended also to expect numerous other types of accounts to be kept—rentals, household accounts, and estate accounts, to name just three. This resulted in sets of annual records concerning the general management of farms which though not in any way standardized across the country contained at least reasonably uniform information from year to year. The type of information included in the archive varied both between owners and according to the function of the farm. Financial accounts were common, as were records of produce supplied to the estate from the farm. A number of owners were interested in the more detailed running of farms on their estates. At Blockley in Gloucestershire, Lord Northwick routinely added questions about the running of the estate farms on the accounts submitted to him by his steward.[29] On the Ford Estate near Coldstream in Northumberland, the steward was required to produce details of the cost of working estate farms for the year 1764.[30] On the Hook Estate in East Sussex records of the treatment of fields and the resultant output from the crop were kept over a number of years.[31] William Marshall is said to have been instrumental in encouraging the keeping of the

---

[29] Worcestershire RO, 705: 66 BA4221/42. [30] Northumberland RO, ZDE 19/4/17.
[31] East Sussex RO, HOOK 16/3, Hook Family Archives.

particularly good set of records on the Drake Estate at Buckland Abbey in Devon in the late eighteenth and early nineteenth centuries.[32]

By their very nature, estate farm records are functional sources of information. The material tends to be well presented, routinely kept, and easy to use. Unfortunately, the information they contain needs careful handling to determine the degree of influence the close association with the estate had on its constituent holdings. A home farm was often managed to reflect the direct needs of the estate, particularly if part of it constituted a landscaped park, so that the system of husbandry was not necessarily the most appropriate for the local environment. Labour records in estate collections often included workers who were partly, or exclusively, employed on the grounds of the estate, and consequently should not be included in a study of farm labour. Capital expenditure on estate farms was not necessarily typical of farms in general. This was especially true during periods when a farm was taken in hand for improvement because efforts to return exhausted soil to fertility would have involved greater than average applications of soil conditioners, manures, and fertilizers. Underdraining and the construction of new buildings sometimes took place during periods when farms were in hand. A number of landowners, particularly in the nineteenth century, used the home farm as a base for agricultural experimentation. The earls of Leicester at Holkham in Norfolk are probably the best known, but many others were also actively involved in the science of agriculture.[33] New crops, plant varieties, seeds, fertilizers, cultivation regimes, watering techniques, and other innovations were tried, and the results recorded in farm records. In 1819, the steward at Rushbrooke Park in Suffolk reported on the experiments into various methods of treating turnips carried out on the home farm.[34] In the 1860s on Lord Bolton's estate in Hampshire, trial plots of named varieties of corn and root crops were grown under carefully recorded rotational conditions using different sowing techniques and fertilizers.[35] All of these factors could have had knock-on effects on output, and this may have an impact on

---

[32] Devon RO, 346 M. This information is taken from 'Survey of Farm Records in Other Repositories and in Private Hands', covering Bedfordshire to Gloucestershire (Reading: Rural History Centre, 1972).

[33] R. A. C. Parker, *Coke of Norfolk: A Financial and Agricultural Study, 1707–1842* (Oxford: Oxford University Press, 1975).

[34] RUL, SUF 5/5/2. These archives and the others noted in this and later chapters are listed in full in University of Reading Library, *Historical Farm Records: A Summary Guide to Manuscripts and Other Material in the University Library and Collected by the Institute of Agricultural History and the Museum of English Rural Life* (Reading: University of Reading, 1973).

[35] Hampshire RO, 11M49/69/1.

the way we interpret data collected from estate records. Farms taken in hand after periods of neglect by their tenants would logically produce yields lower than the local average. Conversely, estate farms which had benefited from inputs in the form of manures, fertilizers, improved techniques, or extra labour would equally logically (though not inevitably) produce higher than average yields.

The potential for the survival of atypical data from estate-run farms has to be offset against the fact that before the mid-eighteenth century these collections are the main source of evidence, although any record, however well kept, should be subject to close scrutiny. In addition, it is sometimes difficult to distinguish estate farm records from non-estate. The inclusion of a set of farm records in an estate archive does not guarantee that the farm was either a home farm or in hand. Occasionally such records were found when a tenanted farm was changing hands, perhaps because the outgoing tenant was not interested enough to take the records with him, or failed to do so because of an oversight. Internal evidence from the documents often provides the only clue to the status of the holding.

The second category of records consists of material generated on farms worked by owner-occupiers or tenants. These archives tend to be more serendipitous in survival, more individualist in the way that they were kept, and — consequently — more difficult to use. They were not, as might be expected, generated purely as a result of the size of the enterprise. It might seem logical that the larger the farm, the greater the farmer's need to keep records. Size suggests complexity, hence the need for a written record of the various operations involved in management. Furthermore, large farms required more capital than smaller ones, and the combination of size, complexity, and capital points towards education, albeit of a general rather than a specifically farming-based nature. It follows that we might anticipate a correlation between farm size and record keeping. One of the best and most complete archives to survive is that of the Brown family of Aldbourne in Wiltshire. The several farms on their holding amounted in total to 1,400 acres or so in the eighteenth and nineteenth centuries, and included in the 106 items in their archive are numerous account books (the earliest from 1727 to 1755), cropping books, diaries, valuations, maps, catalogues, and sundry other papers.[36] Yet size of operation was not necessarily the governing

---

[36] RUL, WIL 11. This collection has been returned to the Brown family but microfilms are retained in Reading, WIL P 482 and WIL P 713.

factor determining the keeping, and hence the survival, of records. An archive which is almost as elaborate as the Aldbourne example survives for a farm of 345 acres at Ash in Kent.[37] Of the twenty-one holdings for which we have been able to establish the total area in cultivation, the largest is Aldbourne and the smallest is Westgate-in-Weardale, in county Durham, with just under 16 acres. The median size is 170 acres and the mean is 309 acres.

Generally there was no obvious reason why tenant farmers and owner-occupiers should keep records. When income tax was introduced from 1799 to 1816, and again from 1842, payment could be based on the rental value rather than on the actual income from a farm, 'the income-tax acts assume the tenant's earnings to be measured by one-half his rent in England. ... it coincides very closely with an average return of nine per cent', and from 1897 the assumption was that net profits amounted to one-third of the annual value of the holding or one-third of the rent. Even farmers losing money seldom kept accounts to demonstrate this loss for use as evidence to adjust the taxes they owed accordingly.[38] This remained true up to, and beyond, the First World War.

Nor did other official developments encourage record keeping. From 1866 farmers were expected to make an annual return, known today as the June returns. As a minimum the June returns recorded details of the acreages of the principal grain, root, and pulse crops, the hay and the pasture, and the numbers of animals—horses, cattle, sheep, and pigs—sometimes broken down into age groups. Unfortunately for our purposes, although collected at the farm level the material was later aggregated to parish level and the original returns destroyed, but in any case the associated requirements in submitting the returns did not suggest the need to keep any form of compulsory record. Where farmers did not keep records, their answers to the annual returns must have relied on memory. Tenant right, that is the right to compensation for unspent improvements on a farm at the end of a lease, did generate some records, particularly after it was codified by the passing of the Agricultural Holdings Act of 1875.[39] A claim from Street Farm, South Warnbrough, in Hampshire illustrates not only the compensation received by the farmer for the use he made of purchased and

---

[37] RUL, KEN 4.
[38] P. G. Craigie, 'Taxation as Affecting the Agricultural Interest', *JRASE* 2nd ser. 14 (1878), 392; *AHEW* viii. 58; S. H. Webb, *The Practical Farmers' Yearly Account Book* (London, 1891), preface by C. S. Read.
[39] J. A. Perkins, 'Tenure, Tenant Right, and Agricultural Progress in Lindsey, 1780–1850', *AgHR* 23 (1975), 1–22; Webb, *Farmers' Account Book* (1883 edn.), preface by C. S. Read.

home-grown feeds and fertilizers, but also the payments owed by the tenant for his sales of hay, straw, and chaff.[40] Before the 1875 Act, some landowners allowed tenants to claim compensation. In 1864, the outgoing tenant from Stoke Wood Farm at Meonstoke, Hampshire, was paid for tillages as well as for items such as spreading chalk in a field and folding sheep on various roots while they were fed oilcake.[41]

The first systematically collected evidence of production at the level of the individual farm was contained in the volumes of material published by the Royal Commission on the Agricultural Depression 1894–6. The Commission produced almost one hundred examples from across the country of farmers' own assessments of their farms, sometimes for isolated years in the early 1890s and sometimes for runs over the preceding twenty years. The material was primarily put together in terms of income and expenditure, profit and loss, and was based on farm accounts. But the fact that it was the 1890s before such records were even requested by Westminster is perhaps an indication of the level of interest displayed in the work of the farming community.

For the most part, those farmers who kept records did so for their own information and recorded data because they thought it might be useful to them. Their archives often took the form of aide-memoires. Unlike the more systematic books kept on estate farms, many from tenanted farms were jottings, notes, and annual lists of important numbers or events. John Kinglake was a prosperous farmer on the Somerset Levels, who kept a careful record of purchases and sales from 1784 until his death in 1809. Unfortunately for our purposes, although his record included rent receipts, loan repayments, and sales agreements on the income side, tax payments, losses on stock, and expenses at fairs on the outgoings side, they contain nothing on the cost of running the farm, such as labour or tools. However, he did note that a domestic servant who left him in 1786 took away a shirt, two blankets, four bolster cloths, one tablecloth, cheese, and 'other things to a considerable value'. After Kinglake died the account book was put to one side, but in 1874 it was reopened by James Day, a farmer at Whitstock, who used it for two years before passing it to a relative who was a butcher and who, in 1886, recorded sales to all his customers.[42]

---

[40] Hampshire RO, 3M48/XIV308.   [41] Hampshire RO, 53M63/43.

[42] Mary Siraut, 'A Somerset Farming Account Book', *Somerset Archaeology and Natural History*, 129 (1984–5), 161–70.

Records kept in this way seem ostensibly to be only semi-coherent, but they are none the less significant, and by about 1750 many farmers were generating at least some written records. In 1858, John Coleman commented that 'the necessity of keeping some kind of accounts is so generally admitted by farmers that it is unnecessary to dwell upon the importance of the subject'.[43] By the end of the nineteenth century the practice of keeping records was widespread.[44] A survey in 1964 of established farming families in the Oxford area revealed that 20 per cent of those that had been actively farming since the 1920s owned records of *historical* interest. One of the more significant conclusions resulting from this survey was the impression that 'prior to 1914 most farmers kept records of one sort or another'.[45]

Farmers were encouraged to keep records by the availability of pre-printed annual farm account books, which included space to keep a complete record of the farming year. An early example, printed by Bye and Law in the first decade of the nineteenth century, took the form of a daily journal.[46] These printed books grew in sophistication. In the early 1830s, J. C. Loudon recommended the use of *Harding's Farmer's Account Book*. This contained forms for a weekly journal of transactions, including the state of labour, of cash accounts, of management of arable land, of pasture land, and of woodland, an account of crops, of the dairy, and of stock.[47] By the mid-nineteenth century, Webb's *Farm Account Book* was widely used. With some annual variations this contained pages dedicated to: an agricultural calendar; produce in hand and estimated value; hiring of servants and wages paid; the inventory and valuation of live and dead stock; crops; weekly labour accounts; weekly expenditure and receipts; information on crops and dairy produce; a summary of produce of corn, hay, and seeds bought, of livestock bought or bred, of livestock sold, of stock sent out or taken in to be fed, of cake, manure, and chalk bought, of tradesmen's bills and other payments, of payments for rent, tithes, rates, and taxes; a general view of cultivation; a summary of labour and cash accounts; a general statement of crops; and the balance sheet for the year. Loudon found these books to 'be useful, by directing the attention of farmers to the particulars of which they should keep an accurate record'.[48]

---

[43] J. Coleman, 'Farm Accounts', *JRASE* 1st ser. 19 (1858), 122.
[44] E. J. T. Collins, 'Historical Farm Records', Archives, 7 (1966), 145–6.
[45] E. J. T. Collins, 'Introduction', in University of Reading, *Historical Farm Records*, pp. x–xi.
[46] A sample page from a Warwickshire farm for 1806 can be found in 'The farmer's daily journal and complete accountant for a farm in Wootton, Warwick, 1806–7': RUL, WAR 1/1/1.
[47] J. C. Loudon, *An Encyclopaedia of Agriculture*, iii (2nd edn. London, 1831), 790–3.
[48] Ibid. iii. 793.

Printed farm books illustrate the variety of information a farmer might have usefully recorded. They included the full range of records we might expect to be generated on the farm, although it was not necessarily in a format that is of real value to historians.

Agricultural 'experts' often disagreed as to the type of records that farmers would find it most useful to keep. For the small, that is non-estate, farm, Loudon recommended a time-book for recording labour payments and use, some memoranda books, a cashbook, and a ledger.[49] John Coleman suggested a pocket memoranda book, a journal/cashbook, a ledger, and a labour book.[50] In the 'Report of the Judges of Book-Keeping' following a prize essay competition of the Royal Agricultural Society of England in the 1880s, only two books, a diary and a farm account book, were recommended. The diary combined a cash account and a daily record of all farm transactions. The farm account book used columns to show all payments and receipts with an annual balance sheet. The judges felt that the two books, 'correctly kept, would enable the occupier of any farm to see the annual results of his farming'. To recommend that small farmers kept other types of records 'might deter them from beginning to keep any'. For those in the habit of keeping records, the report suggested that they continue to use the form to which they were accustomed.[51] Clearly little consensus existed among the experts.

In the first decade of the twentieth century the editor of the professional journal the *Accountant* commissioned a series of handbooks to be published under the heading *The Accountant's Library*. Over fifty titles eventually appeared. These were designed to provide information and advice on 'the most approved methods of keeping accounts in relation to all the leading classes of industry whose books call for more or less specialised treatment'. The volume on agricultural accounts, the fourth in the series, appeared in 1901. The author set himself two tasks, to recommend a system that was as brief as possible, and one which was 'free from complications, and therefore easily understood'. This was to be achieved by just two books of accounts. The first was to be a cashbook, including cash and bank transactions, of which the former included standard receipts and payments and the latter was made up of 'paid into' and 'drawn from'. The second was a tabular ledger,

---

[49] J. C. Loudon, *An Encyclopaedia of Agriculture*, ii. 549.   [50] Coleman, 'Farm Accounts', 141.
[51] C. Randell, W. M. Frankish, and R. A. Warren, 'Report of the Judges of Book-Keeping', *JRASE* 2nd. ser. 19 (1883), 693.

within which were columns for the capital account, accounts for plant, machinery, and other items of capital outlay, and a profit and loss account.[52]

Although farm records are a type of business record, they have one characteristic that distinguishes them from most other business archives. The farm was not only the location of the business but it was also the home of the farmer and his family, and although the records relate to the farm, the family perceived them as personal property. This distinction works against the survival of such records because when families left farms they took their personal possessions with them. It follows that the records of estate-run farms are much more likely to have survived because the documents formed part of a much larger estate archive and a much larger business. Landed families and their estates were relatively enduring and stable. Thus, while there was no guarantee that the estate records would have been kept in anything like ideal conditions, many have survived. Records from tenanted farms were most at risk of being lost. Some families remained on a farm for many generations, but this was not the norm. The death of the head of the family, particularly if the sons were minors or had left and taken farms of their own, often resulted in a family leaving a farm. In times of agricultural depression, many families were unable to continue on the farm. Prosperity also resulted in mobility as the family moved to a larger holding. Each time this happened, any records the family possessed were at risk. Leather-bound books and parchment documents had an intrinsic value that increased their chances of survival, but the majority of records were far less glamorous. Outside the landowning group, an awareness of the historical value of almost any old document is relatively recent. Few farming families had a notion of the potential value of the box of old notebooks and scraps of paper in the cellar or attic. The volumes of annual accounts made good scrapbooks for children or were useful for recording recipes, but otherwise simply took up valuable space.

While information about the farming system, the introduction of innovations, and a plethora of detail of other related issues is scattered throughout farm records, quantitative data relating to output was not always recorded. For the most part, farmers do not appear to have been specifically concerned with keeping detailed records of land productivity. This was especially true of the earlier records where yield data are scarce, and is most commonly to be found for estate-run farms. Some farmers were aware of the growing

[52] T. W. Meats, *Agricultural Accounts* (The Accountants Library 4, London: Gee, 1901).

interest among men of business and industry in the concepts of productivity and production costs. To calculate these requires output details, which, before the common use of printed account books, were not generally recorded. That is not to say output data do not exist. However, the way the information was recorded, and the intention of the farmer in keeping records in the first place, varied from farm to farm. For example, for the most part, indications of livestock weight were found in account books, indicating that the farmer's main concern was in the sale price of the animal rather than in its weight. However, there are a number of documents in which a more analytical approach was taken. James Quantock at South Petherton in Somerset had an ox weighed on a weighbridge at Illminster in 1804. The gross weight of the live animal was 15 hundredweight 2 quarters, or 1,736 pounds. Once dead, each forequarter weighed 284 pounds and each hindquarter 229 pounds, giving a dead weight of 1,026 pounds. Quantock went on to make the simple calculation that the ratio of live to dead weight was 100 to 59.[53] On the Wentworth Estate in Yorkshire seven Scotch heifers were bought in October 1773 and sold a year later. Between them they produced 144 stone 5 pounds of beef at 3s. 6d. per stone, 22 stone 1 pound of tallow at 3s. 9d. per stone, 19 stone 13 pounds of hide at 2s. per stone, and seven calves weighing approximately 4 stone each. The total value of the animals after a year was £36 16s. 2¾d.[54]

The records of some arable farmers demonstrate a very specific interest in comparative annual yields. For example, at Chilgrove Farm in West Dean, Sussex, several generations of farmers between 1753 and 1831 recorded annual grain yields along with occasional comments on harvest conditions, the weather, and the nature of the crop. In 1823, for example, barley was 'much blighted, thin, and ordinary the worst Hill year ever known'. Two years later barley produced 'much straw—but yielded badly'. In the same year wheat was 'a very large crop'.[55] The harvest of 1772 at Hampstead Norreys in Berkshire produced 'a very bad yeald this year but the wheat was very good in Boddy and it carryed a good wait'. In 1777 the 'clover overpowered the barley and was above it in bloom'.[56] At Aldbourne in Wiltshire, the Brown family routinely recorded yields for wheat, barley, and oats between 1831 and 1912. The interest of the son was sufficient that when he took over the holding from his father in 1839, he copied out the information

---

[53] Somerset RO, DD/MR 110. [54] Sheffield AO, WWM A 1380.
[55] West Sussex RO, MP 1477. [56] Berkshire RO, D/Ex 62/2.

collected by his father before beginning his own records.[57] Other farmers were less specifically interested in yield *per se*, but none the less calculated and then recorded crop output. Over time, farmers increasingly reckoned yields as part of their field and/or labour records. At Park End in Northumberland, the Ridley family recorded the crop, the variety, the quantity of seed, and the yield for each field planted between 1851 and 1855.[58] Farmers sometimes completed their annual labour record by dividing the amount of corn threshed by the area harvested, hence giving the crop yield. More often, farmers simply recorded the area harvested and the quantity threshed, leaving the historian to calculate yield. This, along with the large number of records that provide only one part of the data needed to establish yield, is illustrative of the more traditional attitude to production. Farmers knew if they had a good crop, so measuring and recording arable output was the exception rather than the rule.

The Records

With provisos as to both the quantity and quality of the written record, we turn to the various types of material that constitute the national archive of farm records. For this purpose, we can usefully, though somewhat artificially, divide the material into three categories: farm accounts, labour records, and a range of memoranda and related books. The farm accounts primarily record the financial records, such as income and expenditure; the labour books, which include wage and hiring books, provide a record of farm work and payments for labour; and the memoranda books cover a wide spectrum of records including pocket notebooks, farm diaries, weather books, crop and field books, stock books, valuations, and inventories.

Farm Account Books

The keeping of written public accounts was hardly new in the seventeenth century. The annual accounts of Crown revenues, the Pipe Rolls, survive in an almost continuous series from the twelfth century.[59] For private estates the situation was more varied. Over 2,000 accounts are extant for estates in

---

[57] RUL, WIL 11.   [58] Northumberland RO, 414/14.   [59] PRO, Exchequer 372.

Norfolk between 1248 and 1450 representing 219 different demesnes,[60] but elsewhere there are hardly any at all. The key principle that seems to have developed over time was access to information *in absentia*. Just as the Crown was absent from most parts of the kingdom from which revenue had to be raised, so landowners who were away from their estates for long periods were the most likely to require written accounts since they could not be in day-to-day control. However, even on an estate managed by a bailiff or steward who was responsible for its management, and hence answerable to the landowner, written records have often not survived.

These records may never have been numerous, but they required some form of order, and the master and steward system was developed primarily for the better running of large landed estates.[61] It was based on the charge and discharge system of bookkeeping with income on the left, expenditure or outgoings on the right, and a positive or negative balance on the appropriate side of the double page. Not surprisingly, the value of these accounts was limited:

Where all items were entered in monetary form it could, in fact, be looked upon as an elaborate cash account, presented annually and running, on large estates, into tens of thousands of pounds. It also provided a general view of the estate, and since incomes, such as rents, sales of timber, etc., were usually grouped together in some commonsense way, e.g. by manors, unchanged over the years, it also provided a kind of check on the efficiency of the estate, as it was possible to compare from year to year the movement of rents between village and village, or the proportions of rents in arrears, or, on the expenditure side, the salaries or other expenses on the upkeep of the farms.[62]

These accounts could give a few insights into the efficiency of the estate by comparing incomes and outgoings on a yearly basis. However, although it was possible to go beyond simple questions of profit and loss to 'test the profitability of certain enterprises within the estate, or even of certain

---

[60] B. M. S. Campbell and M. Overton, 'A New Perspective on Medieval and Early Modern Agriculture: Six Centuries of Norfolk Farming, *c*.1250–*c*.1850', *Past and Present*, 141 (1993), 38–105.

[61] Contemporary manuals offered advice on the keeping of accounts: G. Jacob, *The Complete Court-Keeper or Land Steward's Assistant* (London, 1713), 477–88; E. Laurence, *The Duty of a Steward to his Lord* (London: John Shuckburgh, 1727), 133–4; J. Mordant, *The Complete Steward; or, The Duty of a Steward to his Lord*, ii (London, 1761), 35–57.

[62] S. Pollard, *The Genesis of Modern Management* (London: Edward Arnold, 1965), 210; see also M. J. Jones, 'The Accounting System of Magdalen College, Oxford, in 1812', *Accounting, Business and Financial History*, 2 (1991), 141–61, who describes the standard charge and discharge system but in the context of an educational institution.

experiments, by and large the system was utterly unsuited to this and was, in fact, very rarely so used'.[63] Nor, indeed, were these accounts designed for such usage in the first place. Basil Yamey's view of the seventeenth and eighteenth centuries was that 'calculation and quantification were less important than the availability of records for ordinary administration'.[64] Barbara English, writing about the Sledmere Estate in East Yorkshire, noted that the master and steward system was 'primarily intended to make sure that the master was not being cheated'.[65]

Estate farm accounts tended therefore to be relatively straightforward. The reasons for this were apparent in the way the accounts were put together. Incomes and appropriate expenditures were not usually grouped together, or entered in a comparable form. Capital expenditure was seen only as part of the outlay in one particular year, 'grossly and fatally distorting any possible conclusions as to efficiency or even profitability'. Similarly, 'the return on investments could not be isolated and calculated, and there was no guidance on stocks, which might be run down or built up without showing in the accounts, yet greatly distorting the "profitability" of the estate'.[66]

Such accounts worked for the landed estate, even if they make little sense in terms of modern accounting methods, and since estate stewards dealt with landowners as well as farmers these were probably the type best known in the countryside. However, these accounts had built-in assumptions. For most landowners the principal concern was cash flow, and statements of assets and liabilities were of little meaning to a life tenant since the major asset—the land—would in all probability have been either inherited or have come into the family through marriage. Therefore, except for any property left out of settlement for the purpose of sale or mortgage, it had no meaningful market value. Consequently the major asset would not have been expected to bear a return in the normal manner.[67]

---

[63] Pollard, *Genesis*, 211.   [64] Quoted in Jones, 'Magdalen', 144.
[65] B. English, 'On the Eve of the Great Depression: The Economy of the Sledmere Estate 1869–1878', *Business History*, 24 (1982), 32. Some landlords went to considerable lengths to ensure loyalty, and therefore trustworthiness, in their stewards. On this point see J. V. Beckett, 'Estate Management in Eighteenth-Century England: The Lowther–Spedding Relationship in Cumberland', in J. Chartres and D. Hey (eds.), *English Rural Society, 1500–1800* (Cambridge: Cambridge University Press, 1990), 55–72.
[66] Pollard, *Genesis*, 211.
[67] C. J. Napier, 'Aristocratic Accounting: The Bute Estate in Glamorgan, 1814–1880', *Accounting and Business Research*, 21 (1991), 163–74; Jones, 'Magdalen', 149.

Farmers keeping accounts were likely to record their ongoing business transactions simply as a written reminder of actions undertaken, and at best some of them noted the arithmetic difference between costs and revenues. In the absence either of bookkeeping education or of clear guidelines as to the best accounting procedures that were available, the quantity and quality of farm accounts vary enormously. At one level we can identify a man like Edward Strutt, who took charge of the 9,000-acre Essex estate of his brother Lord Rayleigh in 1876, and was subsequently to impress Rider Haggard as one of the most skilful farmers in the country. More than anything his success was based on a remarkable business acumen. He kept exact accounts of the profits produced by each field, and he regularly measured the yield of each cow so that unprofitable beasts could be slaughtered and the average yield thereby raised.[68] The importance of keeping records of this quality was recognized by experts. Daniel Hall argued that farmers particularly in the 150–500 acre bracket would find that 'a sound system of bookkeeping would open the farmer's eyes to wastages and mistakes of policy'.[69] Christopher Turnor argued in 1921 that any measure of the effectiveness of scientific methods was hampered by the absence of exact, 'carefully kept accounts'.[70] Yet these were voices crying in the wilderness. Modern historians have concluded that whatever the theory, the practice simply did not compare favourably. Mingay has written that:

Farmers rarely kept the kind of records which would have enabled them to gain an accurate idea of either the extent of their capital investment in the business or the return it earned in the form of profit.... Large numbers in any case lacked the education and expertise to keep any kind of books, and indeed rarely engaged in correspondence.[71]

And, according to Holderness, 'the defects in farm accounting and the infrequency of their survival suggest that many, probably most occupiers knew little of the ratio between profit and spending money'.[72] Consequently, where they survive at all, farm accounts tend to be inconsistent, haphazard, and generally very basic. They were likely to record everyday transactions but not to examine accountancy issues such as cash flow and the return on investment.[73] As the Royal Commission on Agricultural Depression

---

[68] *AHEW* vii. 775.   [69] Hall, *Pilgrimage*, 440.
[70] C. Turnor, *The Land and its Problems* (London: Methuen, 1921), 132.
[71] *AHEW* vii. 788.   [72] Ibid. 912.   [73] Ibid. 923.

concluded in the 1890s, accounts, where they survived at all, were the product only of 'men of exceptional business capacity'.[74]

In view of the lack of formal education, if farmers received any advice at all on accounting practices it is likely to have come from their landlord's agent, and therefore to have reflected the in-built assumptions of estate accounts. This is reflected in two separate methods we can identify from the records.[75] First, many farmers kept chronological records of their income and expenditure. When maintained in this fashion there was generally no attempt to arrange entries into type of transaction or even to distinguish between those related solely to the farm and those essentially of a domestic nature. In his expenditure accounts for October 1845, a farmer at Kirkby Thore in Westmorland included entries for the purchase of pigs, a shorthorn calving cow, seed wheat, a half-year's hiring, and threshing and shearing corn, together with ordinary domestic purchases of tea, sugar, salt, coffee, a pair of new clogs, and a gallon of gin.[76] This intermixture of household domestic expenditure and specifically farm-related expenditure abounds in farm records, and it is often difficult to separate the two.[77] If the keeping of home and business accounts was felt to be non-businesslike behaviour then we might expect to see it discouraged in printed account books, but some of these produced in the nineteenth century continued to include a section for 'Household Expenses'.[78] Chronological accounts provided the farmer with a check on cash flow. Over the course of a year he would be able to check if the holding was in credit and compare this with previous years. Occasionally, particularly once printed account books were in use, an annual inventory of stock was taken so that the actual profitability of the enterprise could be assessed, but this was a rarity and, in any case, farmers did not necessarily incorporate the information into their other records.

---

[74] RC on Agriculture, *Final Report of Her Majesty's Commissioners Appointed to Inquire into the Subject of Agricultural Depression*, BPP, C. 8540, XV (1897), 30, para. 109.

[75] E. Jones, *Accountancy and the British Economy 1840–1980* (London: Batsford, 1981), 23–37, makes the point that new accountancy practices were primarily brought into play during the industrial revolution period by the practical demands of industrial and particularly railway businesses. Estates, and individual farms, did not deploy the scale of capital that required more sophisticated methods of accounting. In this context compare the more modest ambitions proposed by Meats in 1901 for keeping agricultural accounts in *Agricultural Accounts*.

[76] RUL, WES P 242.

[77] For other examples see Somerset RO, DD/X/PPP; RUL, WIL 11.

[78] See for example RUL, BER 20/20/1, Edward Roe's Improved Farmers' Account Book.

The second accounting method commonly used recorded entries according to client or product. John Ellis, of Beaumont Leys, Leicestershire, continued the method used by his father Joseph in the years 1807–13. His accounts were arranged primarily according to the individuals with whom he usually dealt. For example, he sold corn to Thomas Pratt at Ansty Mill, for which he was usually paid in cash. However, on occasions, he was paid in flour or bran. It is possible that in a chronological account such exchanges would go unrecorded. If not, the relationship between the sale and the purchase would be obscured by this barter system.[79] Other interesting relationships become apparent from the accounts. William Smeeton of Sibbertoft regularly purchased cows and cattle from Ellis. This was a long-standing arrangement that began before Ellis took the farm from his father. The accounts show a growing debt owed by Smeeton as well as an apparent loan of £700 advanced by Ellis in 1813. In 1814, the debt, including £31 10s. in interest, was cleared. Part of the payment for the livestock was again in kind. William Smeeton also charged Ellis expenses for taking the cattle to London for sale and occasionally for grazing.[80] Using a system of this sort, a farmer was instantly able to see details of outstanding debts owed to him, the income that had been made from the year's crop sales, and what he was spending on different inputs. More typically, however, the farmer did not include this stock inventory information in his records.

Farmers sometimes used more complex methods. A small number of accounts employ double-entry bookkeeping. This was a system which recognized that each business transaction had a twofold aspect to it in which one party received a payment and another party made the payment. Thus the purchase by a farmer from a blacksmith or other service provider was a

---

[79] Note the same point in S. Wade Martins and T. Williamson, 'Labour and Improvement: Agricultural Change in East Anglia, c.1750–c.1870', *Labour History Review*, 62 (1997), 277. In their case this was in the context of the labour services provided by and for neighbours in which payment was in kind or in reciprocal labour services.

[80] RUL, LEI 4/1/2. John Ellis (1789–1862) farmed 375 acres at Beaumont Leys 1814–49. He was a witness to the parliamentary Select Committee of 1836 on the State of Agriculture. In answer to the opening questions about where he lived and what he did he explained that he lived in Beaumont Leys, near Leicester, where 'Until the last three years I have been a farmer entirely; I still occupy about 400 acres of land, but I have other business now; I am a coal and lime-merchant, and I am concerned in the cultivation of a farm at Chatmoss, in Lancashire': *Third Report from the Select Committee Appointed to Inquire into the State of Agriculture with the Minutes of Evidence, Appendix and Index*, BPP, VIII, part 2 (1836), Qs. 10696–9. Subsequently he was MP for Leicester (1848–52) and Chairman of the Midland Railway (1849–58), L. Stephen and S. Lee (eds.), *Dictionary of National Biography*, vi (Oxford: Oxford University Press, 1921–2). Additional material on Ellis is in Leicestershire, Leicester, and Rutland RO, 28D64.

credit to the blacksmith, and a debit to the farmer. The sale of some crops or animals or anything else by the farmer to a buyer was a credit to the farmer and a debit to the buyer. Consequently double-entry bookkeeping involved parallel pages of debit and credit. Accounts from Bocking in Essex and Beaumont Leys in Leicestershire both appear to contain elements of double-entry accounting. In the latter the farmer set off each of his debts (or debits) with an equivalent amount of goods that he sold and vice versa. Each sale or purchase was credited or debited as many times as it took to pay off the debt or retrieve the credit.[81] Those contemporary observers who advised farmers on this point differed in their opinions. Loudon argued, 'In fact, there is no correct mode of keeping accounts but by the principles of double entry.' John Coleman, on the other hand, believed that the method tended to be more sophisticated than was either necessary or practical for use on a farm.[82] The problems involved in using a double-entry system at farm level are obvious. Each of the inputs would have been assigned to a single product. Thus, a fertilizer or payment for labour on turnips would be assigned to that crop when, in reality, it would benefit the entire rotation as well as any stock fed on the crop. Other farmers used various systems that combined chronological and thematic account books. A number of books—a daybook, a cashbook, a journal, and a ledger—were sometimes advocated and occasionally used. At Goss Hall Farm in Kent cash- or daybooks and ledgers were used concurrently.[83] Other account books indicate that the transactions from one book were transferred into another volume, but the latter has since been separated from the archive or lost. An eighteenth-century account book from Tudeley, Kent, notes that the accounts were transferred into a cashbook, but this is no longer with the archive.[84] An advantage of more complex accounting lay in the potential both to track the profitability of the entire holding over time and to check on the financial well-being of each element. At a glance the farmer could assess his income and expenditure and identify late payments, unprofitable crops, and so forth. Printed farm account books expedited this by providing pages for an annual valuation and inventory, as well as for weekly cash accounts, and for more detailed recording by type of produce and expenditure.

Such was the development of account keeping, and of printed account books. We have to ask whether, from the viewpoint of a farmer, such

[81] RUL, ESS 11/1.
[82] Loudon, *Encyclopaedia*, iii. 793; Coleman, 'Farm Accounts', 133-5.
[83] RUL, KEN 4/3.
[84] RUL, KEN 13/1/1.

developments made them better farmers. In general, the reality is that they did not keep their accounts in such a way that we now know might have helped them to understand more clearly the nature of any problems they faced. As P. J. Perry has argued, contemporary understanding of the causes of the great agricultural depression at the end of the nineteenth century was hampered by farmers' inadequate accounting skills: 'even [Edward] Strutt had to admit that no man could keep cost accounts and yarded cattle except a dealer; the problem was as much technical as personal.'[85] This general conclusion is borne out in detail by the reports of assistant commissioners submitted to the Royal Commission on Agricultural Depression in the 1890s. Dr W. Fream, reporting on Andover and Maidstone, generally thought that farmers kept accounts that were sufficient for their own use, and that the best accounts were kept by those with some existing experience in commerce.[86] Jabez Turner, reporting on Frome and Stratford-upon-Avon, stressed the unwillingness of farmers to keep accurate accounts.[87] Other assistant commissioners noted a gulf between large and small farmers, among them Mr R. Hunter Pringle, reporting on Essex, and Mr A. Wilson Fox, reporting on Lincolnshire, Suffolk, and Garstang and Glendale.[88] The Royal Commission concluded that only the substantial farmers kept good accounts, and that a representative cross-section could not be found. Clearly, the financial records kept by most farmers simply did not match up to the high standards expected by the men who sat on government commissions.

The truth, plainly enough, was that the farm accounts were not primarily a measure of accountability. Formalized accounts were of limited use to those farmers who were not accountable to others for the running of their farms. The records they kept may have been influenced by estate record-keeping practices and the information about accounting methods was available to farmers if they chose to seek it out, but they were not preparing documents for scrutiny by others, whether for the Inland Revenue, or simply for their

---

[85] P. J. Perry, *British Farming in the Great Depression, 1870–1914* (Newton Abbot: David & Charles, 1974), p. xxiv.

[86] RC on Agriculture, *Andover and Maidstone: Report by Assistant Comissioner Dr W. Fream*, BPP, C. 7365, XVI, part 1 (1894), 10.

[87] Id., *Frome and Stratford-on-Avon: Report by Assistant Commissioner Mr Jabez Turner*, BPP, C. 7372, XVI, part 1 (1894), 28, para. 32.

[88] Id., *Isle of Axholme and Essex: Report by Assistant Commissioner Mr R. Hunter Pringle*, BPP, C. 7374, XVI, part 1 (1894), 53, para. 108; *Lincolnshire: Report by Assistant Commissioner Mr A. Wilson Fox*, BPP, C. 7671, XVI, part 1 (1894), 63, para. 105; *Suffolk: Report by Assistant Commissioner Mr A. Wilson Fox*, BPP, C. 7755, XVI, part 1 (1894), 63–4, para. 82; *Garstang and Glendale: Reports by Assistant Commissioner Mr A. Wilson Fox*, BPP, C. 7334, XVI, part 1 (1894), 84, para. 20.

landlords or landlords' agents. The use of double-entry accounting terminology—*cr, dr, to, by*—suggests that many farmers had some knowledge of the principles of accounting. However, these men had limited time and resources. Simple pragmatism resulted in farmers recording what they found useful rather than what they were told by experts to note. Their understanding of modern accountancy methods was weak, and so, like all other farm records, accounts were individualistic, sometimes to the point of being idiosyncratic. Their value lies in what they reveal about the mentality of the farmer, his farming, and his relationship to his environment. In general they do not offer much by way of deep insights into his financial solvency.

Labour Records

Documents relating to the use and payment of farm labour are a second general category of farm records. These include *hiring books, wage books*, and *labour books*. The last two, in particular, are important records for the agricultural historian. Where the farmer kept no formalized labour records, the payments he made to his workforce, particularly for task work, and notes regarding the terms and conditions of the annual hiring, were often scattered through pocket notebooks and other memoranda books. Details regarding wages, especially the weekly totals of his wage bill, were often included in account books. As labour payments grew more complex, it became increasingly common to use dedicated volumes to record details of labour and wages. The presentation of these records tended to be more consistent than other types of farm records. This can be deceptive, particularly in the case of *wage books*. These contain lists of labourers and their weekly payments. Superficially they appear to be an excellent source from which to calculate the annual labour cost of a farm, and the distribution of labour use throughout the year, as well as the relative pay of each worker and the number of weeks worked each year. The difficulty is to be sure that they accurately distinguish between full- and part-time or occasional workers. The wage book for the Oakes farm, at Norton in Derbyshire, may well be such a record. On a weekly basis it lists the labourers and their weekly wages and their per diem rate. A select few of the labourers were paid on a four-weekly cycle, and for an equally select few there are indications of the nature of their work. Since this appears to be a full list of the labouring year we can distinguish between the full-time or regular, and part-time or occasional, workers. Therefore in 1772, there were five full-time labourers who each worked a six-day week, more or less, and therefore each accumulated about 300 days. These five workers represented around 42 per cent of the total days worked on the farm, as recorded in this record (just over 3,600), and they received about 53 per cent of the recorded labour bill. Only three other labourers worked for more than 100 days in a year. Most labourers worked odd days and half-days, often on a seasonal basis. This was a rhythm that can be picked up in other years in the 1770s. Women, for example, were employed in late April spreading muck and gathering sticks and stones, and again in the first week of October when, together with their children, they were to be found shearing corn.[89]

There are ways in which the material from wage records is problematical, and has to be treated with care. They often only record payments of weekly or daily wages, whereas not all work was waged. Many jobs—threshing, hedging, and harvesting for example—were paid by the task.[90] Payments for such work were sometimes entered into farm account books but often omitted from the wage book. This may easily be the case with the Oakes farm. Therefore a reasonable precaution when using wage books is to check the payments at busy times of the year, especially at harvest, to see if the expected increase in labour payments and number of people employed did occur, which indeed was the case with the Oakes farm. Although this is no guarantee of accuracy, it does begin to give clues to the use that the farmer made of the wage book as well as to his use of casual labour or of workers from other farms. Other payments associated with wages were also made and often not recorded in the wage book. Perquisites were detailed in annual contracts and the costs of such benefits were occasionally noted in other records such as account books. Such expenditure was rarely entered in the wage book. Sometimes labourers on an annual contract received a lump sum payment in addition to the weekly wage. Thus, any attempt to compare the wages of different workers, as recorded in a single volume of accounts, may be seriously flawed. Using the wage book to determine relative wages is further complicated by the custom of including wages of some family members, particularly those who were irregularly employed, in the wage of the head of the family. Such entries might mistakenly be reckoned to represent

---

[89] Sheffield AO, OD 1518. The same archive has been used by G. Clark and Y. Van Der Werf, 'Work in Progress? The Industrious Revolution', *JEH* 58 (1998), 830–43, esp. 837. In our view their use of this archive is flawed because they failed fully to accommodate the distinction between full- and part-time work. A related issue is the problem of accounting for the labour of the farmer and his own family.

[90] Wade Martins and Williamson, 'Labour and Improvement', 277.

particularly high wages paid for some kind of unspecified special skill. Where they exist, annual contracts provide the details of the work to be provided by the family members of a regular employee.

A. Wilson Fox conducted a series of detailed examinations of farm labour and wages in the late nineteenth century. He made a particular point of itemizing the various perquisites and payments in kind associated with agricultural employment, all of which raise questions and present problems in modern studies of agricultural labour when the farm records do not record such details. Thus:

Men in charge of animals, in addition to higher wages than ordinary labourers, usually get more payments in kind. The married men generally have cottages and gardens free, and frequently have potato ground. Among the other allowances sometimes given are straw for pigs, coal, milk, vegetables, and certain allowances for food, beer and cider. They also often receive some extra cash payments, such as lamb money, journey money, and Michaelmas money.[91]

Unlike wage books, which give only names and payments, other records include more detail of the work that was performed on the farm. They range from haphazard lists found in memoranda books to detailed volumes of labour books kept specifically for the purpose. The complexity of these detailed volumes developed over time. By the mid-nineteenth century it was common to list daily tasks performed by employees along with payments and the number of days, or half-days, worked in a week. Pages set aside for the weekly record of labour constitute the bulk of the printed farm account books. When an accurate, daily record of the work of each employee on the farm was maintained, labour accounts were the most time-consuming of all records kept by farmers, and consequently they are a particularly useful source for details of farm labour and farming systems. With this information, it is often possible to determine whether a job was paid at a daily rate, by the task, or by a combination of task and daily payments. While these volumes tend to contain more information, and are a less ambiguous source than the wage books, some of the problems encountered when using wage books are repeated. For example, it is difficult to determine without supporting documents whether the payments that were made to a labourer accounted for his or her total income. Fortunately, because details of the jobs that were

---

[91] A. Wilson Fox, 'Agricultural Wages in England and Wales during the Last Half Century', *JRSS* 66 (1903), 283. See also his *Second Report by Mr A. Wilson Fox on the Wages Earnings and Conditions of Employment of the Agricultural Labourers in the United Kingdom*, BPP, Cd. 2376 (1913), esp. 18–23.

performed are included, it is generally possible to assess whether other family members were involved in the work, through the payments for task work such as threshing, and the use of casual labour.

Labour records can also be used for other purposes. Take, for example, farm output. Harvesting was often paid on the basis of the area that was harvested, while threshing was task work paid on the basis of the bushel or quarter threshed. In those cases where the area of cropped ground was stated and the entire crop was threshed, the crop yield can be calculated simply by dividing the total quantity of grain threshed by the total area harvested. There would have been more examples if we could have been sure that the whole of the crop that was harvested had subsequently been threshed. The records from Moccas in Herefordshire, for example, give the total area harvested at one point, and then later the quantity of corn threshed. Thus the wheat crop grown on 34.5 acres in 1801 produced 569.5 bushels of threshed grain, yielding 16.5 bushels to the acre. It was not possible to make this calculation in all years because sometimes the threshing record made no distinction between old and new grain.[92] At Milstead in Kent another complication arises. Oats were routinely left in the field for the horses during harvest. Fortunately the farmer provided details of the average number of bushels the uncarted sheaves would otherwise have produced, thereby making it possible to calculate the oats yield.[93]

Another use of labour books is to investigate different farming systems. For example, information about the cultivation and use of home-grown feeds is often difficult to find. Since this kind of feed was not sold off the farm it failed to generate a direct income. Similarly, when seed was recycled, only the labour cost of sowing it was a direct expenditure. Other records may indicate if a feed crop was folded or lifted and carted, but from the labour books it is possible to reconstruct even the timing of the process as well as the fact that it took place at all. For example, at Colworth Farm in Sharnbrook, Bedfordshire, in December 1805, one of the workers, William Norman, spent almost every day pulling or carting turnips for which he was paid 1*s*. 4*d*. per day or 8*s*. per week. Other workers on the farm were threshing and dressing wheat and barley, ploughing ashes and blood, spreading mould, feeding cattle, and undertaking a range of other jobs.[94]

---

[92] Herefordshire RO, J56/III/116 and 117.
[93] Centre for Kentish Studies, U593/A3.
[94] RUL, BED 5/2/1.

Memoranda Books

Memoranda books, the third generic category, are some of the earliest farm records to have survived. The records in this category tend to be the most individualistic and idiosyncratic of all farm records. Amongst the best known of these records are *farm diaries*. The narrative descriptions of the farming calendar found in some diaries have warranted their publication as complete volumes or in abridged formats.[95] They can provide insights into the focus, concerns, and strategies of the farmers who kept them as they planned, undertook, and finally examined the results both of their routine practices and of the introduction of changes and innovations. At the same time, diaries generally lack the structured data available in many other farm records. Documents included in this generic category were often highly individualistic. As such they are not easy to use systematically in the analysis of farm production.

The memoranda book type of records was often employed to provide a means of comparing events or information across seasons. By keeping a record of what each field contained, farmers were able to see at a glance the order of crops over time and across the farm. *Cropping books* provide just this information. At Houndhill, Yorkshire, for example, the rotation of the arable was often very intensive. In the sixteen seasons from 1802 to 1817 a field of just over 6 acres was planted in wheat every other year with just one exception, 1812, when peas and oats were followed by a year of fallow before wheat was grown again. However, generally wheat was followed alternately by a fallow or a legume such as beans, clover, or peas. During the same period a number of fields appear to have been converted from arable to pasture. First the land was planted to 'seed', or rotational grasses and clovers. Then after four or five years under this medium-term ley, the field was said to be in 'grass'. The change in nomenclature suggests that the ley was considered more or less as permanent pasture. The intensity of the growth on the arable along with the conversion of arable land to pasture begs questions about the motive for the change. Was it a response to the altered economic conditions created by the end of high grain prices at the close of the French wars, or was it the result of soil exhaustion caused by over-cropping?

---

[95] Among those which have been published are J. Stovin (ed.), *Journals of a Methodist Farmer, 1871–1875* (London: Croom Helm, 1982), and S. Wade Martins and T. Williamson (eds.), *The Farming Journal of Randall Burroughes (1794–1797)* (Norwich: Norfolk Record Society 58, 1995).

Unfortunately the records tell us only what happened. They generally do not record the decision-making process.[96]

More detailed and elaborate records were the *field books*. These were used to detail the treatment of the field, often throughout the planting year. Typically, they included seeding rates, use of soil conditioners, the application of manure and fertilizer, and details of one-off improvements such as field draining. Occasionally a farmer would also include information on the results of the year's planting. When this occurred, they provide an excellent account of the system of farming on a holding. Inputs and the resultant outputs are found in conjunction with the treatment of the field in previous years and the field books also record the farmer's assessment of the crop, and the reasons why he thought that a harvest was particularly good or bad. Most of the early field books were kept on estate farms. On the Danny Estate in East Sussex in the mid-eighteenth century, for example, seed, or rotational grass, was followed by wheat planted on ridges after four ploughings. Following the wheat, the field was 'mended' and manured with both livestock and pigeon dung before being planted in the spring with barley.[97] By the second quarter of the nineteenth century, as field treatment became more involved, detailed field books became increasingly common on tenanted farms. Several particularly good examples of these records come from Bocking in Essex. The Tabor family kept one on their farm between 1847 and 1882. In 1860, Barn Field, Spring Farm, was partially pipe-drained then put down to fallow. The following spring it was planted with oats and undersown with red clover. Both the oats and the clover were reported to have been good crops. Next the ground was treated with burnt earth and sown with Halletts wheat, and this yielded 5.5 quarters, or 44 bushels, per acre. After the wheat, in 1863/4, part of the field was ridged for mangles, which yielded 7 tons per acre, and part of the field was sown with beans. In 1864 wheat was grown again and it produced another good yield. In 1865 and 1866, further draining was carried out. This time deep drains of pipes and wood were used. The record continues with similar details for both the arable and pasture on the holding.[98]

Information on arable farming can be found in many other types of memoranda book. *Granary books* record the use of various grains for animal

---

[96] Sheffield AO, EM342. See also the use of cropping books in A. D. M. Phillips, 'Agricultural Land Use and Cropping in Cheshire around 1840: Some Evidence from Cropping Books', *Transactions of the Lancashire and Cheshire Antiquarian Society*, 84 (1987), 46–63.
[97] East Sussex RO, Dan 2201. [98] Essex RO, D/Dta A16.

feed and human consumption.[99] They are more commonly found on estates and larger farms where size prompted such records. *Winnowing books* record the dressing, and also often the disposal and use, of corn crops. These are particularly useful as a source for home consumption. Arable farmers commonly kept *weather diaries*. These often detail the effect of rain, drought, frost, and wind on the progress of production during both the growing and the harvest seasons. *Daybooks* were often small bound volumes used daily by the farmer for recording informal notes about all aspects of the farm. The small, 'pocketbook' size of many suggests that they were often carried around the farm. The information in these is usually arranged randomly. One such notebook has survived for John Twynam of Whitchurch in Hampshire. He is well known for his part in establishing the Oxford Down sheep breed from Hampshire Down/Cotswold crosses. His notebook of 1845–6 gives details of his sheep-breeding activities, and also contains comments on the newly arrived railway, as well as the use he made of fertilizers, including the mixture of sulphuric acid, bones, and ash.[100]

In comparison with the arable sector the survival of farm records for livestock farms is not nearly so rich. Fewer livestock records seem to have been kept in organized volumes. Marketing details occasionally provide useful information on the size of animals, and on the length of time they were kept on a farm, and the general structure of the livestock sector as animals were bought and sold and moved around the country. In April 1755 on a farm at Milton Abbot in Devon, the farmer bought a heifer calf at Lesant which he kept for 34 weeks and sold on at a sale and resale profit of £2 2s. He also bought a heifer at Milton Damerell which he kept for 20 weeks and 4 days at £1 15s. 6d. A heifer he purchased at Hartland he kept for 21 weeks and sold at a profit of 19s. 6d.[101] At Saltmarshe in the East Riding of Yorkshire, a local farmer routinely recorded the date and place his cattle were purchased and sold. In May 1843, for example, he purchased two beasts from Weighton for £10 and £10 10s. respectively. Fourteen months later he shipped them by rail

---

[99] This is of importance when it comes to a true assessment of the total output from a farm. The grain and other home-produced products consumed by animals realized their value when the livestock and livestock products were sold, but the consumption by the farmer, his family, and his workforce of home-produced grain and other products was generally not accounted for in the general run of farm accounts. On this point see also G. A. Lee and R. H. Osborne, 'The Account Book of a Derbyshire Farm of the Eighteenth Century', *Accounting, Business and Financial History*, 4 (1994), 159.

[100] RUL, HAN 9/1/1. On his sheep-breeding activities see also G. G. S. Bowie, 'New Sheep for Old: Changes in Sheep Farming in Hampshire, 1792–1879', *AgHR* 35 (1987), 17.

[101] Devon RO, 2168 M/E2.

to Wakefield where they were sold for a total of £47.[102] This type of marketing information is relatively scarce. However, one of the best-documented details of livestock production is the weight of the animals at slaughter. Occasionally, especially when the animal was slaughtered for home consumption, this information can be found in memoranda books, but it was more often recorded in sales accounts.

A more common record of the memoranda book type relating to animals dealt with aspects of livestock breeding. A number of complete breeding histories of individual cows, sows, and mares have survived in notebooks. These are useful for demonstrating the age at commencement of breeding, the mortality rates of both the mothers and their offspring retained by the dairy, and the weight of the calves sold out of the herd. For example, from the calving book at Bloxham Grove in Oxfordshire it is possible to trace the breeding career of individual cows and the destiny of their calves. In May 1827, Julia was born of Jesse bulled by North Star. In November 1829 she was first bulled and produced a cow calf which was sold in calf in April 1833. She produced a bull calf in 1831 which was sold fat in December 1834 at 120 stone for £22, and she had another in 1832 that was sold as a calf in the same year. In June of 1832 she was again put to the bull and produced a cow calf in 1833. Shortly after this birth, Julia and her calf were sold.[103] These breeding books contain similar information for pigs. Between 1791 and 1838 Thomas Wood at Didsbury recorded the date of the birth of piglets (known as the 'brawning'), the size of litter, and the number of piglets born alive and dead, as well as the number reared.[104] Sheep breeding was rarely recorded for individual animals but it was relatively common to record information for whole flocks. On such occasions there are usually details of the date the ram was put to the ewes and the number of ewes put in with each ram, the tupping rate for the ram, the expected yeaning (birth) date, the actual date of lambing, the number of lambs born as well as the number castrated and tailed, the mortality rate in the breeding flock, and the cause of death.[105] *Sheep accounts* are useful for information about flock size and structure. The Brown family farmed near Aldbourne on the chalks of Wiltshire, and kept sheep flocks for commercial breeding. The sheep accounts for 1883/4 show that the policy on the holding was to maintain a fairly constant age structure of their flocks by

---

[102] East Riding of Yorkshire RO, DDSA 1203/6.   [103] Warwickshire RO, CR 1635/135.
[104] Manchester Local Studies Unit Archives, City Library, M62/1/ 2 and 3.
[105] Warwickshire RO, 1635/125; Somerset RO, DDX/PPP; Hampshire RO, 35M63/90; Wiltshire RO, 1787/1.

regularly culling the older animals. In January 1884, there were 231 six-tooth, 280 four-tooth, and 275 two-tooth ewes. Additionally, there were 300 tegs that had been bred on the farm the previous year. During the year, all the older ewes were sold or died, and 279 of the tegs were taken into the flock. There were also six rams in the flock. The simple purchase/resale profit of the sheep and wool during the year was £1,874 18s. 10d.[106]

Livestock feeding is another common topic found in notebooks. Many entries deal with the use of corn or roots as feeds, or they refer to purchased feeds for livestock. Sometimes this information is included in notebooks relating to the disposal of the crops grown on a holding. When this was done, it provides a means of assessing the proportion of a crop consumed on the farm. Unfortunately, the number of animals fed is rarely noted. One of the most important parameters determining or influencing the output of the livestock sector was the acreage needed to rear, maintain, and fatten an animal. While it is sometimes possible to determine the total number of animals on a holding and also the size of the farm, it is not possible to assess the proportion of the land used to feed the livestock. This is further complicated by the change during the course of a year in the number of animals on a holding. Even on those infrequent occasions when the number of animals on a given acreage was noted, the total supply of feed including purchased feeds, and feeds diverted from any arable on the holding, is not always recorded.

Other records relating to the feeding regime of animals do exist, often as part of more general notebooks or even as simple sheets of paper. Occasionally these give sufficient detail to assess the quantity and cost of feeding the livestock over a specified period, and a sense of the conversion ratio of feed into fattened animal. On 23 November 1864 a farmer from Kingston Deverill in Wiltshire purchased 100 Down lambs at Marlborough Fair at a cost of £215. These animals consumed 2 bushels of corn and 9 hundredweight of mangolds, per 100 sheep, per day. They were sold in three lots: twenty on the following 28 January; six on 22 February; and seventy-three on 6 March. One of the sheep was ill. These sales realized £280 7s. 6d., and the cost of feeding the animals was £41 14s. 9d., making a simple profit (but not including any rent or labour or other charges in keeping the sheep) of £23 12s. 9d.[107] In Northumberland in the 1830s it was estimated that 220 acres of grass would maintain a flock of 400 ewes along with ten yearling

[106] RUL, WIL 11.   [107] RUL, WIL 6/2/2.

beasts for a year; while 100 acres of two- to three-year-old grass would maintain 24 cows with 20 young cattle of two to three years along with 200 gimmers (female sheep after the first clip).[108] In October 1855 the farmer at Histon in Cambridgeshire detailed the cost of keeping four cows. Between them they consumed 8 bushels of corn per week for five months at a total cost of £30; 41 pounds of cake at £12 10s.; 141 pounds of trefoil at £13; and 24 hundredweight of chaff at £24. Thus the total feed bill was £79 10s. In addition, the cost of labour in attending the four cows came to £5. The farmer paid £12 10s. for each cow and then sold them on for £25 each, but the production costs of feed and labour amounted to £21 2s. 6d. per cow. Therefore if this farmer was fattening heifers for the London or any other market then the arithmetic suggests that he was running at a net loss of over £8 per animal. Possibly he bought cows in calf and made his profit from the initial milk sales and the resulting sale of the calves. There is also the value of the manure to be considered. Conversely he might also have sold on the cows as cows in calf. The records bring out these interesting farm practice questions, but without always providing the answers.[109] At Swallowfield in Berkshire in 1808, 21.75 quarters of beans, peas, and barley were converted between October and January by three hogs and a porker into a final combined weight of 993 pounds.[110]

Records concerning the agistment of livestock give details of the number of animals sent out or taken in for feeding and the period involved, but they rarely indicate the acreage on which the animals were being fed. At Dunster Castle in Somerset in the 1750s and 1760s a farmer took into his farm cows, heifers, and oxen and fed them on grass, for which he charged at the rate of 1s. 3d. per week. For sheep he charged at 3d. per week. In March 1766 at Forestall Farm, at Burmarsh in Kent, the farmer kept thirteen calves at a rate of 4d. per week, one heifer at 1s. 3d. per week, and two others at 12d. He also kept 55 sheep at 1d. for eight weeks and 38 horned ewes at $1\frac{1}{2}d.$ for four weeks.[111]

The division of farm records into the three categories, though useful, is artificial, because most records were a combination of two, or often all three, groups. The 'farm accounts' kept by the brothers Richard (for the period 1817–35) and Henry Angier (for 1830–55) at Histon and Coton in Cambridgeshire, for example, contained weekly wage material and harvest costs as well as annual crop yields.[112] This was also true of the accounts kept

---

[108] Northumberland RO, ZSI 82. [109] Cambridgeshire RO, R51/1/17/57.
[110] RUL, BER P 322. [111] Somerset RO, DD/L 1/5/16; RUL, KEN 19/1/1.
[112] Cambridgeshire RO, R62/36; R 58/9/4/3.

by a farmer at Charlton Marshall in Dorset 1784–90.[113] He regularly assessed the costs involved in the growth of his crops and then calculated the yield. Memoranda books, and particularly the pocket diaries or notebooks, generally contain details of sales and expenses along with labour payments, hiring agreements, lease conditions, and similar information.

Conclusion

In this chapter we have looked at the farmer and we have looked at his records. With the exception of those farmers who were required by a landowner or some other person to keep accounts, they kept only those records that they found useful. More formalized records may have been advocated in the literature but surviving examples are unusual. Evidence of a pragmatic attitude of farmers to their records can be found in the printed account books. While these provided space for all aspects of agricultural production, many farmers felt no obligation to fill in the columns and pages. Thus, even these printed volumes demonstrate the fact that farm records are highly individualistic documents and hence difficult to use in a systematic fashion.

Yet there are a great many such records. In our work we examined no fewer than 2,743 records. Of these, 979 records (from 281 collections) contained data we were able to use (Table 2.1). We located at least one collection and one usable record for every English county, although with a bias towards the south and east. Best represented are Essex with twenty-two collections and Hampshire with twenty-one (Fig. 2.1).[114] The bias arises from farming methods and collection policy. Arable and mixed farming proved to be more likely to generate the type of records useful in a survey of farm production, while both Essex and Hampshire have excellent archive collections which are well indexed and relatively easily searched. Less comprehensive indexing may mean we have overlooked some records in other repositories. The chronological distribution of the archives (Fig. 2.2)[115]

---

[113] Dorset RO, D15339.

[114] Yorkshire has nineteen collections, but we have not disaggregated the county by Ridings for the purpose of this table.

[115] The figures vary from M. E. Turner, J. V. Beckett, and B. Afton, 'Taking Stock: Farmers, Farm Records, and Agricultural Output in England, 1700–1850', *AgHR* 44 (1996), 21–34. A full list of the collections consulted and records used can be found in Appendix 1.

Table 2.1 Geographical distribution of archives used, 1700–1914

County	Number of collections	Number of records
Bedfordshire	4	6
Berkshire	8	38
Buckinghamshire	6	10
Cambridgeshire	8	57
Cheshire	4	4
Cornwall	3	4
Cumberland	4	5
Derbyshire	3	7
Devon	11	43
Dorset	9	21
Durham	5	7
Essex	22	71
Gloucestershire	9	22
Hampshire	21	165
Herefordshire	1	2
Hertfordshire	3	4
Huntingdonshire	1	1
Kent	6	18
Lancashire	9	19
Leicestershire	5	8
Lincolnshire	8	100
Norfolk	16	105
Northamptonshire	6	7
Northumberland	12	34
Nottinghamshire	8	16
Oxfordshire	10	27
Rutland	1	6
Shropshire	6	9

Table 2.1 *Continued*

County	Number of collections	Number of records
Somerset	9	31
Staffordshire	3	4
Suffolk	15	20
Surrey	2	3
Sussex	5	15
Warwickshire	4	5
Westmorland	1	2
Wiltshire	8	44
Worcestershire	6	8
Yorkshire	19	31
Total	281	979

Source Appendix 1.

reflects both our own bias in concentrating on the period before the June returns were introduced in 1866, the natural process whereby more archives tend to survive the closer we come to the present, and the pressure brought to bear on farmers through time to keep records.

A simple statistical aggregation gives little idea of the scope of the material. A collection could vary from a single document to the entire archive of a farm, a large estate, a firm of solicitors, or some other business. Some collections contain annual records such as diaries and accounts, so that the number of documents from a single archive can heavily bias the overall total. The archive from Langham Farm in Norfolk consists of eighty-two diaries, and the West collection in Lincolnshire has thirty annual volumes of accounts. Other collections are more varied. The Scorer Farm collection in Lincolnshire has fifty-seven volumes containing accounts, daybooks, labour and wage records, cropping books and stock books, dating between 1836 and 1901. On the other hand quantity of records was no guide to quality, and sometimes useful material was found in a single volume. The surviving

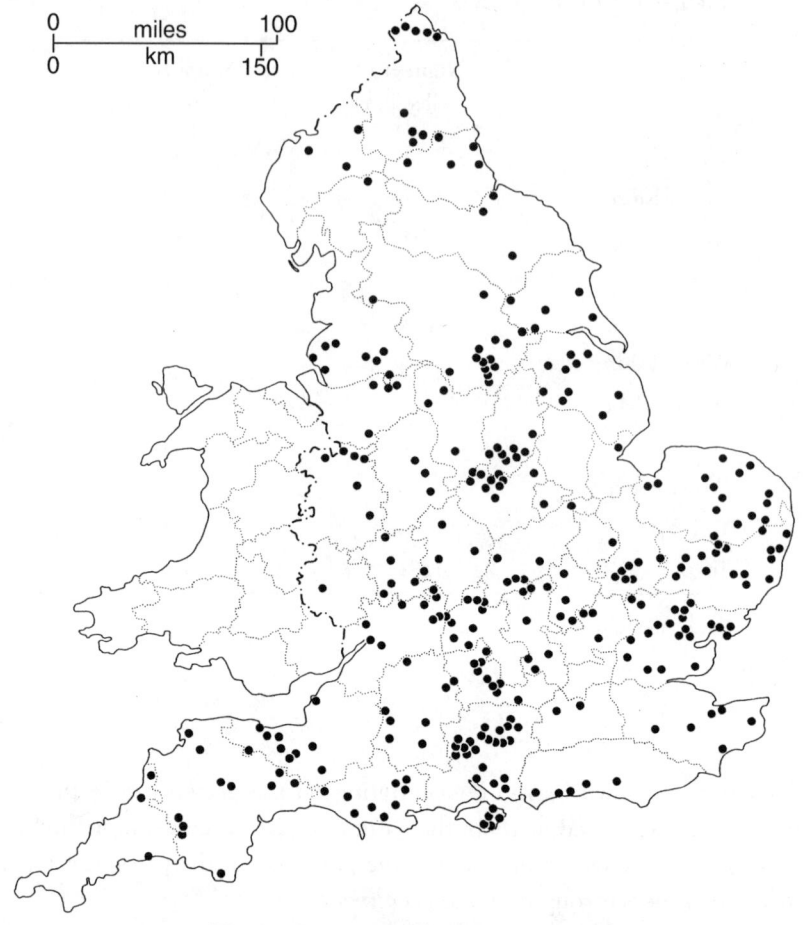

Fig. 2.1 Location of farms, 1700–1914

archive for Coton Hall Farm near Bridgnorth in Shropshire is one volume covering 1744–69, but it is an invaluable record containing annual valuations, labour records, and detailed income and expenditure accounts.

Agrarian historians have used individual data from particular farm records, whether memoranda books, accounts, or diaries, but in the chapters which follow we make the first attempt to use the records in a systematic fashion to discuss questions about farming practice, production, and output. We recognize the limitations of the material. In the case of labour records, for example, they provide considerable evidence about crop yields and carcass weights,

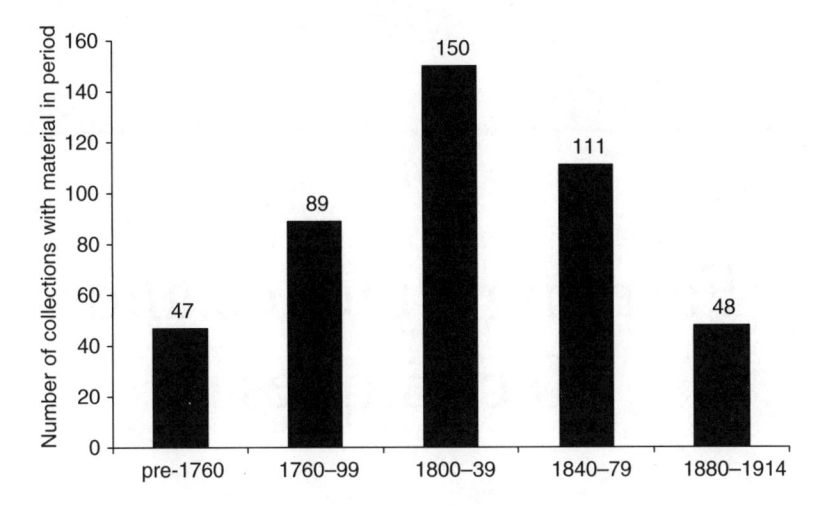

Fig. 2.2 Chronological distribution of archive collections, 1700–1914

but insufficiently consistent data from which satisfactorily to assess labour productivity. Historians usually work with incomplete data, and farm records are no exception, but they contain a great deal of data which sheds significant light on the development of agriculture in England during the period conventionally known as the industrial revolution.

CHAPTER 3

Farming Practice and Techniques

Farm records tell us a great deal about farming. This may seem obvious, but a simple demonstration of their reliability as a source is that we should be able to use them to trace the course of agricultural work, whether in the form of established practices or innovatory methods, and thereby test some of the commonly held assumptions about the business of farming. Most farmers, most of the time, farmed according to custom and tradition, but many were prepared to innovate, whether by planting new crops, experimenting with new animal breeds, replacing labour with machinery, or altering the terms under which they took their produce to market. The records allow us to examine both inputs and outputs at the farm gate level and to seek links between the two. In this chapter we use evidence from the farm records to look at the traditional practices of the working farmer, and also to seek out substantive examples of farmers innovating on individual holdings.

The Farming Cycle

Apart from those upland areas where only pastoral farming was possible, over much of the country in 1700 farmers combined the production of crops with the rearing of animals. Usually they farmed using two, three, or more courses. Perhaps the best-known traditional method of farming was the three-course rotation of (1) wheat, rye, or barley; (2) beans, peas, oats, or barley in one or more combinations; and (3) fallow. Wheat was generally the chief cash crop, and was grown every third year. Variants on this system

Table 3.1 Principal arable crops growing on English farms, 1700–1914 (% of farms)

	Wheat	Barley	Oats	Rye	Peas	Beans	Vetches	Turnips	Swedes	Mangolds	Clover	Trefoil	Sainfoin	Grasses
Pre-1760	94	94	82	35	71	41	18	53	0	0	71	29	18	59
1760–99	97	100	92	21	74	54	36	67	0	0	79	33	15	67
1800–39	100	92	97	26	70	62	58	82	26	8	86	35	12	65
1840–79	98	94	94	23	65	60	54	81	50	33	77	27	15	56
1880–1914	91	87	91	39	56	48	52	74	52	52	83	22	22	61

Source Appendix 1.

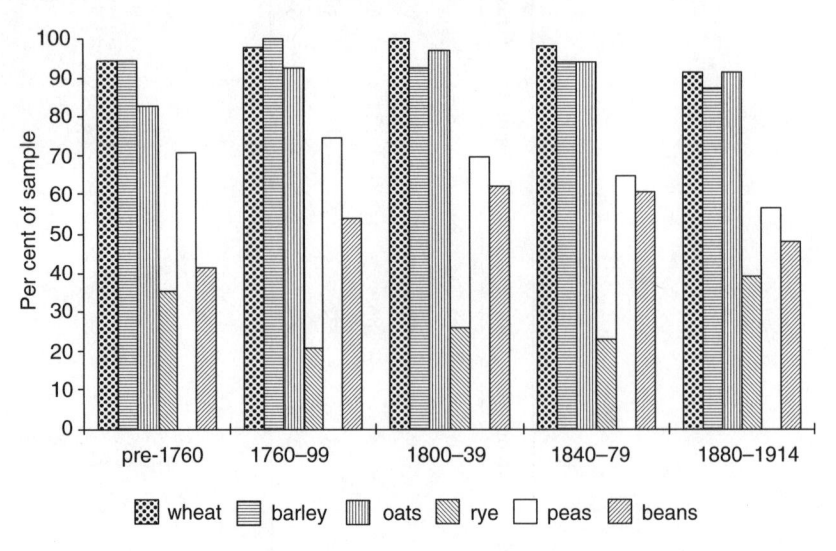

Fig. 3.1 Traditional crops grown on English farms, 1700–1914

were found across the country. Table 3.1 gives a broad indication of the arable crops grown in the eighteenth and nineteenth centuries on the farms in our sample. No attempt has been made to break the material down by region (Fig. 3.1). Wheat and barley were grown at all times on almost all the arable farms for which the evidence is clear. Only in the agricultural depression towards the end of the nineteenth century is there a suggestion that some farmers were abandoning these staples in the search for better returns on alternative crops or animal husbandry.[1] In fact, given the severity of conditions in these years, the figures for wheat, barley, and oats are remarkably high. Barley was still required both for malting and as an animal feed crop. The figures for oats may also have reflected the demand for animal feed. Rye was less widely grown, and was in decline at least until the final quarter of the nineteenth century.[2] Like oats, rye was cultivated on poorer soils often as an adjunct to livestock systems (particularly in upland areas for feed) rather than as a cash crop for the market. Rye was often grown on mossy land where oats were inappropriate. Both crops, together with barley, could be

---

[1] J. Thirsk, *Alternative Agriculture: From the Black Death to the Present Day* (Oxford: Oxford University Press, 1997), 147–61; B. Afton, 'The Great Agricultural Depression on the English Chalklands: The Hampshire Experience', *AgHR* 44 (1996), 191–205.

[2] E. J. T. Collins, 'Dietary Change and Cereal Consumption in Britain in the Nineteenth Century', *AgHR* 23 (1975), 97–115. Collins shows, among other things, that rye was no longer used in bread production.

eaten young, particularly when planted in the winter as a catch crop. This may account for the increased popularity of rye 1896–1914.

All mixed farms grew grain crops, and many also cultivated peas and beans (legumes), normally in the second year of the cycle. Beans and peas are nitrogen-fixing crops, and the improvement in yield that they helped to bring about had long been recognized. By the end of the thirteenth century legumes accounted for as much as one-fifth or even one-quarter of the cropped acreage in parts of East Anglia and the south-east.[3] Between 56 and 74 per cent of farmers grew peas throughout our period, but there was more of a trend with beans, rising from 41 per cent at the beginning of the eighteenth century to 62 per cent in the early years of the nineteenth century, and then showing a gradual downward slope thereafter. Peas and beans remained part of the cycle even though they were less effective as nitrogen fixers than small-seeded forage legumes such as clover.[4]

In the third year of the traditional open-field rotation the land was left fallow. This was the regenerative and cleaning period for the land. During this year the field was grazed by livestock, especially sheep. The dung and urine added fertility to the field and at the same time the weed growth was checked by the feeding animals. The fallow was the period of the rotation during which the field could be cleared of perennial weeds both by grazing and through repeated ploughing. This allowed the land to be brought into a fine tilth for the grain crop that followed. With no pressure to crop the field, even on heavy clay soils the land could be ploughed often enough to break down the earth and form a good seedbed. This was important, although in fact the notion of the benefit of the fallow was based on the mistaken belief, held well into the nineteenth century, that the earth was itself a source of plant nutrition. It was believed that there were four basic elements, earth, fire, water, and air, and that the soil—rather than the elements (nitrogen, phosphorus, potassium, and so on) it is now known to contain—was taken into the plant through the roots. In order for this to happen, the earth had to be finely pulverized. Hence the belief in the importance of the bare fallow.[5]

---

[3] R. S. Shiel, 'Improving Soil Productivity in the Pre-fertiliser Era', in Campbell and Overton (1991), 51–77.
[4] G. P. H. Chorley, 'The Agricultural Revolution in Northern Europe, 1750–1880: Nitrogen, Legumes and Crop Production', *EcHR* 34 (1981), 71–93. Nitrogen fixation by pulses is about half that fixed by forage crops such as clover: Shiel, 'Improving', 76.
[5] E. J. Russell, *Soil Conditions and Plant Growth* (11th edn. ed. A. Wild, Harlow: Longman, 1988), 5; G. E. Fussell, *Crop Nutrition: Science and Practice before Liebig* (Lawrence, Ks. Colorado Press, 1971), 113–49.

While the fallow gave the farmer the chance to produce a fine tilth, on many soils it could actually reduce the fertility of the land as nutrients were leached out of the topsoil. In those areas where livestock feeding was central to the farming system, alternative rotations were introduced. The use of a grass ley rather than a bare fallow was already widespread by the eighteenth century: in the period to 1760, 71 per cent of farmers grew clover and 59 per cent temporary grasses (Fig. 3.2). Typically a rotation would include two years of grain crops followed by a ley of two or more years. This rotation often meant that a fallow was unnecessary, but one could be included if the farmers found it necessary. Importantly, it provided livestock feed.[6] On poor soils and as part of an infield/outfield system the grasses could be left down for many years as medium- and long-term leys. The ley was often sown in the spring under a corn crop. When the corn was harvested the ley was already well established. This practice further limited the possibility of nutrient loss through leaching.

One of the most important features of the ley was the use of small-seeded forage legumes. Among these were clovers (*Trifolium*) and trefoils (*Medicago* and *Lotus*) and on long-term leys sainfoin (*Onobrychis*) and lucerne (*Medicago*). Cultivated red clover is believed to have been introduced as a farm crop by Sir Richard Weston in 1633. By the 1660s clover seed was 'commonlie sold in Exeter, and other easterne [Devon] marketts',[7] and it subsequently came into widespread use. The inclusion of clover and trefoil seed in the ley had the effect of increasing the nutrient supply in the field through atmospheric nitrogen fixation. Leys can be identified from farm records although their exact nature cannot always be determined because 'grass seed' was often used generically to mean any of a number of grasses and legumes, and 'clover' to mean any of a number of leguminous forage plants. In the 1720s on the Grafton Estate in Northamptonshire the steward often recommended using hayseed. On one occasion he noted that the 'most proper grass seed to be sown there are Trefoil and ryegrass together'. At other times he suggested clover or sainfoin.[8] The importance of clovers, trefoils, and ley grasses is clear from Table 3.1 and Fig. 3.2.

Farmers anxious to improve the output of their land looked at ways of altering the existing farming systems to make it more productive. The

---

[6] Some of the misconceptions about fallowing, which were only understood as science progressed in the 19th century, are explained in Russell, *Soil Conditions*, ed. Wild, 110–12, 658–9.

[7] Royal Society, 'Georgicall Account of Devonshire and Cornewall', *Philosophical Transactions of the Royal Society*, 10 (3)/12 (1667), unpaginated.

[8] Northamptonshire RO, G 3883.

FARMING PRACTICE 71

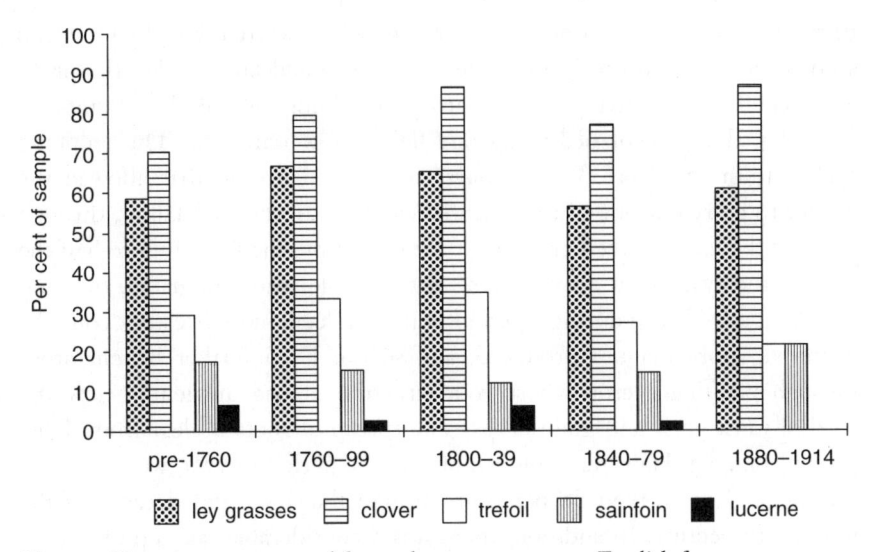

Fig. 3.2 Temporary grasses and forage legumes grown on English farms, 1700–1914

Norfolk rotation of wheat, roots, barley or oats, clover and grass ley, is perhaps the best known of the innovations. Named after its place of origin, the system, together with numerous variants, spread to many parts of the country through the eighteenth and into the nineteenth centuries. It was best suited to soils sufficiently light to be worked and fed throughout most of the year, hence its development in Norfolk where areas previously in pasture or only occasionally cropped could be brought into more regular cultivation. The Norfolk rotation reduced the acreage under wheat, but it allowed a second cash crop (barley or oats) without compromising the supply of animal feeds, and it probably increased the overall yield of corn crops. It included a grass and legume ley with the combined benefits of producing a field crop and including nitrogen-fixing plants. The major development was the introduction of a root course.

The innovative feature of the Norfolk four-course was the introduction of the cropped fallow as an essential part of the rotation, and this was where the turnip (and later swedes and mangolds) was vital. Except where a fallow period was necessary to clean the field or to break down heavy clay soils, leaving the land uncropped was inefficient. At Heath and Reach near Leighton Buzzard in Bedfordshire, those with rights in the open and common fields decided in 1814 to change the three-field system, which since 'time immemorial' had been regularly cropped in a succession of wheat, beans, and fallow, to a four-field system in which anyone could sow turnip seed or potatoes on the fallow. The three fields had not 'yielded the profit and benefit' they were thought capable of doing, hence the alteration.[9] However, the concept of leaving one field fallow remained. The cropping cycle within the Norfolk four-course introduced a crop that allowed the farmer to carry out one of the essential elements of the bare fallow, through clearing the land of weed growth. At the same time the loss of up to half the land in the system (in a two-course, for example) was eliminated, the land produced a feed crop during a period of limited feed, more livestock could be wintered, more manure produced, and so on. Even further benefit arose through planting a crop whose root structure tapped nutrients below the level of most grass and grain crops. The spread of roots is clear from Table 3.1 and Fig. 3.3. Half the farms sampled in the period to 1760 were growing turnips, and more than 80 per cent through the first eight decades of the nineteenth century. In addition, there was a considerable take-up of swedes and mangolds from the early nineteenth century, partly because they extended the feeding potential of the rotation. Swedes provide feed later into the winter, and mangolds suit heavier soils. From the 1880s the number of farmers growing turnips declined, possibly in response to higher labour costs.

The Norfolk four-course combined cash and feed crops, and cleaning and restorative crops. It was too demanding for some areas of the country, but it could be modified to suit local conditions. The use of temporary leys increased the flexibility of the rotation. When left for a second year, the ley was likely to be fed rather than mowed as a hay crop, and this increased the amount of muck on the field. Extending the rotation to five years by including a second year in temporary ley may have produced a gain in yield of up to 7 per cent, although because the acreage devoted to grain was reduced it would have brought a slight fall in total output.[10] Alternative systems were also developed. One of the most sophisticated was found in Wessex in the third quarter of the nineteenth century. It included a Norfolk rotation of wheat, turnips, barley, and ley, followed by wheat, barley, ley, and ley. A further modification produced a rotation of turnips, swedes, wheat, barley, ley, ley, wheat, barley.[11]

The impact of both the ley rotations and the Norfolk-type farming systems is clear from Table 3.1. Prior to 1760 both turnips and leguminous ley crops were being used on at least half of the farms in our sample, and these

---

[9] Bedfordshire and Luton Archives and Record Service, BO/1334.
[10] Shiel, 'Improving', 75.   [11] Afton, 'The Great Agricultural Depression', 196–7.

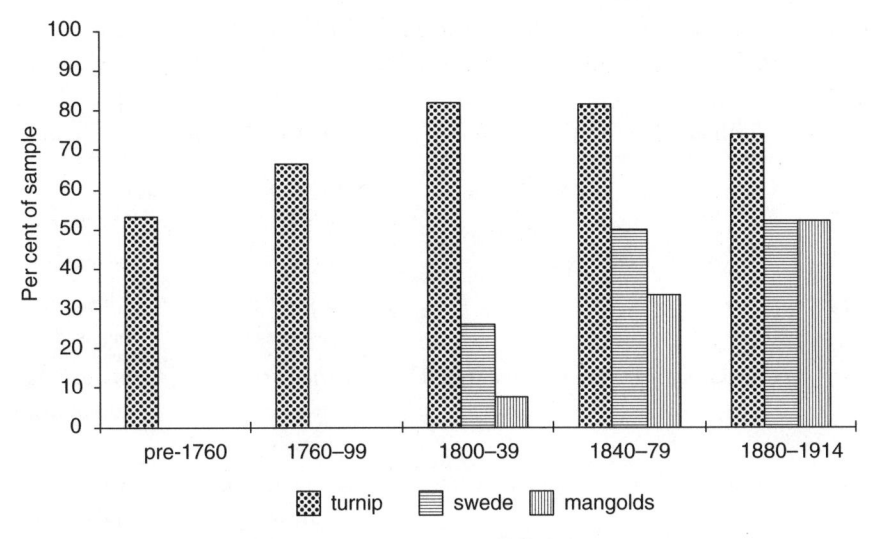

Fig. 3.3 Root crops grown on English farms, 1700–1914

were spread through twelve separate English counties from Lancashire in the north, through the Midland counties of Staffordshire, Nottinghamshire, and Leicestershire, to Somerset and Dorset in the south-west and Hampshire and Kent in the south-east. In the last county East Sutton Court Farm was using a Norfolk-type rotation in the period 1727–46. The farm contained 103 acres of arable land divided into eleven lfields, of which four are given in Table 3.2. The rotations included crops typical of the Norfolk system, while at the same time several fields were left in longer-term leys, either a seed mix or lucerne. This was the case with Pound Field, illustrated here. The accounts show that sainfoin was purchased on occasion but none is recorded in the rotational list. The extent to which new techniques had been introduced points to a farm on which many of the methods that we usually associate with the agricultural revolution were already in place well before 1750 in a county not normally considered to be in the forefront of the agricultural revolution.[12]

---

[12] Centre for Kentish Studies, U120. Our example may be a particularly early one, but Wade Martins and Williamson have recently suggested that the Norfolk four-course was widely used in the county of its origin by the mid-18th century: *Roots of Change: Farming and the Landscape in East Anglia, c.1700–1870* (Exeter: British Agricultural History Society, 1999), 99–115. They question the findings of B.M.S. Campbell and M. Overton, 'A New Perspective on Medieval and Early Modern Agriculture: Six Centuries of Norfolk Farming, c.1250–c.1850', *Past and Present*, 141 (1993), 54–66 that it was rather later.

Table 3.2 Rotations in four fields at Sutton Court Farm, East Sutton, Kent, 1727–1746

	North East Court	North West Court	Lower Bramble	Pound
1727		wheat	barley	oats and clover
1728	oats and clover	clover	clover	beans
1729	clover	wheat	buckwheat, peas, oats	wheat and beans
1730	buckwheat	oats and trefoil	barley, trefoil, clover, ryegrass	barley, wheat, and beans
1731	wheat	trefoil	clover, trefoil, ryegrass	wheat, peas, and beans
1732	oats, clover, and trefoil	oats and trefoil	wheat, oats	wheat
1733	wheat	turnips	oats	lucerne
1734	oats	barley, clover, and trefoil	fallow and oats	lucerne
1735	fallow	buckwheat, trefoil, and clover	wheat	lucerne
1736	turnips	wheat	oats	lucerne
1737	barley	barley, buckwheat, and clover	oats and clover	lucerne
1738	clover	clover	turnips and tares	lucerne
1739	wheat	wheat	barley and clover	lucerne
1740	oats	oats and ryegrass	clover	lucerne
1741	turnips	turnips	wheat and peas	lucerne
1742	barley	barley	peas, tares, and oats	lucerne
1743	clover	clover	fallow and oats	lucerne
1744	wheat	clover	wheat and turnips	lucerne
1745	oats and clover	wheat	spring corn and clover	meadow
1746	clover	oats and clover	clover	

Source Centre for Kentish Studies, U120/A17.

The adoption of a Norfolk-type rotation did not mean the end of fallowing, although it is not always easy to tell what a farmer meant when he used the term. Some clearly thought of it as a period when cleaning crops were grown, but for others it was a time when no crops were growing in the field. The period between the wheat harvest and the planting of turnips, beans, or whatever followed, could be called a fallow, but it was not necessarily a year-long period. At Doggetts Hall Farm in Essex a fallow was sometimes a full year, and at other times was simply a short period between crops.[13] William Smythe of Little Houghton Farm in Northamptonshire generally used the term to mean a bare fallow, but he could also be referring to a short period between wheat and barley, and even to a cropped fallow.[14] On the heavy soils of Harlow in Essex the farmer made a note of his rotations for the period 1798–1831. According to the cropping routine fields were often fallow. Fortunately, the cropping table is in a diary, from which we know that the fallow was routinely cropped with turnips, vetches, cabbage, and coleseed.[15]

Allowing for these difficulties, we can still locate many examples of farmers fallowing land on a regular basis. In the 1820s on two farms at Brunton in Northumberland, on a medium to heavy soil, the farmer used a fallow and three-course rotation of wheat, spring corn, fallow, alternating with wheat, clover, and fallow.[16] This was not a particularly innovative rotation, but during the same decade at Medmenham in Buckinghamshire, the farmer came close to following a Norfolk rotation, but still occasionally had a bare fallow in some of his fields. He regulated his farming by extending the output of grass, clover, and sainfoin as required, and also by undersowing spring corn with grass and clover.[17] During the 1840s a farmer at Deanshanger in Northamptonshire employed a regular Norfolk four-course in some of his fields, with wheat, turnips, barley, and clover in the rotation. Some of his fields either could not sustain this regime or else the soil was too heavy. The rotation in one field was clover, wheat, fallow, wheat, beans, wheat, beans, and fallow. Legumes were grown in each of his fields at some point in the cycle, and despite the obvious importance of the newer ones such as clover and trefoil, peas and beans continued to play a significant role in his farming system. Here was a farmer obviously prepared to devise a farming system tailored to each of his ten fields depending upon soil conditions and his experience of production (Table 3.3).[18] Bare fallowing never entirely died out.

---

[13] RUL, ESS 18/3/1. [14] Northamptonshire RO, ML 1226.
[15] Essex RO, D/DU676/1. [16] Berwick-upon-Tweed RO, 1978/3/1.
[17] Buckinghamshire RO, D85/13/1. [18] Northamptonshire RO, XYZ 1336.

Table 3.3 Rotations at Deanshanger, Northamptonshire, 1843–1850

Field size (acres)	12	6	9	7	13	10	6	4	16	4
1843	fallow	barley	wheat	wheat	wheat	turnips	peas	vetches	oats	wheat
1844	wheat	wheat	turnips	vetches	peas	barley	wheat	wheat	vetches/clover	clover
1845	clover	peas	barley	wheat	fallow	clover	vetches/turnips	beans	wheat	wheat
1846	oats	turnips	clover	barley	wheat	wheat	barley	wheat	peas	fallow
1847	fallow	barley	wheat	wheat	peas	vetches/turnips	peas	beans	wheat	oats
1848	wheat	oats	oats	fallow	wheat	barley	turnips	turnips	clover	clover
1849	beans	peas	turnips	wheat	oats	peas	barley	wheat	wheat	turnips
1850	wheat	turnips	barley	beans	fallow	wheat	clover	clover	beans	wheat

Source Northampton RO, XYZ 1336.

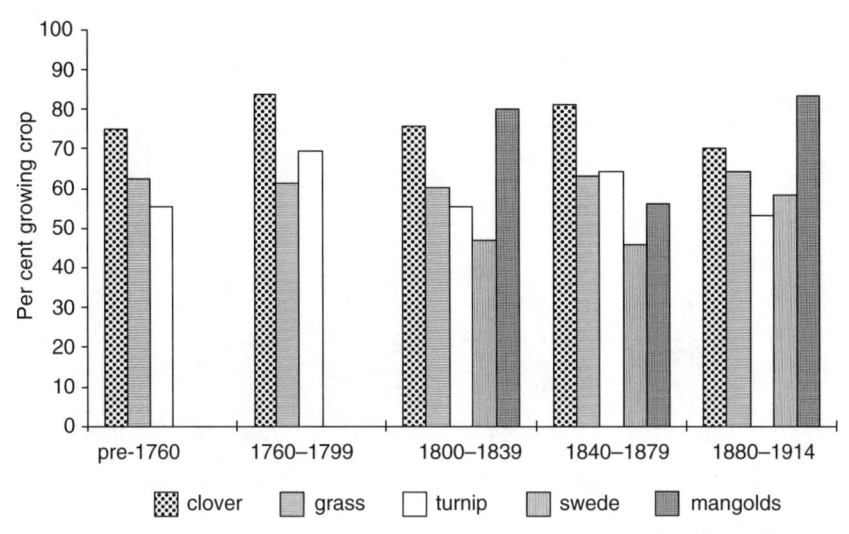

Fig. 3.4a Seeds of recently introduced crops purchased or sold on English farms, 1700–1914

During the process of tithe commutation in the 1840s 'bare fallows' were noted by assistant tithe commissioners in various parts of the country,[19] and as late as 1889 there were said still to be 'advocates of the system of bare fallowing... on very heavy clay soils, especially suited for wheat cultivation'.[20]

Along with the changes to the cropping patterns we need also to consider seed. Unlike the traditional crops, including the various corns and beans and peas, the newer crops of small-seed forage legumes, grasses, and roots were harvested or eaten before seed was produced. Thus, farmers often found it convenient to purchase or trade the seeds (Fig. 3.4a). Of farmers planting clover pre-1760, 75 per cent purchased seed on at least one occasion. The acquisition of grass seed was the next most frequent with 63 per cent of farmers recording the use of purchased seed. Some of this was bought locally but often it came from greater distances. On the Grafton Estate in Northamptonshire some of the ryegrass seed could be purchased in the locality, but clover often could not be found near the estate. In 1730 a ton of clover seed was obtained from Reading and Finchampstead. Often hayseed was used

---

[19] R. Kain, *An Atlas and Index of the Tithe Files of Mid-Nineteenth-Century England and Wales* (Cambridge: Cambridge University Press, 1986); J. V. Beckett and J. E. Heath, *Derbyshire Tithe Files 1836–50* (Chesterfield: Derbyshire Record Society 22, 1995).
[20] J. Wrightson, *Fallow and Fodder Crops* (London: Chapman Hall, 1889), 13.

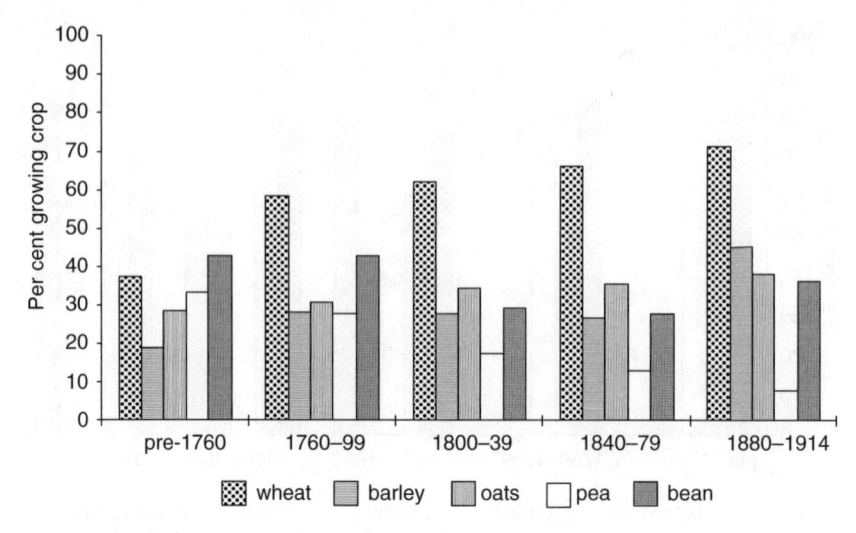

Fig. 3.4b Seeds for traditional crops purchased or sold on English farms, 1700–1914

when there was sufficient quantity.[21] A century later on the Bridgewater Estate in Shropshire, the steward commented that red and white clover seed were available locally but other grass seeds 'are imperfectly known to farmers and also local seedsmen'.[22] The recorded use of clover seed, often a generic term for any of the small-seed forage legumes, never fell below 70 per cent and of grasses below 60 per cent. Turnip seed was recorded as having been purchased by 56 per cent of those planting the crop before 1760. This rose to 69 per cent 1760–99, and then fell to between 53 and 64 per cent.

With corn, beans, and peas the position was different (Fig. 3.4b). Seed could easily be saved from the previous crop, and therefore purchase was not really required. However, there was a general if mistaken belief that a better crop could be grown if seed from a distance was used.[23] George Boswell, a Dorset farmer who corresponded regularly with the renowned Northumberland livestock breeder George Culley, noted in a letter of 9 September 1787, 'Much has been said relative to the advantage of change of seed, from a distance. Although I am not convinced by any argument I've heard; yet wishing not to dissent from every received opinion I subscribe

---

[21] Northamptonshire RO, G 3883.
[22] Shropshire Records and Research Centre, 212/bundle 361.
[23] J. R. Walton, 'Varietal Innovation and the Competitiveness of the British Cereals Sector, 1760–1930', *AgHR* 47 (1999), 32–3.

Table 3.4 Examples of crop varieties noted in the farm records

Langham, Norfolk	Tartarian oats 1770; White Loaf turnips 1815; Red Globe turnips 1855
Somersham, Huntingdonshire	Rivett's wheat 1776; Fen wheat 1771; Poland oats 1776; Fen oats 1776; Dutch clover 1776
High Legh, Cheshire	Red clover 1790; White clover 1790
Didsbury, Lancashire	Red Champion potatoes 1794; Ox Moble potatoes 1792; Marrow Fat peas 1792; Superfine Early peas 1792; Turkey long pod beans 1792; Early Sugar Loaf cabbage 1792
Upminster, Essex	Rivett's wheat 1794
Kings Somborne, Hampshire	Marlborough Grees peas 1800; Grey Partridge peas 1800; White Hedge wheat 1800; Cobham wheat 1805; Taunton wheat 1805; Kentish wheat 1805; White Dutch oats 1801; Sparrowbill White oats 1801; Yellow barley 1800; White Thanet barley 1801; White Round turnips 1805; Tankard turnips 1805; Black Siberian oats 1805; American wheat 1810
Charminster, Dorset	Golden Drop wheat 1836; Talavera wheat 1814; Broad clover 1784; Dutch clover 1784; Churchill oats 1807; Potato oats 1808; Devon Ryegrass 1803; Norfolk turnips 1815; Yellow turnips 1812; Chevalier barley 1833; Norfolk barley 1831
Swallowfield, Berkshire	Golden Swan wheat 1806; Taunton wheat 1806; Winterslow wheat 1808
Beaumont Leys, Leicestershire	Poland oats 1806; Potato oats 1808
Clenchwarton, Norfolk	Burwell wheat 1811; Globe turnip 1811; White Essex wheat 1816; Red Harty wheat 1816
Sandford, Devon	Red, White, and Dutch clover 1821
Nettleton, Lincolnshire	Tartarian oats 1821; Polish oats 1824
Histon, Cambridgeshire	Dutch clover 1823; White clover 1823; Tankard turnips 1824
Bloxham Grove, Oxfordshire	Tawneys barley 1827

Table 3.4 *Continued*

Nether Wallop, Hampshire	Taunton wheat 1831; Red Chevalier wheat 1838; Red Lammas wheat 1838; White Chevalier wheat 1838; Whitington wheat 1838; Golden Drop wheat 1840; Trump wheat 1854; Britannia wheat 1850; Juliana wheat 1850
Searby, Lincolnshire	Poland oats 1831; Early Anglesea oats 1835; Golden Drop wheat 1840
Shelford, Nottinghamshire	Chevalier barley 1836
Sompting, Sussex	Globe turnips 1845; Tankard turnips 1845; White Round turnips 1846; Dutch clover 1846; Red clover 1861
Stokenchurch, Buckinghamshire	Red Northumberland wheat 1850; Victoria wheat 1850
Thornbrough, Northumberland	Potato oats 1861; Tartarian oats 1861
Dunholme, Lincolnshire	Hallett's Pedigree barley 1870
Towcester, Northamptonshire	Rivett's wheat 1874
Preston, Rutland	Little Fois swede 1898
Shelley, Yorkshire	Hartley Short Tap swede 1902; Purple Top swede 1902

Source Appendix 1.

to it.'[24] Boswell's instinctive reaction was correct. As understanding of the theory of selective breeding advanced, it was recognized that selection rather than distance was important. However, until farmers began to buy seeds rather than to use farm-produced seed, it was not possible to introduce new varieties. The use of purchased wheat seed increased from 38 per cent pre-1760 to 71 per cent by the end of the nineteenth century.

The regular introduction of unfamiliar varieties, together with the selection, multiplication, and distribution of superior specimens from existing stock, needed conditions which really only came into play around the mid-eighteenth century. Table 3.4 shows a sample of named crop varieties appearing in farm records. For wheat, a major impetus came from the flow of grain coming into Britain once the grain surplus of the early eighteenth century turned into a

[24] Northumberland RO, ZCU 12–27.

deficit from the 1760s. The availability of new seed varieties increased from the 1830s, and numerous experiments were tried in the nineteenth century. The introduction of wheat varieties appears to have been yield led. Unfortunately, many of the wheat varieties were high yielding but did not have good bread-making qualities, and the desire to improve yields had costly consequences for the English farmer when higher-quality wheat imports arrived in the late nineteenth century. Oat production was geared to livestock feed. For this reason emphasis was not necessarily placed on high yield but on straw length and quality. The demands of malting meant that barley varieties which produced a good malting sample rather than a high yield were often preferred.[25]

Tending the Soil

One of the more important changes in farming practice during the eighteenth and nineteenth centuries was in the use of soil conditioners, and manures and fertilizers. Soil conditioners, particularly lime, were recognized by the mid-seventeenth century as 'one of the mainstays of soil fertility'.[26] Farmers found that they increased the output of the land. They discovered by trial and error that soil acidity could be corrected, or at least countered, by adding a conditioner, which was then mixed into the soil in the normal course of ploughing. They believed that lime was a fertilizer, although in fact it was principally a soil conditioner. Lime and other soil conditioners were bulky and subject to significant transport cost constraints, but the advantages they brought, especially on heavy soils, were such that farmers were often prepared to carry them considerable distances, particularly as they were applied only occasionally.

Figure 3.5 shows that lime was already used on a considerable number of farms by 1760 and that its deployment increased subsequently. At East Sutton in Kent in 1722, thirteen loads of lime at £17s. 6d. a load (105 baskets—of 6 gallons each—to the load) was delivered 'upon the ground' and spread at the rate of one load per acre. Over the next five years the farmer used another forty-four loads at a total cost exceeding £78.[27] The use of lime peaked in the

---

[25] Walton, 'Varietal Innovation', 47–50. The impact of new seed varieties for yields are discussed in more detail in Chapters 4 and 5 below.
[26] C. Spedding (ed.), *Fream's Agriculture* (16th edn. London: John Murray, 1983), 178; M. A. Havinden, 'Lime as a Means of Agricultural Improvement: The Devon Example', in C. W. Chalklin and M. A. Havinden (eds.), *Rural Change and Urban Growth, 1500–1800* (London: Longman, 1974), 104–34.
[27] Centre of Kentish Studies, U120 A 17.

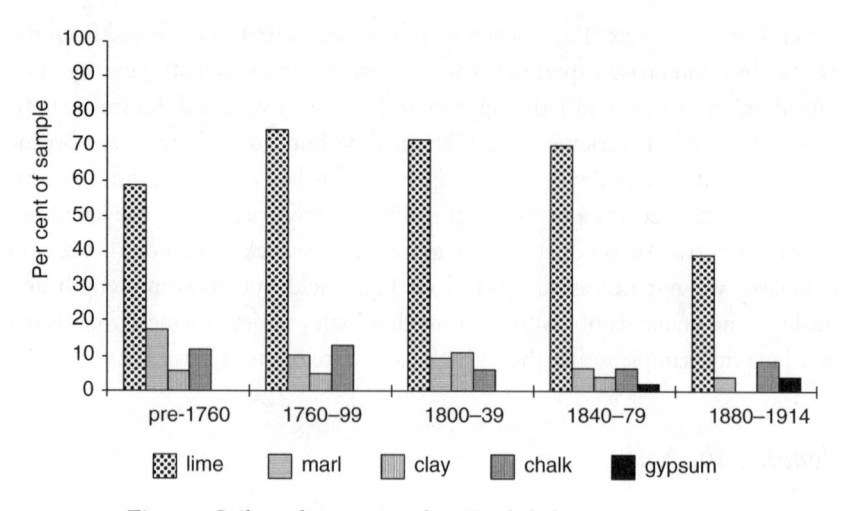

Fig. 3.5 Soil conditioners used on English farms, 1700–1914

late eighteenth century and during the French wars when high prices offered immediate returns on outlay.[28] Its use declined gently from a high of 74 per cent 1760–99 to 72 per cent 1800–39, and to 70 per cent in 1840–79. On the London Clays at Rochford in Essex, 74 acres in five fields were limed between 1851 and 1867. Generally the lime was mixed with earth and applied at a rate of twelve loads per acre. In 1866 and 1867 the lime was specifically used to control slugs.[29] The use of lime declined in the closing decades of the nineteenth century. From the early nineteenth century farmers began to neglect lime through a misplaced trust in artificial manures.[30] It was deployed on only 39 per cent of farms 1880–14, but it appears in farm records as late as 1900 at Aldbourne in Wiltshire, and in 1901 at Shelley in Yorkshire.[31] As Robinson suggested in 1943, farmers 'found by experiment that they got a better return for money spent on basic slag, superphosphate, and other artificial fertilisers',[32] although a good dressing can last for decades, so that the impact of liming would have been felt long after the application ceased.[33]

Figure 3.5 also shows the application of clay and marl. It is difficult always to distinguish the kind of clay that was applied. If it was marl, a clay containing carbonate of lime, it acted on the soil much like lime, chalk, and gypsum.

---

[28] *AHEW* vi. 280.   [29] RUL, ESS 18/3/1.

[30] H. W. Gardner and H. V. Gardner, *The Use of Lime in British Agriculture* (London: Farmer and Stock-Breeder Publications, 1953), 23.

[31] RUL, WIL 11, YOR 8.   [32] G. W. Robinson, 'The Use of Lime', *JRASE* 104 (1943), 136.

[33] *AHEW* vii. 539.

The soil was improved particularly if it was acidic or low in organic matter, by making it more workable, by improving the availability of nutrients, and by reducing acidity so that crops such as clover, wheat, barley, mangles, sugar beet, and sainfoin would grow better. These conditioners also aided pastures by increasing bulk and improving grass quality, by helping to suppress weeds and pests, and by preventing certain diseases such as finger and toe in root crops.[34] Clays and marls gave retentiveness to light soils.[35] Like liming, marl was well established in English farming by the seventeenth century. Walter of Henley had commented on its use in the thirteenth century. Marl was absolutely vital to the agricultural revolution in Norfolk.[36] Before 1760 marl was used on 18 per cent of the farms in the sample, by 1760–99 this had declined to 10 per cent, and the trend continued downwards thereafter. By the end of the nineteenth century marl was no longer appearing in farm records. In the 1800–39 period generic clays at least rivalled the use of marl, but as the nineteenth century proceeded soil conditioners came to be neglected in favour of artificials. Gypsum was a late and somewhat marginal addition to the list of conditioners, partly reflecting the development of gypsum mining in the second half of the nineteenth century.

Conditioning the soil improved the uptake of nutrients, but apart from supplying calcium and occasionally traces of other elements, it did not directly increase fertility. The use of various manures and fertilizers helped to supply nutrients naturally deficient in a soil and to replace those nutrients removed by crops growing in a field. From time out of mind farmers recycled natural waste products. Walter Blith in 1652 and J. Worlidge in 1687 suggested the use of seaweed, oyster shells, fish, pigeon, hen, and goose dung, ash, soot, rags, hair, malt dust, straw stubble, bone, horn, tree bark, blood, and urine along with soil conditioners including chalk, lime, marl, fuller's earth, and clay.[37] In the 1660s at the time of the Georgicall Inquiries undertaken by the Royal Society, farmers in Devon and Cornwall were found to be routinely using dung, seaweed, and ashes as well as sea sand (containing calcium from shells), marl, and lime. Figure 3.6 shows the

---

[34] W. M. Mathew, 'Marling in British Agriculture: A Case of Partial Identity', *AgHR* 41 (1993), 97–110.
[35] Gardner and Gardner, *Use of Lime*, 14.
[36] Wade Martins and Williamson, *Roots of Change*, 55–60.
[37] W. Blith, *The English Improver Improved* (London: J. Wright, 1652), 144–51; J. W. Worlidge, *Systema Agriculturae* (4th edn., London, 1687), 65–84.

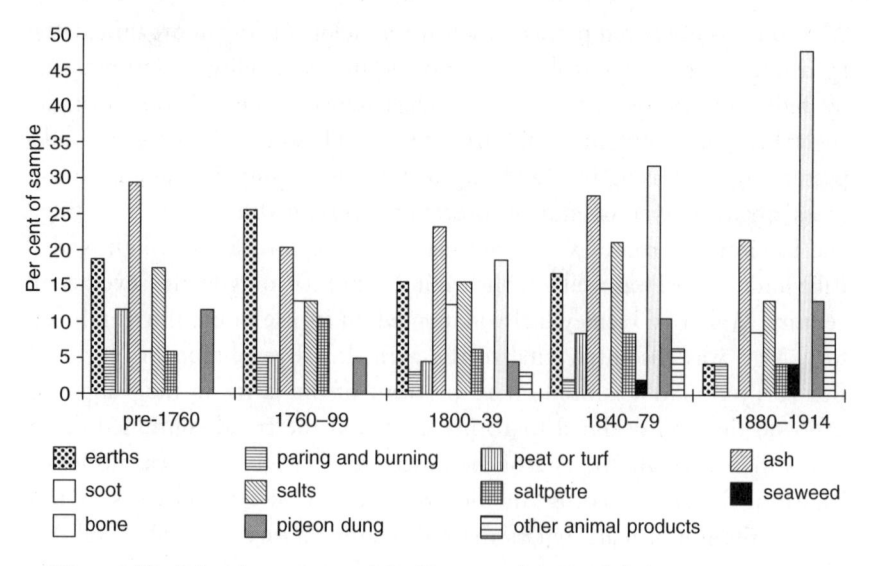

Fig. 3.6 Traditional manures and fertilizers used on English farms, 1700–1914

use of traditional manures and fertilizers 1700–1914. Ash was the most consistently used by farmers. Generic 'salts', routinely employed by between 10 and 20 per cent of farmers, could have been a reference to any number of nutrients from ammoniac salts providing nitrogen to potash salts. Some farmers may even have been using the term to refer to saltpetre. Until the end of the nineteenth century, road scrapings, earths, moulds, and the like were used. Often they were mixed with either dung or soil conditioners, as at Rochford in Essex. The use of bone and other animal products grew in importance during the nineteenth century. The main nutrient in bone and in pigeon dung is phosphate. This promoted root development and so it was important for growing turnips and other root crops.

Throughout the period dung and urine from farm livestock was by far the most important source of fertility available to farmers. Every farm had livestock, if only in the form of working animals, and farmers were usually prevented by covenants in their leases from selling dung off the farm. Farmyard manure is an organic manure and a source of all the plant nutrients. The actual nutrient value of the manure was dependent both on the diet of the animal and on the way the manure was collected, stored, and applied.[38]

[38] Russell, *Soil Conditions*, ed. Wild, 101.

Farmyard manure was often mixed with straw or other bedding material that added to its value. Folding was another way to control the distribution of both dung and urine. Often straw was put into the field to supplement the feed, to provide bedding, and the like. Direct evidence of folding from the records probably understates its use, but between 30 and 40 per cent of farmers recorded the use of the fold. Numbers peaked 1800–39 when 39.4 per cent of farmers folded livestock, particularly sheep, on arable fields. Both spreading farmyard manure and the use of the fold were more efficient means of conserving the nutrients than random grazing, but even this was an important source of both nutrients and organic material. Farmers expended considerable energy collecting, carting, and spreading dung. At Woodstreet Farm, Wool, Dorset, in October, 1758, four horses were being used to plough in preparation for wheat sowing while five horses carted dung into the fields.[39] The arable fields at East Sutton Court Farm were regularly manured with animal dung. In the 1850s and 1860s, at Rochford in Essex, dung continued to be an important source of fertility despite the growing use of artificial fertilizers. Over time, with various alternatives becoming available, some commentators thought that farmers neglected the natural waste products available on their farms. Christopher Turnor complained in 1911 that

> the great majority of farmers do not realise the very great value of the liquid from their manure ... it is even sometimes purposely allowed to escape in the ordinary drains. It is all the more regrettable that this carelessness exists as the larger the amount of money that is spent on cake in feeding the animals the greater is the relative richness of the liquid excreta.... Manure should be regarded as capital.[40]

The increasing use of forage legumes was one of the most significant factors in the improvement of nutrient levels during the eighteenth and nineteenth centuries. Beans and peas are nitrogen-fixing plants, but their root systems are restricted. Up to 90 per cent of the nitrogen in the nodules associated with fixing nitrogen is transferred to the tops of the plants, so that these crops are not particularly good at restoring nitrogen in the soil. By contrast, the root system of forage legumes like clover, trefoil, sainfoin, and lucerne is more extensive and active for a longer period, and consequently these crops are more efficient at increasing the soil nitrogen content. This increases the supply of nitrogen available both to the plants growing with

---

[39] Dorset RO, 406/1.
[40] Christopher Turnor, *Land Problems and National Welfare* (London: The Bodley Head, 1911), 75. Turnor was far from alone: *AHEW*, vii. 535–9.

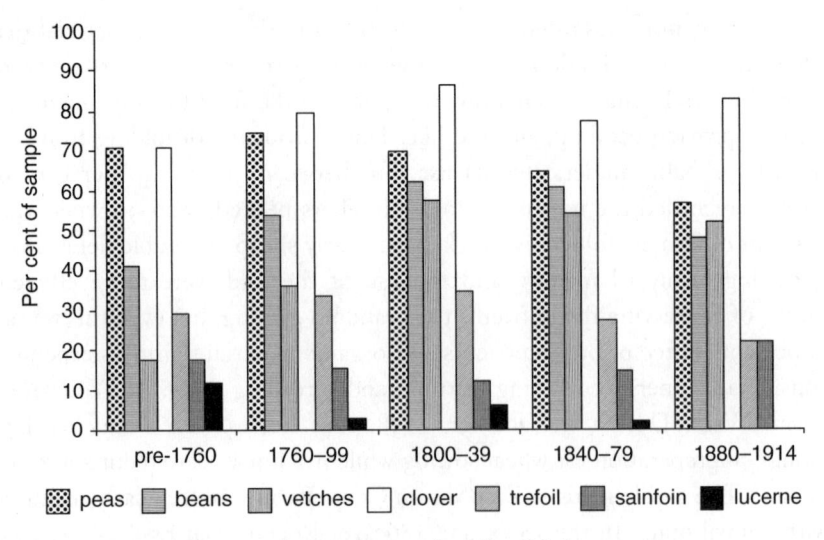

Fig. 3.7 Nitrogen-fixing crops used on English farms, 1700–1914

them, such as barley and oats, as well as to the plants that follow in the rotation.[41] Figure 3.7 illustrates the use of beans, peas, and vetches as well as the more efficient clovers, trefoil, lucerne, and sainfoin. The use of all legumes apart from peas and sainfoin peaked 1800–39.

Measured purely in statistical terms, it is relatively easy to argue the case for a significant intensification of the use of agricultural inputs in the 1830s and 1840s.[42] Fertilizers retained in the United Kingdom increased from 26,000 tons 1810–14 to 781,000 tons 1872–6. These figures suggest that despite the traditional conservatism of farmers, by the 1830s and 1840s many of them were moving away from the simple recycling of organic matter which made the farm self-reliant but which only maintained fertility, towards a more business-orientated vision of farming. In this brave new world inputs were purchased from off the farm, in the belief that they would go beyond the mere maintenance of fertility and actually help to raise output. Traditional manures were supplemented by an increasing array of new, often straight (single-nutrient) fertilizers (Fig. 3.8). Imports of guano (excrement of sea fowl) from Peru began in the 1840s, and during the same decade the process for producing superphosphate was discovered.[43] Oilcake, one of the 'new'

[41] Russell, *Soil Conditions*, ed. Wild, 549.
[42] F. M. L. Thompson, 'The Second Agricultural Revolution, 1815–80', *EcHR* 21 (1968), 62–77.
[43] See for example W. M. Mathew, 'Peru and the British Guano Market, 1840–70', *EcHR* 23 (1970), 112–28.

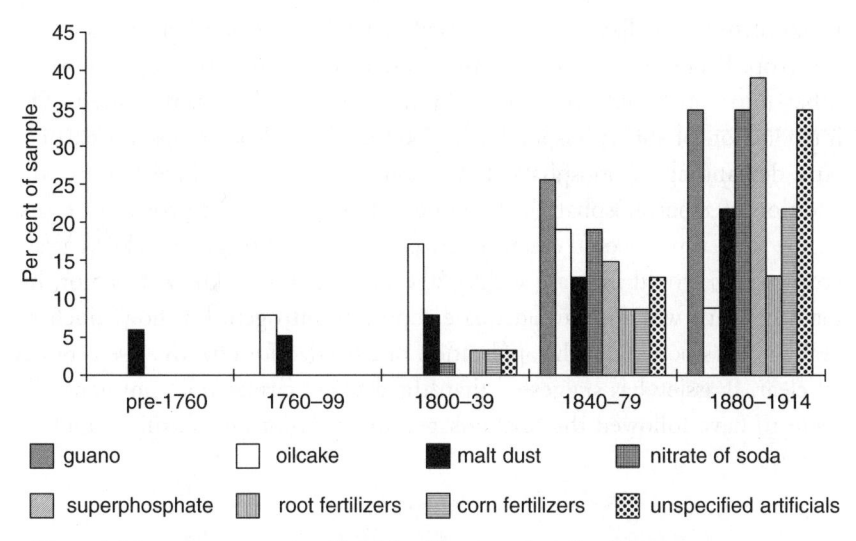

Fig. 3.8 New manures and artificial fertilizers used on English farms, 1700–1914

artificial fertilizers, was ground and spread on the field, but this technique decreased in popularity when it was found that feeding cake to livestock and then spreading the resultant manure was more efficacious. A range of generic artificials also came into use. Some of these were crop specific, such as turnip manure and wheat fertilizer. They reflected the growing importance of straight, often inorganic fertilizers, which came into use from the 1840s.

The increasing sophistication of the choices available to farmers can be seen at Saltmarshe on the River Ouse in the East Riding of Yorkshire. The farmer regularly bought manure and fertilizers which reached him by boat from Hull, York, and sometimes further afield. His purchases included 'Hull manure' in 1802, ashes in 1815, whale blubber in 1817, rape dust in 1827, bone dust in 1832, linseed dressing between 1827 and 1834, 'wheat manure' in 1838, nitrate of soda in 1843, and guano in 1844. In addition cow and horse manure was regularly brought from York.[44]

The enthusiasm for the new inputs undoubtedly led to misplaced application. For generations lease covenants had prohibited the removal of dung, straw, and the like from the farm. Now leases were altered to allow sales if certain quantities of (often unspecified) artificial fertilizers were purchased. Unfortunately, understanding of the way these fertilizers worked was as yet

[44] East Riding of Yorkshire AO, DDSA 1203/1–6.

incomplete.[45] Fertilizers had to be used in a balance to suit both the soil and the crop. If one particular nutrient was deficient, additional applications of other nutrients could not be used by the plants, and so were wasted. The introduction of superphosphate is a good example. Root crops are particularly demanding of phosphate. If the soil was deficient, as many were, the addition of superphosphate had a noticeable impact on the growth of roots. At the same time it could create a new potential for nitrogen uptake by grain crops which would increase yield. Whether the farmer knew it or not, his main concern was with an adequate supply of nitrogen, but how much he really understood about the application rate needed for effective use is not at all clear. Brassley has suggested that 'the average farmer does not generally seem to have followed the textbook recommendations on fertiliser application rates'.[46]

Enclosure, Drainage, and Machinery

By the early nineteenth century other developments were under way which also had an impact on farm output. The most obvious example is enclosure. Contemporaries seem to have been in little doubt that enclosure was a key element in raising output. Arthur Young, ever the optimist, never ceased to regale his readers with stories of major increases in output, and to compare them with the uncertain yields achieved in open fields.[47] This is contested ground since modern historians have claimed that the open fields were not nearly so inflexible as was once believed.[48] Supporting evidence from farm records shows that some farmers introduced innovative practices in the open fields where local husbandry rules allowed. This is clear from the surviving diaries from Langham Farm in north-west Norfolk. In its unenclosed state,

---

[45] G. W. Cooke, 'Advice on Using Fertilisers, 1861–1967', *JRASE* 128 (1967), 107–24.

[46] *AHEW* vii. 543. Brassley argues that the use of artificials increased because farmers had insufficient farmyard manure and because of falling relative prices, but they could have been adopted on a greater scale than was the case: 544–7.

[47] For extracts from a number of Arthur Young's writings see G. E. Mingay (ed.), *Arthur Young and his Times* (London: Macmillan Press, 1975), 98–111. For a specific reference to the Vale of Aylesbury, a location where he was particularly scathing of open-field agriculture when he said that 'The landlords have fourteen shillings where they might have thirty and the tenants reap bushels, where they ought to have quarters,' see A. Young, *The Farmer's Tour through the East of England*, i (London: W. Strahan, 1771), 20.

[48] 'Progress' in open fields was first suggested by M. A. Havinden, 'Agricultural Progress in Open-Field Oxfordshire', *AgHR* 9 (1961), 73–83.

the farm grew a variety of crops typical of a Norfolk four-course rotation, including wheat, barley, oats, peas, vetches, clover, trefoil, ryegrass, and turnips. The fields were regularly limed and dunged, particularly for the turnip crop, some of which was sold to be grazed by sheep, and some to be fed to cattle. Livestock included cattle and pigs, and by 1775 sheep had been introduced. Along with Scottish cattle they were sent to market to London. Therefore, before enclosure the farm had the characteristics of a well-run innovative enterprise employing up-to-the-minute techniques. The parish was enclosed in 1815 after which the farming system changed in response to new methods as they became available. Oilcake was introduced for feed as well as for manuring. Swedes were added to turnips in the rotation, and by 1835 greater prominence was placed on arable production, with the application of more sophisticated fertilizers, including 'pigeon manure' and, by 1845, nitrate of soda and guano.[49] This may be a good illustration of Thomas Davis's dictum that enclosure made a good farmer better, although he added that it might make a bad one worse.[50] Among historians there is no real consensus as to the significance of enclosure for yields.[51]

A second significant change took the form of land drainage. Heavy land needed to be drained to combat acidity, to make it more easily workable (particularly in wet weather), and to encourage crop growth. Traditionally, particularly on the Midland clays, water was run off ill-draining soils by ridge and furrow methods. Farmers recognized this to be an inefficient way of draining land, and experimented with various methods whereby they dug trenches and filled them with stones or vegetable matter. Many eighteenth-century records refer to ditching. At East Sutton Court Farm in 1722 the ditches were located at a depth of 2 feet 6 inches, and they were 2 feet across at the bottom. Waterfurrows were mentioned in a number of records, and

---

[49] Norfolk RO, LC 120/1–82; B. Afton, 'Investigating Agricultural Production and Land Productivity: Opportunities and Methodologies Using English Farm Records', paper for the Twelfth International Economic History Congress Session C54 'Production et productivité agricoles dans le monde occidental (XIIe–XXe siècles)', Madrid, 24 to 28 Aug. 1998.

[50] Thomas Davis, Jr., *General View of the Agriculture of Wiltshire* (London, 1811), 46; M. E. Turner, 'English Open Fields and Enclosures: Retardation or Productivity Improvements', *JEH* 46 (1986), 687.

[51] Turner, 'English Open Fields and Enclosures', 669–92; J. R. Wordie, 'The Chronology of English Enclosure, 1500–1914' *EcHR* 36 (1983), 504–5; D. N. McCloskey, 'The Economics of Enclosure: A Market Analysis', in W. N. Parker and E. L. Jones (eds.), *European Peasants and their Markets* (Princeton: Princeton University Press, 1975), 73–119; R. C. Allen, 'The Efficiency and Distributional Consequences of Eighteenth-Century Enclosures', *Economic Journal*, 92 (1982) 937–53; R. C. Allen and Cormac Ó Gráda, 'On the Road Again with Arthur Young: English, Irish and French Agriculture during the Industrial Revolution', *JEH* 48 (1988), 98–116.

by the early years of the nineteenth century farmers were experimenting with underdraining at Garboldisham in Norfolk (1805), Bradfield in Yorkshire (1806), and Searby in Lincolnshire (1831). Pipes and tiles were in use at Histon, Cambridgeshire, 1818–24. At Coton in the same county a major drainage scheme between 1838 and 1842 involved tiles and bricks. Water pipes were laid at Clenchwarton in Norfolk in 1817, and at Chilwell in Nottinghamshire in 1832 tiles were laid in furrows 16 inches deep.[52] Farmers experimented to find the system most suited to their needs. On heavy soils in parts of East Anglia 'effective systems of field drainage were widespread . . . by the end of the eighteenth century'.[53] Randall Burroughes of Wymondham in Norfolk undertook a considerable amount of under-drainage on his land 1794–9. In his journal he noted the following:

Early in November [1794] began to prepare for draining first a close at Inton call'd Feeks Ten Acre. Two men . . . undertook to drain it at one halfpenny per yard or $3\frac{1}{2}$ d per rod of 7 yards about 28 inches deep the main drains & 24 or 26 the crop drains, all materials to be brought to them. According to custom the drainers should find their own tools but as draining is an expensive improvement in order to receive no inconvenience from the use of defective tools and as an encouragement to the men to execute their undertaking in the completest manner I save them the expense by procuring them at my expense. The alders are brought from a carr belonging to Mr Bernard at Wymondham about four miles off. One man will cut & bring out of the carr at this time of the year and lay them on sound land for the waggons to take them about one load & a half per day & I expect it will take about five loads of alder for six acres drain'd regularly at 12 yards distance from drain to drain besides the main drains.

This work, hampered by heavy frosts in December, took until mid-February 1795 to complete. An entry from 28 February commented 'Measured up underdraining of Feeks 10 acre at Sutton which has been finished about two weeks. Number of rod at 7 yards was 650 rod which cost £9 9s. 7d. at $3\frac{1}{2}d$.'[54]

Perhaps because it was in the 1840s that land drainage became a major concern of agricultural experts, it has sometimes been argued that effective drainage methods only date from the introduction on a substantial scale of

---

[52] Centre for Kentish Studies, U120, A17–19; Norfolk RO, BR 149/1, MS 576; Sheffield AO, MD 3518; Lincolnshire AO, Dixon 5/1/1–64; Cambridgeshire RO, R62/36, R58/9/4/3; Nottinghamshire AO, DD 1089/5.

[53] P. Pusey, 'On the Progress of Agricultural Knowledge during the Last Four Years', *JRASE* 1st ser. 3 (1842), 177; P. Pusey, 'Evidence on the Antiquity, Cheapness and Efficacy of Thorough-Draining, or Land-Ditching', *JRASE* 1st ser. 4 (1843), 23–49; Wade Martins and Williamson, *Roots of Change*, 61.

[54] Norfolk RO, MC 216/1.

manufactured cylindrical pipes.[55] Clearly this was not the case, although the switch to pipes is reflected in the records dating from the 1840s. At Doggetts Hall Farm, Rochford, Essex, ten fields totalling 126.5 acres were drained in the 1850s. The pipes were placed 2 rods apart and 3 feet deep. In 1862–3 another 62 acres were drained, this time using a mole plough. By 1873 ten of these acres needed further drainage work.[56] While it is generally accepted that drainage had a positive impact on ouput, it is also recognized that 'no single measure of the effect of drainage is possible'.[57]

The introduction of ploughs, harrows, horse-hoes, seed drills, and similar machinery affected land productivity positively. Farm records occasionally mention ploughs, implying that equipment was either being purchased new or was already in use. Ransomes ploughs are mentioned at Histon, Cambridgeshire, in 1817, and Beaumont Leys, Leicestershire, in 1825.[58] During a period of especially wet weather in May 1807 William Barnard of Harlow, Essex, noted that 'I have been plowing Old Field; I began with a cast iron plow on the 6th and am in great hopes it will answer.'[59] Ploughs constructed of cast iron were widespread in East Anglia by the 1820s,[60] and a farmer at Clayhidon in Devon was using an iron plough in 1828.[61]

Seed drills, and horse-drawn hoes, in one form or another were invented in the seventeenth century, but they were adopted relatively slowly. Broadcast sowing or dibbling (the practice of making holes in the soil and dropping in the required number of seeds) were both inefficient relative to drilling seed. For wheat it was necessary in the spring to harrow-out weeds which were growing with the crop. With the seed drill the sowing rate could be reduced. Drilling seed meant that it was planted in straight rows, allowing weeding to take place between the rows with a hoe (by hand or horse). It was most effective in terms of controlling weeds, which competed with plants for the limited amounts of nitrogen available in the soil, and which could damage the crop. Even so, drills were not widely adopted, at least prior to the 1780s. Contemporary evidence suggests that there was considerable experimentation with drill husbandry by the end of the eighteenth century, but that drilling was only common in Norfolk and Suffolk, Northumberland and Durham. Indeed, there were still so few machines

---

[55] A. D. M. Phillips, *The Underdraining of Farmland in England during the Nineteenth Century* (Cambridge: Cambridge University Press, 1989); Shiel, 'Improving', 53.
[56] RUL, ESS 18/3/1. [57] *AHEW* vii. 514–21, esp. 520.
[58] Cambridgeshire RO, R62/36; RUL, LEI 4/1/1–2. [59] Essex RO, D/DU676/1.
[60] Wade Martins and Williamson, *Roots of Change*, 115. [61] Devon RO, 1061A/ZB3 (V).

available that contract drillers were found in some areas, and it was not until the 1830s and 1840s that drills came into common usage.[62] Unfortunately the farm records throw up only a few examples of the adoption of drills: they were in use at Eathorpe in Warwickshire from at least 1803; at Riby in Lincolnshire in 1805; at Saltmarshe in Yorkshire in 1806; at Beaumont Leys in Leicestershire in 1813; at Braunton in Devon in 1825; and at Barton in Cambridgeshire in 1830.[63] Other equipment mentioned at different dates in the records includes a chaff sieve (1806), a haymaking machine (1822), seedlips (1816, 1856–8), mangle drills (1858, 1898), a turnip cutter (1864), wheat reaper (1867), winnowing fan (1759), cake crusher (1883), and even 'tools for underdrainage' at Saltmarshe in Yorkshire in 1833.[64]

Some new tools made little or no difference to land productivity but were important labour-saving innovations. The gradual introduction of hand tools such as the scythe, the reap hook, and the bagging hook represented considerable labour-saving opportunities. In 1750 most wheat was cut with the sickle, but by 1850 the scythe or bagging hook was the common method of harvesting, the sickle having largely been abandoned by this time. A farmer at Upminster in Essex recorded a scythe in 1794, another at Swallowfield in Berkshire had a reap hook and chaff sieve by 1806, and a farmer at Sandford in Devon had a reap hook in 1822.[65] By the 1870s the use of the scythe and bagging hook was widespread and the principal means of harvesting. Together they helped to raise the productivity of the harvest labour force, since a scythesman could cut over an acre of wheat daily whereas a labourer using a sickle could reap only about one-third of an acre in the same time.[66]

---

[62] R. Wilkes, 'The Diffusion of Drill Husbandry, 1731–1850', in W. Minchinton (ed.), *Agricultural Improvement: Medieval and Modern* (Exeter: University of Exeter Press, 1981).
[63] Lincolnshire AO, Dixon 4/1–3; East Riding of Yorkshire AO, DD SA 1203/1–6; RUL, LEI 4/1/1–2; Cambridgeshire RO, R56/15/1. Farm records are not ideal as a source for mechanization. Some drills were probably recorded only when they were purchased. Even so, the fact that we have not found more evidence is interesting in the light of recent arguments that suggest that drills were of considerable importance for raising crop yields: L. Brunt, 'Nature or Nurture? Explaining English Wheat Yields in the Agricultural Revolution', *University of Oxford Discussion Paper in Economic and Social History*, 19 (1997), 21. On farm implements more generally see G. E. Fussell, *The Farmer's Tools: The History of British Farm Implements, Tools and Machinery, AD 1500–1900* (London: Bloomsbury, 1985).
[64] RUL, BER P322/1–3, BUC P321/2–4, HERT 4/1/1, DOR 5/1/3–4, NORTHUMB 2/3/1–2; Isle of Wight RO, WHT 1–2; Cambridgeshire RO, R56/15/1; Lincolnshire AO, Dixon 4/1–3, Scorer Farm 1/16–19; East Riding of Yorkshire AO, DD SA 1203/1–6. J. R. Walton, 'Mechanisation in Agriculture: A Study of the Adoption Process', in H. S. A. Fox and R. A. Butlin (eds.), *Change in the Countryside: Essays on Rural England 1500–1900* (London: Institute of British Geographers, Special Publication no. 10, 1979), 23–42.
[65] Essex RO, D/DJn E5; RUL, BER P322/1–3; Devon RO 1283/M/E1.
[66] E. J. T. Collins, 'Harvest Technology and Labour Supply in Britain, 1790–1870', *EcHR* 22 (1969), 453–73.

Much more common as 'new technology' were threshing machines.[67] Andrew Meikle invented the first practical threshing machine in 1786, and its successors seem to have been coming into widespread use by the early nineteenth century. Examples can be found from Yorkshire in 1801, Suffolk in 1815, Gloucestershire in 1823, Warwickshire in 1839, Essex in 1841, Buckinghamshire in 1850, Surrey in 1852, and Dorset in 1865.[68] Threshing machines may be less prominent in the records than other tools simply because they were often hired (as at Coton, Cambridgeshire, in 1837) rather than purchased. Portable machines were often employed on cost-sharing principles, and what was known as the renting of 'itinerant' machines.[69]

Towards the end of the nineteenth century changes in motive power and developments in the steam engine started to influence agriculture. Steam cultivation was in use at Dunholme, Lincolnshire, by 1870. At Easton, Hampshire, in 1872 a farmer recorded hiring a steam thresher, and at Whitchurch in the same county a steam thresher was in use by 1907. Other 'new' equipment included reaper-binders at Thornbrough in Northumberland in 1884 and also at Hitchin in Hertfordshire in 1898. Reapers are recorded at Preston in Rutland in 1888, and at Bingham in Nottinghamshire in 1893. The Rutland farmer also had a grass mower in 1888 and a winnower in 1897.[70]

Whatever the pace of innovation and mechanization, farmers remained subject to the vagaries of the weather, as well as to weeds, pests, and diseases. A great deal was known of them all, but much less about how to counter them. Throughout our period weeding and hoeing was regularly used in an effort to keep grain crops free of weed. To counter pests and disease rats were trapped, mice were poisoned, birds were scared, and sheep were salved. Vitriol was occasionally sprayed on wheat from the 1840s. A farmer at

---

[67] S. Macdonald, 'The Progress of the Early Threshing Machine', *AgHR* 23 (1975), 63–77. See also the debate about the introduction and diffusion of the threshing machine that ensued from this article in N. E. Fox, 'The Spread of the Threshing Machine in Central Southern England', and S. Macdonald, 'Further Progress with the Early Threshing Machine: A Rejoinder', in *AgHR* 26 (1978), 26–8, 29–32. See also E. J. T. Collins, 'The Diffusion of the Threshing Machine in Britain, 1790–1880', *Tools and Tillage*, 2 (1972), 16–33. Wade Martins and Williamson, *Roots of Change*, 118, argue that machine threshing 'seems to have had relatively little impact in East Anglia for at least the first two decades of the nineteenth century', largely due to labour abundance.

[68] East Riding of Yorkshire AO, DD SA 1203/1–6; East Suffolk RO, HA/2/A3/185; RUL, GLO 1/2/1–9, ESS P300, BUC 7/1/1, SUR 3/1/2, DOR 5/1/3–4, WAR 1/1/1.

[69] Cambridgeshire RO, R58/9/4/3; *AHEW* vii. 787; Macdonald, 'The Progress of the Early Threshing Machine', 76.

[70] Lincolnshire AO, WEST 1–32; Hampshire RO, 60M77/F2, 83M76/8215; RUL, NORTHUM 2/3/1–2, HERT 4/1/1; Leicestershire, Leicester, and Rutland RO, DE 3858/6; Nottinghamshire AO, DD 1405/2–3.

Atherton in Lancashire wrote extensively in his records of the impact of wire worm. In 1827 he sowed the upper Highfield with oats, but 'the crop was of no value having been completely ruined by the wire worm'. The following year the same land was well ploughed and manured before being planted with potatoes, but the result was the same: 'this crop was also much injured by the wire worm.' To try to counter the problem, he bought in and spread seven loads of salt on the field, and then sowed it with wheat. This seemed to have worked, but in 1829 the same problem recurred and he reported that he:

[found] the wheat again destroyed in the Upper Highfield, and seeing that the Field was becoming very foul, I determined to plough up what little wheat there was growing, and plant it with potatoes without manure, in the hope that by the frequent stirring of the land I might weaken this host of wire worms. At the same time I sowed upon the new ploughed ground a quantity of brimstone (sulphur) but nothing seemed to have any effect.

Nor was this a great success. The wheat he had sown soon proved to be 'much injured', so he ploughed up half of the field and resowed it with vetches. This produced 'a very fair crop', but the wheat that he had left to go full term 'was a very indifferent crop, indeed there was more weed than wheat'. The following year he came up with a new idea for the Crowbank field. He planted it with early potatoes, and because these were also infested with wire worm he hired 'a number of lads' to go in front of his labourers, to turn over the clods 'and pick by hand the wire worms which were here very numerous'. He was delighted to find a good use for the wire worms: 'the poultry eat them with great avidity'.[71]

Although there is occasional reference in the contemporary literature to farmers trying experiments to rid their crops of disease, it needed the potato blight of 1845–6 to focus attention on the devastating damage that crops could sustain under truly adverse conditions. Yet the scientific study of the extreme affliction plants could sustain was still in its infancy, and the methods available for the control of insect and fungal attacks on crops did not change much before the twentieth century. Cornelius Stovin spoke for many farmers when he reflected in 1872 on the golden-winged fly that 'these tiny insects still baffle the farmer's skill notwithstanding the

---

[71] Lancashire RO, DD Li box 92. Cornelius Stovin of Binbrook on the Lincolnshire Wolds also complained about wire worm: J. Stovin (ed.), *Journals of a Methodist Farmer* (London: Croom Helm, 1982), 77.

triumphant conquests already achieved over the natural world'. For the most part farmers deployed cultural rather than scientific methods in their struggle with the insect world, and chemical control methods were still in their infancy at the beginning of the twentieth century.[72]

Livestock

The relative emphasis among agricultural historians on arable culture, and particularly on wheat, has tended to underplay the role of livestock in the agricultural cycle.[73] In earlier generations sheep had been integral to the agricultural economy, producing the wool which drove the prosperity of medieval and early modern England. During the agricultural revolution, as the emphasis shifted towards increasing arable output, the animal world was relatively neglected, despite the importance of sheep and cattle, to say nothing of horses and oxen, within the mixed farming system, and the simple fact that the sale of stock was a vital element in the farmer's financial solvency. Yet, within the mixed farming economy, it was well into the nineteenth century before farmers began to concentrate on the profitability of livestock production rather than viewing animals as little more than a necessary adjunct to cereal cropping.[74]

Livestock farming systems adapted to different stages in the long-run agricultural production process. Until the eighteenth century much of the emphasis was on the production of wool, but subsequently different areas of the country came to concentrate on breeding, or on store cattle, or on fattening. Long-term animal improvement required selective breeding, the creation of new breeds, or the development of native varieties by crossing with new varieties. The idea of crossing animals to improve breeds was not new, since the principles were essentially those which had long been used in

---

[72] *Journals of a Methodist Farmer*, 117. Despite Brassley's hopes (*AHEW* vii. 554), that 'further analysis of farm diaries and account books' will shed fresh light on the whole question of pest control, there is little in the records to suggest that his rather gloomy conclusions on the issue are misplaced.

[73] Wade Martins and Williamson, *Roots of Change*, 119. Recent exceptions include B. M. S. Campbell and M. Overton, 'Norfolk Livestock Farming, 1250–1740', *Journal of Historical Geography*, 18 (1992), 377–96; J. R. Walton, 'Pedigree and the National Cattle Herd, circa 1750–1950', *AgHR* 34 (1986), 149–70; and M. E. Turner, 'Counting Sheep: Waking up to New Estimates of Livestock Numbers in England, c.1800', *AgHR* 46 (1998), 142–61.

[74] T. Davis, Sr., *General View of the Agriculture of the County of Wiltshire* (London, 1794), 20–1; C. Hillyard, *Practical Farming and Grazing* (4th edn. London, 1844), 207–8; Pusey, 'On the Progress of Agricultural Knowledge', 205; P. H. Frere, 'On the Feeding of Stock', *JRASE* 1st ser. 21 (1860), 233–4.

racehorse breeding.[75] Farmers had long experimented in the hope of improving the weight and quality of their animals. Although there is a widely held belief that they had made little progress by the eighteenth century, our evidence suggests that farmers who specialized in animal production had achieved some significant breakthroughs. At Burmarsh in Kent, in the early years of the eighteenth century, one farmer worked an unspecialized combination of dairy, fatting, and breeding. He ran sheep on his pasture in Romney Marsh, both for wool and to sell fat animals on the London market. In 1703 he had more than 4,000 sheep, valued at £1,644, and 172 cattle valued at £575. On a farm in the same area by the 1770s the emphasis had shifted towards feeding and fattening. Animals were both bred on the farm and purchased, and the farmer was using some specialist breeds such as Dorset sheep. Other animals were taken in from neighbouring farmers and agisted, or grazed.[76]

Until the eighteenth century the place of origin of an animal largely determined its breed.[77] Each region had its native cow, pig, sheep, and so on. A full understanding of the rudiments of selective breeding awaited the experiments of men like Robert Bakewell at Dishley, near Loughborough, in the 1740s.[78] Where Bakewell led others were following by the 1780s, although it took time before the importance of quality in livestock spread from the progressive few to the greater mass of farmers.[79] Regional specialization altered over time, partly as interest in breeding resulted in the spread of particularly successful breeds, such as the shorthorn cattle and the Leicester sheep.[80] Just occasionally individual motives are known. William Barnard of Harlowbury Farm, Harlow, Essex, observed his animals closely and acted according to experience. In June 1807 he noted that 'I have a great inclination to part with my stock of sheep and to replace them with South Downs. I used to

[75] M. J. Huggins, 'Thoroughbred Breeding in the North and East Ridings of Yorkshire in the Nineteenth Century', *AgHR* 42 (1994), 115–25.

[76] RUL, KEN 19/1/1. Although the data for 1703 and the 1770s is taken from the same farm accounts, the records are for separate farms.

[77] R. Brown, *The Compleat Farmer* (London: J. Coote, 1759), 33; W. Ellis, *The Modern Husbandman*, ii (London: T. Osborne, M. Cooper, 1744), 143–4.

[78] N. Russell, *Like Engend'ring Like: Heredity and Animal Breeding in Early Modern England* (Cambridge: Cambridge University Press, 1986); S. Macdonald, 'The Diffusion of Knowledge among Northumberland Farmers, 1780–1815', *AgHR* 27 (1979), 30–9.

[79] R. Trow-Smith, *The History of British Livestock Husbandry, 1700–1900* (London: Routledge & Kegan Paul, 1959).

[80] On breeding in general see Russell, *Like Engend'ring Like*, and esp. 210–12 for cattle and sheep breeds/breeding, and on the dissemination of ideas from Robert Bakewell to the Culleys of Northumberland. See also D. J. Rowe, 'The Culleys, Northumberland Farmers 1767–1813', *AgHR* 19 (1971), 156–74; S. Macdonald. 'The Role of George Culley of Fenton in the Development of Northumberland Agriculture', *Archaeologia Aeliana*, 3 (1974), 131–41.

lose many of my large cows, [but] since I have had Welch ones I have lost very few.' A year later he was regretting the decision to change: 'I am sick of my South Downs, the lambs are so very small.' However the following month, July 1808, a prolonged drought caused him to change his mind yet again: 'South Downs appear to do much better in this sharp time than the horned sheep.'[81] Barnard's understanding of the difference between breeds was shared with many other farmers, partly because animals were driven great distances from Scotland, Wales, and Ireland to the fattening pastures of England. This trade was a result of demand for meat from London and the growing Victorian cities, but it brought a greater appreciation of different breeds.[82] Table 3.5 provides examples of breeds mentioned in farm records.

William Marshall thought that the motives of early breeders were over-dominated by hopes of attracting high prices in the market place. He observed that stock that showed the earliest signs that they might fatten were in fact not saved for breeding but instead were the first to be slaughtered. This prevented the natural improvement of stock, and breed improvement had therefore to be a conscious decision on the part of farmers.[83] John Ellis of Beaumont Leys in Leicestershire is a case in point. By 1815–16 he was buying in animals, mostly locally but often from further afield, feeding them, and selling them on as fat animals. Some went to London. He also bought in-calf cows for dairy and calf production, and fattened pigs from the by-products of the dairy,[84] but despite his proximity to Dishley he did not always show much appreciation of breeds. His accounts usually mention only heifers and cows, without distinction as to type, although in 1817 in the list of cows bulled he added some descriptions: grey, poled, Side bag'd cow, Little Black spotted cow, Scot cow, Southport cow. These comments were not accompanied by any reflections on breeds. The effect of breed on weight is well documented,[85] but from our point of view it is unfortunate

---

[81] Essex RO, D/DU676/1.

[82] R. J. Moore-Colyer, *The Welsh Cattle Drovers: Agriculture and the Welsh Cattle Trade before and during the Nineteenth Century* (Cardiff: University of Wales, 1976). The earlier example involving the Scottish trade to England is discussed in D. Woodward, 'A Comparative Study of the Irish and Scottish Livestock Trade in the Seventeenth Century', in L. M. Cullen and T. C. Smout (eds.), *Comparative Aspects of Scottish and Irish Economic and Social History 1600–1900* (Edinburgh: University of Edinburgh Press, 1977), 147–64. The different breeds commonly found in late 19th-century England are discussed in *AHEW* vii. 555–69.

[83] W. Marshall, *The Rural Economy of the West of England* (London: G. Nicol, 1796), 244–5.

[84] RUL, LEI 4/1/2.

[85] G. E. Fussell and C. Goodman, 'Eighteenth-Century Estimates of British Sheep and Wool Production', *Agricultural History*, 4 (1930), 131–51; G. E. Fussell, 'The Size of English Cattle in the Eighteenth Century', *Agricultural History*, 3 (1929), 160–81.

Table 3.5 Examples of livestock breeds noted in the farm records

	Sheep	Cattle	Pigs
East Sutton, Kent	Hampshire (1722); Marsh (1722); Horned (1732)	Welsh (1732)	
Quainton, Buckinghamshire	Wiltshire (1748); Dorset (1748)		
Bridgnorth, Shropshire	Welsh (1758); Dorsetshire (1758)	Scotch (1748); Black (1758)	
Burmarsh, Kent	West Country (1766); Dorset (1766)	South Wales (1766)	
Swallowfield, Berkshire	Dorset (1791); Down (1808); Southdown (1813)	Berkshire (1791)	
Garboldisham, Norfolk	Southdown (1807); half-bred Leicesters (1807)	Scotch (1808)	
Bradfield, Yorkshire			
Charminster, Dorset	Portland (1808); Dorset Down (1824)		
Chippenham, Cambridgeshire	Norfolk (1810); Leicester (1810); Lincoln (1810); Southdown (1810)	Suffolk (1810)	
Beaumont Leys, Leicestershire	Scotch (1815)	Scotch (1806); Longhorn (1808); Yorkshire (1808); Welsh (1817); Hereford (1817); Irish (1817); Shropshire (1817); Denbighshire (1817); Shorthorn (1817)	
Preston, Rutland		Longhorn (1818); red (1818); Shorthorn (1818)	
Thornbrough, Northumberland	Cheviot (1856); grey faced (1861)	Irish (1884)	
Adderbury, Oxfordshire		Hereford (1868); Shorthorn (1868)	
Hitchin, Hertfordshire	Down (1904); Masham (1904); Cheviot (1907); Suffolk (1909)	Irish (1910)	Suffolk (1807)

Source Appendix 1.

that references to breeds, or even places of origin, often received no mention at all when animals were sent for slaughter.

The act of breeding brought considerable advantages to the farmer. A good example is the Hampshire Down and later the Oxford Down sheep breed in the nineteenth century. The animal went from a long-legged, woolly creature to a large compact animal by the end of the century. It was especially well suited to arable folding, it fattened easily, and matured early. With high feeding it became normal to slaughter it at nine months. On a less intensive feeding regime the animal was fat at about eighteen months. It lambed early in the year, a characteristic derived from the feeding combination of arable and water meadows that ensured a good food supply in late winter and early spring. The lambs became sufficiently physically mature for breeding in their first year.[86] At South Wonston in Hampshire there was a typical mid-nineteenth-century commercial sheep-breeding farm. What typified it was a regular culling system. Ewes were removed from the system at between three and four years, and replaced by ewe lambs.[87] Wether and ram lambs were sold, although the farmer retained some wethers on another holding that he possessed.[88] This was a specialist farming system. One of the main sources of income of John Twynam of Whitchurch in Hampshire was ram breeding, but many of his animals were not sold on. Instead he let them out for a season at a time. His clients came from as far afield as Northamptonshire, Shropshire, and Norfolk.[89]

A farmer at Kingston Deverill in neighbouring Wiltshire in the 1860s was selling lambs out of his breeding flocks. They were moved first to grazing and then on to fatting areas. He fed the animals through the winter months and then sold them on. He was also dairying in western Wiltshire, and produced calves, pigs, cheese, and butter. Such was the care with which he produced his accounts that he was able to calculate the return per cow from his dairy.[90]

---

[86] Afton, 'The Great Agricultural Depression', 195, 203; B. Afton, 'Mixed Farming on the Hampshire Downs, 1837–1914' (University of Reading, Ph.D. thesis, 1993), 88 ff.; G. G. S. Bowie, 'New Sheep for Old: Changes in Sheep Farming in Hampshire, 1792–1879', *AgHR* 35 (1987), 151–8; and id., 'Northern Wolds and Wessex Downlands: Contrast in Sheep Husbandry and Farming Practice, 1770–1850', *AgHR* 38 (1990), 117–26.

[87] RHC, D93/10.

[88] In a flock breeding top-quality pedigree sheep older ewes were not regularly culled. In these flocks, ewes that were producing good lambs were more likely to be kept until they were barren.

[89] RUL, HAN 9/1/1; A. K. Copus, 'Changing Markets and the Development of Sheep Breeds in southern England, 1750–1900', *AgHR* 37 (1989), 36–51.

[90] RUL, WIL 6/2/2.

Farmers entered animals in their records as calves, heifers, oxen, bullocks, cows, and so on, rather than simply as cattle. This separation of uses was partly related to breed. It was recognized that cattle breeds best suited to meat production were often inferior milk producers, and that where oxen were used for motive power they would be less productive for the quality of their meat relative to their acquired skills. The 'experience' of an older animal was of greater value than rapid turnover with the butcher. When horse power generally replaced oxen the choice for the end use of cattle was more simply divided between meat and dairy production.[91] This was also true for sheep in terms of their value as wool producers relative to providers of manure or meat. For pigs this was less often the case, probably because the rate of growth of the animal was relatively fast and turnover was correspondingly rapid. In the case of pigs the slaughter weight was determined more by the intended end product—whether as a porker, hog, or pig—rather than by age. To a significant degree breeding affected end use. Long-wool sheep were less suited to folding than short wool, partly because the length of wool made the animals too heavy in wet weather, and they were also more susceptible to foot rot. In contrast, it was the breeding of short-wool sheep, particularly the Southdown crosses with native sheep, that produced excellent folding sheep. The increased tendency to meat eating through the period also had an influence on the type and size of animals produced.[92]

In time, specialist livestock production systems developed—breeding, store stock feeding, and fattening systems. By the mid-nineteenth century English farmers looked more to breed as a means of improving outputs than did their European neighbours, who emphasized the need to improve the nutritive quality of the animal diet through feed. The breed versus feed debate was long-running.[93]

---

[91] T. Wedge, *General View of the Agriculture of the County Palatine of Cheshire* (London, 1794), 30: T. H. Horne, *The Complete Grazier* (3rd edn. London: B. Crosby & Co., 1808), 15; H. Evershed, 'The Agriculture of Staffordshire', *JRASE* 2nd ser. 5 (1869), esp. 269–86; L. G. L. Guilhaud de Lavergne, *The Rural Economy of England, Scotland and Ireland* (Edinburgh, 1855), 36; G. D. Amery, 'The Writings of Arthur Young', *JRASE* 3rd ser. 85 (1924), 182–3.

[92] W. Marshall, *The Rural Economy of Norfolk*, i (London, 1781), 332; *Farmers Magazine*, 2 (1801), 133; J. C. Curwen, *Hints on Agricultural Subjects* (2nd edn. London: J. Johnson, B. Crosby & Co., 1809), 131.

[93] T. Brown, *General View of the Agriculture of the County of Derbyshire* (London, 1794), 21–2; W. James and J. Malcolm, *General View of the Agriculture of the County of Surrey* (London, 1794), 28; C. Webster, 'On the Farming of Westmorland', *JRASE* 2nd ser. 4 (1868), 12; J. B. Lawes and J. H. Gilbert, 'The Feeding of Animals for the Production of Meat, Milk and Manure, and for the Exercise of Force', *JRASE* 3rd ser. 6 (1895), 49.

The development of production systems had implications for feed requirements, and for the condition of the animals measured through age at slaughter and carcass weights, as well as for market turnover. From the eighteenth century grass and clover leys were becoming the norm in many arable systems. The hay from additional grass helped farmers to keep animals through what were otherwise famine feed periods, and in addition it had an impact on arable fertility. The increased use of forage crop legumes pushed up the supply of nitrogen, hence protein, and had a correspondingly beneficial impact on both crop yields and livestock feed supplies. The quality of the supply also improved. The traditional winter/early spring fodder famine was solved as new arable crops entered rotations. Some became main rotational crops, like turnips and swedes. Some were catch crops taken during fallow periods between main crop harvesting and new planting. Figure 3.9*a* gives an indication of the range of traditional arable crops grown on the arable primarily with animal feed in view, and Fig. 3.9*b* shows new crops grown for the same purpose. Oats and turnips were the two most important feed crops, and barley grew in importance from the 1840s. The increasing use of new feed crops, including swedes, mangolds, cabbage, cole, and rape, was reflected in their introduction into the rotation, often as catch crops. This is particularly noticeable from Fig. 3.9*b*. Each dead period in the feeding calendar that could be filled improved the output of feed and had significant knock-on effects for livestock output, measured in terms both of quality and quantity. All of the new crops filled a special niche during the annual cycle.[94] Malden's depiction of the integration of sheep rearing with arable cultivation is helpful. In the winter, from November to February, there were available swedes, white turnips, cabbage, kohlrabi, grass, turnip tops, and hay; in early spring, March and April, mangolds, kale, tares, winter grains, water meadows, turnip tops, rape, and hay were used; in late spring and through the summer from May to August there were clover and seed, grass, trifolium, early cabbage and rape, vetches, sainfoin, and kale; and finally in the autumn in September and October there were white turnip, cabbage, late rape, the stubble, grass, young seeds, hay, early kale, kohlrabi, sainfoin, and mustard.[95]

---

[94] The literature on these processes is immense. Mid-19th-century contemporary thought can be found in C. S. Read, 'On the Farming of Oxfordshire', *JRASE* 1st ser. 15 (1854), 268; S. Jonas, 'On the Farming of Cambridgeshire', *JRASE* 1st ser. 7 (1846), 46–7.
[95] W. J. Malden, *Sheep Raising and Shepherding* (London: Gill, 1899), 93; *AHEW* vii. 570–3 discusses the different forage and root crops available as animal feed.

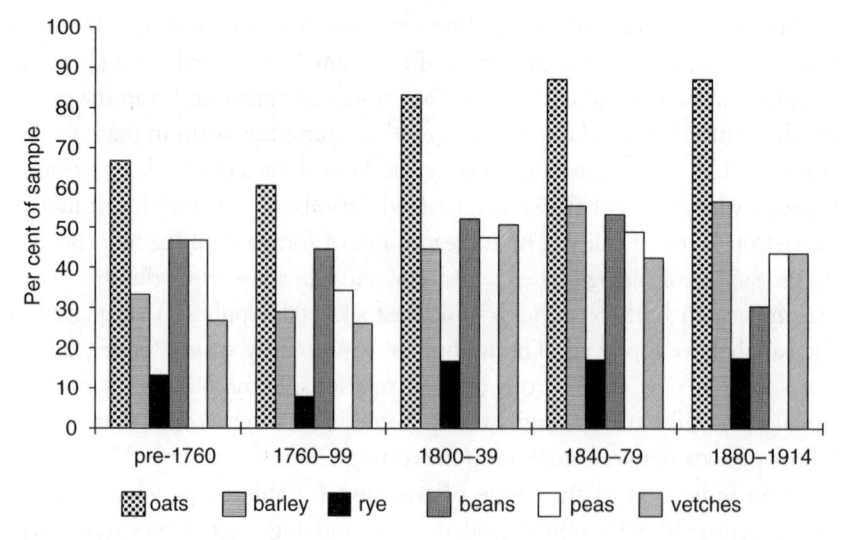

Fig. 3.9a Traditional arable crops used for livestock feeds, 1700–1914

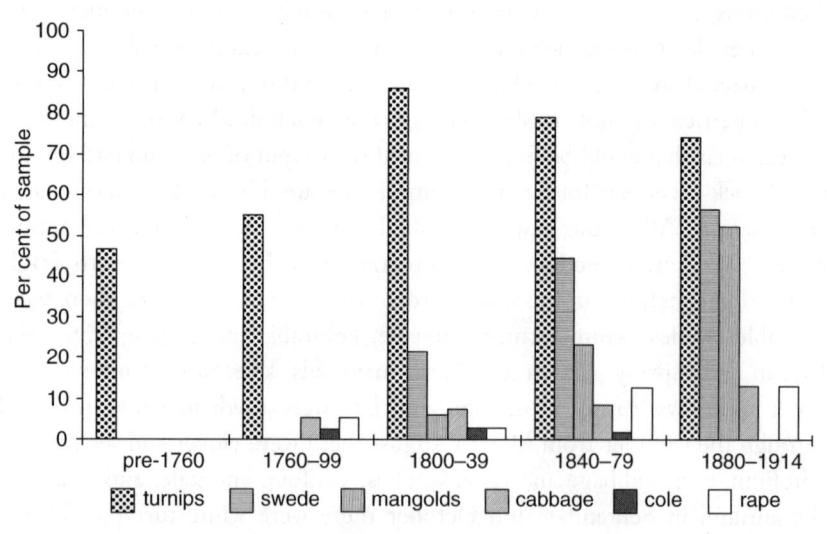

Fig. 3.9b Non-traditional arable crops used for livestock feeds, 1700–1914

A crucial change came with the purchase of off-farm feeds, including cake (used as a feed), maize, and corn (Fig. 3.10). These were supplements rather than substitutes for on-farm feed, although in times of severe shortage they could act as substitutes. In turn, the plant nutrients from the feeds were beneficial to the farm when the resulting manure from those animals feeding on

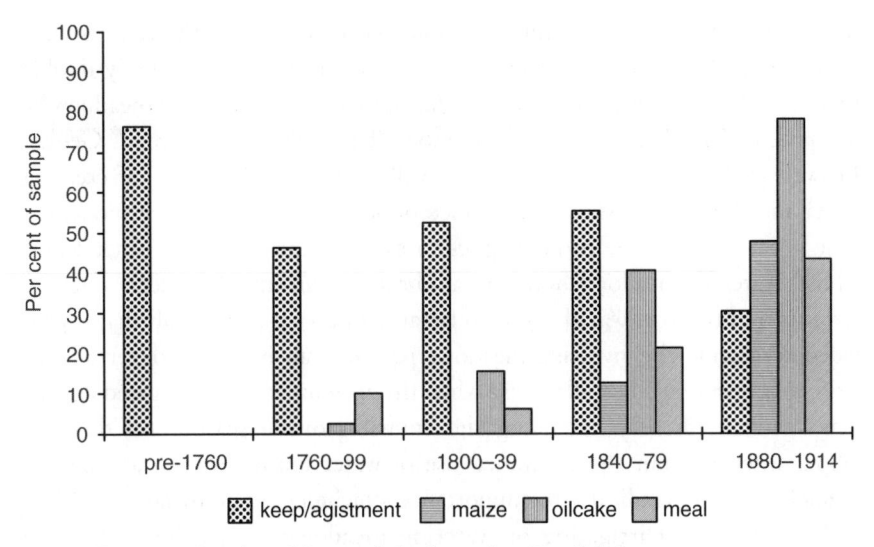

Fig. 3.10 Purchased livestock feeds used on English farms, 1700–1914

the artificial feeds was applied to the land. A farmer at Bungay in Suffolk kept a record of 'the quantities and value of oil cake, corn etc used for bullocks, sheep and cows in each year from Michaelmas 1837 to Michaelmas 1847'. He offset some of his costs against the sales of hay. In 1837–8 he bought 2 tons of oilcake for £18, about 43 per cent of his outlay on feed. The quantity he bought increased to 5 tons in each of 1838–9 and 1839–40, 7 tons in 1840–1, and 9 tons in 1841–2. His purchases declined over the next two years, but in 1844–5 he bought 8 tons of oilcake, and 13 tons in each of 1845–6 and 1846–7. His renewed application of cake may have been encouraged by a fall in the price, from £9 or £10 in the late 1830s to about £8 16s. by the late 1840s.[96] On John West's Lincolnshire farm £398 was spent on purchased feeds in 1860–1 (10.5 per cent of total expenditure) and £1,103 in 1888–9 (18.5 per cent of expenditure).[97]

Over a long period, transcending the introduction of new crops and bought-in feeds, there were developments in the management of the land. One of these was irrigation through a system of water meadows. Floating of water meadows ensured that the hay and grass output for animals was improved by the irrigation, providing spring growth four to six weeks ahead of dryland meadows, a reliable hay crop in July, and further growth during

[96] Suffolk RO, Lowestoft, 1057/2/6.   [97] Lincolnshire AO, West 29.

the late summer months. While the capital costs were high, the returns were impressive. Hay production from good water meadows was roughly double that of dryland meadows, and where a farmer had access to water meadows he was potentially able to increase the number of animals overwintered.[98] George Boswell told George Culley that 'You will remember the work [of creating water meadows] is expensive, but once done, afterwards it's very moderate. One great thing is, the produce goes to assist in the manuring other lands, whilst watered meadows want none brought on them, except water.'[99] Irrigation of this sort was also valued because of the suspended silt that it provided. Watering the riverside meadows provided an insulation layer against frost, and encouraged capillary rise when the air temperature mitigated against plant growth, thereby extending the annual growing season. All this was labour intensive, but the continued use of water meadows even at times of rising labour costs indicates the importance attached to them by farmers.[100]

The practice of irrigating or watering meadowland is believed to have originated in the medieval period, and was widespread, particularly on the chalk streams of Hampshire, Wiltshire, and Dorset, by the beginning of the eighteenth century.[101] Irrigation schemes are documented in a number of farm records from 1714 through to 1848, with examples in Cheshire, Gloucestershire, Norfolk, Warwickshire, and Yorkshire, as well as in Dorset and Hampshire.[102] They were not successful everywhere. A common complaint made by contemporaries about watermeadows was that the technique produced a poor sward. In the Vale of Aylesbury irrigation was easily possible but was found to reduce the value of a hay crop by a reported 250 per cent. In Rutland, water meadows were said to become so coarse after several years of irrigation that the sward was unpalatable. The rank growth associated with water meadows hindered haymaking, the grass had to be cut at an earlier stage than on dry meadow. Even in dry years it was hard to dry the crop evenly.[103]

Other details of grassland management are more difficult to isolate in farm records. Possibly the most important calculation for pasture management

---

[98] J. Wilkinson, 'The Farming of Hampshire', *JRASE* 1st ser. 22 (1861), 290.
[99] Northumberland RO, ZCU/12-27.   [100] Afton, 'Mixed Farming' 120-1.
[101] *AHEW* iii. 315-16.
[102] RUL, LIN P323; Lincolnshire AO, DIXON 4/1-3; Hampshire RO, 102M 71/T133; ibid., 18M54 coffer 6 Box H. See also G. Bowie, 'Watermeadows in Wessex: A Re-evaluation of the Period 1640-1750', *AgHR* 35 (1987), 151-8; S. Wade Martins and T. Williamson, 'Floated Water-Meadows in Norfolk: A Misplaced Innovation', *AgHR* 42 (1994), 20-37; Afton, 'Mixed Farming,', 109-14.
[103] Afton, 'Mixed Farming', 117.

involved a farmer in arriving at the correct stocking rate, but we have been unable to calculate these from the records.[104] Too many or too few animals reduce the quality of the sward. Ploughing up and reseeding was one solution. Landowners recognized the importance of old grassland, and the need to prevent it being ploughed for arable, so they often took control of the process. In 1727 on the Grafton Estate the steward suggested to the duke that the Cow Common should be let for four years with the tenant having the liberty to plough it and take three crops of corn, and then to lay it down with 'proper grass seeds'.[105] In 1736 it was agreed that 400 acres of Micheldever Down on the Russell Estate in Hampshire should be broken up and put into tillage. In 1755 the land was returned to pasture.[106] At Castle Hill in Devon between 1762 and 1769 land was broken up, planted with turnips then barley, and then returned to grass, specifically to improve the quality and value of the land.[107] In the Loughborough area of Leicestershire tenancy agreements on the Prestwold Hall Estate in 1774 made provision for specific fields of grass to be converted to temporary tillage for several years and then to be replanted with 'a proper quantity of grass seed'.[108] Between 1808 and 1810 stipulations were laid down for land at Breedon, Tonge, and Wilson, also in Leicestershire, to be improved by summer fallowing and then laid down to grass.[109]

Another important improvement was the draining of meadow and pasture. A survey of land at Owston, Leicestershire, indicated that much of the pasture needed underdraining. At Breedon draining and 'soughing' (trenching) was undertaken between 1808 and 1810.[110] And, once drained, a field needed proper care and attention to keep the drains clear. After inspecting the Bridgewater Estate in 1827 the steward found that many of the drains were not being cleared properly and on some farms more needed to be spent on drainage. By 1837, much of the required work in the Whitchurch area had been carried out using undertiles.[111]

Manuring and liming of meadow and pasture seems to have been widely advocated. A number of examples, particularly from the later eighteenth century, can be found in farm records. At Castle Hill between 1762 and

---

[104] Brassley, in *AHEW* vii. 535, quotes evidence from Somerset in the 1850s that stocking rates increased from 20 to 33 dairy cows per 100 cultivated acres.
[105] Northamptonshire RO, G 3883. [106] Hampshire RO, 149M89/R4/6063 and 6040.
[107] Devon RO, 1262M/E1/27. [108] Leicestershire, Leicester, and Rutland RO, DE 258/D/3.
[109] Ibid. DE41/1/197/6. [110] Ibid.DE41/1/197/6.
[111] Shropshire Records and Research Centre, 212/bundle 361.

1769 both meadow and pasture were limed and dunged.[112] At Somersham, Huntingdonshire, in 1771, and at Bramhall in Cheshire in 1785, muck was spread on meadows.[113] On Castle Park Farm at Castle Hedingham, Essex, pond mud, stable dung, and wood ash were used on land laid down to grass in the 1780s and 1790s.[114] At Chulmleigh in Devon leases made provision for the use of dung on meadows in the late 1790s.[115] Meadow and pasture land was earthed and limed at Upminster in Essex between 1795 and 1799.[116] On the Bridgewater Estate in the 1830s many farmers were given an allowance for manure, most commonly for bone manuring. On Middle Wood Farm at Houlston no tenant could be found for the 109-acre farm at a rent of £90. It was decided to spend £60 on bones to be applied on 'such land as is now fit to receive it'. At Price's Farm, Brandwood, a yearly allowance of £30 on a rent of £165 for five years was made for boning.[117] In October 1803 at Didsbury in Lancashire five loads of dung were put on the meadow. A load, in this case, was just over 29 hundredweight.[118] Remarks and observations made on the Bridgewater Estate at Whitchurch in Shropshire in 1829 indicate that considerable effort needed to be made to improve the drainage on both meadow and pasture, and by 1837 this was being undertaken.[119] Just how representative these examples are is unclear. A letter of 1787 concerned with improvements to estates of the Earl of Stamford and Warrington at Breedon in Leicestershire noted that the land was very much out of condition and needed liming.[120]

One area where change was slow and halting was in animal health. Just as farmers still understood relatively little about crops and pests, so veterinary science had as yet made little progress. In 1747–8 at Moreton Say in Shropshire, the farmer noted that he had distemper in his cattle, even though there were no other cases within 4 miles of his farm. He had no idea how it had been contracted. The almanac in which he recorded this information also included numerous recipes for salves and other medication.[121] Farm records sometimes refer to the control of disease by the application of ointment, lotion, dip, 'drink', worm medicine, sheep powder, and sheep

---

[112] Devon RO, 1262M/E1/27.   [113] Norfolk RO, MC 64/19; RUL, CHE1/1/1.
[114] Essex RO, D/DMh/E61.   [115] Devon RO, 1262M/E3/7.
[116] Essex RO, D/D Jn E5.   [117] Shropshire Records and Research Centre, 611/bundle 337.
[118] Manchester Local Studies Unit Archives, City Library, 62/1/3.
[119] Shropshire Records and Research Centre, 611/bundle 337.
[120] Leicestershire, Leicester, and Rutland RO, DE41/1/197/4.
[121] Shropshire Records and Research Centre, 2125/1.

dip. In 1852-3 a farmer at Shirley in Surrey engaged in rat and mole catching, and paid for sheep ointment and cow medicine. At Harlow, Essex, in 1814 sheep, particularly Leicesters, were plagued with scab. The farmer noted the cause as being wet weather, as well as neglect by his shepherd.[122] Sheep rot (liverfluke) was a major problem for farmers. In a letter of 1787, George Boswell noted 'the problem of infestation of water meadows', and he recommended that new ones should only be stocked with a few sheep initially, with one being killed in September to see if the liver was infected. If so, his recommendation was that the rest should be sent to market immediately.[123] Autumn feeding of sheep on water meadows, whether watered or not, would cause sheep rot. By the end of the nineteenth century the development both of veterinary knowledge and the veterinary profession was bringing significant breakthroughs in the treatment of disease. In the years just prior to the First World War, the *Journal of the Royal Agricultural Society of England* carried articles on parasites in sheep and cattle (1906), tuberculosis in cows (1910), foot and mouth disease, and grassland infections (both in 1914).

Four Farms

Finally in this chapter, we move away from the aggregated data collected from the records, to look at specific cases of farmers adapting to change. East Sutton Court was the home farm for the Filmer Estate at East Sutton, near Maidstone, in Kent. As the 'Garden of England' this area is particularly well known for hop and fruit growing. We have made reference to this farm previously, but as a relatively early example of improved farming it is worth examining in greater detail. The farm consisted of 103 acres of arable, 9 acres of meadow, 8 acres of woodland, and 80 acres of plantations. Altogether there were fifteen fields, of which the largest was 11.4 acres, and the smallest only 3 acres. From 1740 one of the fields was split, and sown as two separate fields of about 6 and 4.5 acres respectively. As a result, the average field size changed from just under 7 acres prior to 1740 to slightly over 6.5 acres. The meadow land was divided into four pieces, but the individual sizes are unknown. There was also a hop garden.

---

[122] Essex RO, D/D4/676.   [123] Northumberland RO, ZCU 12-27.

The arable fields were regularly conditioned with lime. Soot and road mould were also used on them. Considerable emphasis was placed on nitrogen-fixing crops, including clover, trefoil, sainfoin, lucerne, peas, beans, vetches, and tares. If we assume that the grassland and meadows also contained clovers and other legumes, then 56 per cent of the holding benefited from the nitrogen produced by such crops. When these plants are left for more than one year, nitrogen fixing is enhanced, and it was common practice on this farm to undersow spring corn with clover/trefoil leys that remained growing on the field for a second year. The crops were regularly weeded and turnips were hoed, but apart from references to mole catching and one to digging out anthills, nothing else was noted about pest control. Sheep were fed on the arable, specifically on the turnips, and the leys were also used as pasture. There were sheep, dairy and fatting cattle, pigs, horses, and ducks on the farm. Manure from the animals was routinely recycled.

Inputs, essentially seed and livestock, were bought locally and from further afield. Seed was traded, but also bought at local markets and purchased from a seed merchant in London. Livestock were brought from as far away as Andover in Hampshire. Specific varieties of turnip, peas, and clovers are recorded in the accounts, as well as a number of different sheep breeds. This was a progressive farm by any standards, and particularly for the period.[124]

William Clare was a tenant farmer at Coton Hall Farm, near Bridgnorth in Shropshire during the middle years of the eighteenth century.[125] His farm was essentially a pastoral enterprise. An inventory of his farm stock taken at Lady Day in 1765 is reproduced in Table 3.6. Almost 63 per cent of the total value of his stock came from livestock with just over 26 per cent from arable, including crops in the ground. This is reflected in his accounts with his average income from livestock and livestock product sales between 1765–6 and 1767–8 at over 56 per cent (Table 3.7). Of this, almost nine-tenths was fat stock, with the rest coming from occasional sales of store stock and horses and routine sales of butter, cheese, and wool. Arable sales accounted for slightly over 39 per cent of his sales including wheat, barley, and mixed corn. The arable also grew oats, peas, vetches, and temporary clover and ryegrass leys. These, along with pasture and meadowland, were used for feeding the livestock.

Clare's livestock was an important part of his farming. His expenditure on purchased livestock feeds was over 29 per cent of his total expenditure. This

[124] Centre for Kentish Studies, U120.   [125] RUL, SAL 5/1.

Table 3.6 Inventory of farm stock taken Lady Day 1765, Coton Hall Farm, Bridgnorth, Shropshire

			£	s	p
Fat Cattle					
Oxen	Turk & Tocan £17. Tag & Trulove £19. Darling £9 Captain & Carver £21. Swan & Sweeling £21. Mir & Mahomet £25. Mark & Merryman £20. Buck & Ballface £20. Duke £11. Little Duke £5:10		168	10	—
Cows	Wotton £8. Harris £8. Beauty £6. Veal £4:10. Tomlinson £5:10. Brecknell £5:5. Jordan £5 Williams £4:10. Dolittle £4.		50	15	—
Store Cattle	Ludlow and Calf £7:7. Molly and B.Calf £7:7. Willis milch £6. Willis incalf £6:6		27	—	—
3 years old	Redfinch £6. Grizzle £5:5		11	5	—
	Veals 2		10	—	—
2 years old	£4 £10		14	—	—
	Yearling Molly		3	3	—

	Sheep	£	s	d			
Fat Welch 53 at s9		23:	17:	—			
6 Couples £2:8 a dry Ewe s6		2:	14	—	35	—	6
23 Yearlings s6 each £6:18 6 Rams £1:11:6		8:	9:	6			

	Pigs				
7 Stores £4:11. 2 Dº £1:4		5	15	—	

Horses

		£	s	p
ay Colt £7 ... Chesnut Gelding £7:7. Strong brown Horse coming 5 £6:6 lame ... dark bay Gelding coming 2 £5. Yearling Fillys £8:847		11	—	—
Cart	2 bay Geldings £6. Lion £2. Duke £3. Heriot £3. black Mare £5. Curtain £5. brown Mare £3.	27	—	—
Waggons, Turnbrels, Plows, horse & Oxgearing & other Implements of Husband[x]		63	15	—

	£	s	d			
Wool	4	10	—			
Wheat 169 Bushels at s6	50	14	—			
Mixed Corn 150 Bushels at s6	45	—	—	133	1	—
Barley 88? Bushels at s3 ... 60 Dº 20 Dº at s3	25:	4:	6			
Oats 29 Bushels at s2:6	3:	12:	6			
Peas 32 Bushels at s2:6	4:	—	—			

Total Inventory £596:15:6

		£	s	p
Horse Lesow Corn 20 Acres at £2 20 Bushels own M:Corn sowd		40	—	—
Stables &c 1764		10	4	10 ½

Source RUL, SAL 5/1.

Table 3.7 Expenditure and income, Coton Hall Farm, Bridgnorth, Shropshire, Lady Day 1765 to Lady Day 1768

(a) Expenditure

Product	% of total
Store stock	21.0
store cattle (13.9%)	
store sheep (4.0%)	
store pigs (1.2%)	
horses (2.0%)	
Feed	29.3
oats and beans (14.4%)	
oil cake (14.9%)	
Manure	2.2
Seed	2.4
Labour	27.4
general labour (7.3%)	
hay harvest (0.7%)	
corn harvest (1.0%)	
threshing (2.4%)	
hedging/ditching (0.3%)	
servants (2.9%)	
craftsmen (4.8%)	
thatching (0.4%)	
sundries (7.6%)	
Rent	15.9
Rates and taxes	1.8

(b) Income

Product	% of total
Fat stock	49.3
Fat cattle (38.6%)	
fat pigs (5.1%)	
fat sheep (5.6%)	

Table 3.7 *Continued*

(b) Income

Product	% of total
Store stock	5.2
Store cattle (2.0%)	
store pigs (0.7%)	
store sheep (1.5%)	
horse (0.9%)	
Other animal products	1.9
wool (0.6%)	
dairy (1.3%)	
Arable crops	39.1
wheat (12.0%)	
barley (14.9%)	
mixed corn (11.8%)	
peas (0.5%)	
Other income	4.5

Source RUL, SAL 5/1.

was more than he spent on new stock (21 per cent), labour (27.4 per cent), or rent (15.9 per cent). Approximately half was used to buy oats and occasionally beans. During the three years he also purchased 66 tons 500 pounds of oilcake.

Although Clare was primarily interested in the livestock side of his farm, his crops were not neglected. He regularly purchased wheat, rye, clover, and ryegrass seed along with lime, soot, horn shavings, and muck. His implements of husbandry included both ox and horse ploughs.

In the nineteenth century John West of Dunholme, Lincolnshire, in the arable east of England, was one farmer who saw the need to move with the times.[126] He had a mixed farm of about 1,000 acres. During the High Farming period he used a four-course rotation of turnips or swedes, barley, short-term ley, and wheat. His livestock included cattle and sheep. Sheep breeding was considered the most profitable of the livestock enterprises. Both cattle and

[126] Lincolnshire AO, West 1–32.

Table 3.8 Products sold from John West's holdings at Dunholme and Washingborough, Lincolnshire (% of total income)

	1870–1	1885–6
Arable		
wheat	28.53	18.38
barley	15.95	35.76
peas		2.08
potatoes		0.72
straw	0.04	
Total	44.52	56.94
Livestock		
sheep	32.18	26.34
wool	6.84	4.20
cattle	15.80	11.80
horses	0.66	0.72
Total	55.48	43.06

Source Lincolnshire AO, WEST 11 and 26.

sheep were fed corn and cake during the winter. In addition to manure recycled from his livestock, West purchased guano, cake, blood, superphosphate, bone dust, salt, lime, ash, and night soil. In 1870 he spent £536 on fertilizers and seed, and £670 on oilcake and other livestock feeds.[127] By the 1880s West was faced with the problem of adapting to the depression in order to remain in business. Table 3.8 shows the proportion of his farm income from the sale of various products, and how it changed as the depression proceeded. The balance of crops in the arable rotation was reversed between the early 1870s and the 1880s. Sales of wheat accounted for almost 29 per cent of total income in 1870–1 but less than 19 per cent in 1885–6. High-quality malting barley became West's principal product by 1885–6, and a considerable proportion of

[127] Ibid., West 11, 26.

the crop was sold to Bass breweries. Malting barley from the east of England commanded a premium price well above that paid for feed barley. So important was Lincolnshire production to Bass that in 1901 the company built malting premises in Sleaford.[128] A relatively low nitrogen level in the soil was crucial for the production of a good sample of malting barley, and this had implications for the manure requirements. By concentrating on this crop, and using less wheat in the rotation, West's manure bill was significantly reduced. In 1885–6 the bill for fertilizer and seed fell to £312 of which nearly £60 was spent on seed barley. At the same time £1,160 was spent purchasing cake and feed corn. West increased his livestock, with a change in emphasis from breeding to meat production. By altering the system and reallocating resources from the arable to the livestock sector, West was able to tap into the rapidly growing meat market by producing fat cattle and sheep for sale in the towns of the industrial Midlands and north of England. West was a lively, entrepreneurially minded farmer, who adapted to changing circumstances.[129]

Upton Farm at Sompting in West Sussex was a 340-acre mixed farm on the lightlands of the south coast of England.[130] In the 1840s and 1850s sheep, together with wheat, barley, and oats, provided more than 85 per cent of the value of the products sold from the farm (Table 3.9). In the late 1840s, a rail link with London was opened, making it possible to send perishable goods to the large urban market.[131] Between 1860 and 1864 the sheep flock was sold and the farmer, William Barker, replaced it with a larger dairy herd and a fat cattle herd. The shift in emphasis towards greater livestock production was heightened by the reintroduction of a sheep flock in the late 1870s. This is reflected in the increasing importance of purchased feeds and the relative decline in the purchase of corn seed and manures. Adaptation on this farm was already under way well before the price changes that came about in the 1870s and 1880s. The depression reinforced the messages from the changing marketing opportunities that had already been adopted, and dairy

---

[128] C. C. Owen, *'The Greatest Brewery in the World': A History of Bass, Ratcliff & Gretton* (Chesterfield: Derbyshire Record Society 19, 1992); Jonathan Brown, *Agriculture in England: A Survey of Farming 1870–1947* (Manchester: Manchester University Press, 1987), 38.

[129] West also overcame domestic adversity. Cornelius Stovin noted in 1872 that his wife had died in childbed leaving him a widower with nine children to bring up: *Journals of a Methodist Farmer*, 117.

[130] West Sussex RO, Add. MSS 22769–73.

[131] On milk generally see David Taylor, 'The English Dairy Industry, 1860–1930', *EcHR* 29 (1976), 585–601. On London's retail milk trade see P. J. Atkins, 'The Retail Milk Trade in London, c.1790–1914', *EcHR* 33 (1980), 522–37; id., 'The Growth of London's Railway Milk Trade, c.1845–1914', *Journal of Transport History*, 4 (1978), 208–26.

Table 3.9 Average annual income and expenditure from Upton Farm, Sompting, Sussex (% of total)

(a) Income (from Michaelmas)

	1840s	1850s	1860s	1870s	1880s
Wheat	32.6	34.5	34.9	29.2	18.9
Barley[a]	13.3	7.6	5.4		
Oats, peas, tares, etc.	5.3	13.5	21.3	21.9	13.5
Sheep[b]	34.5	32.2	7.0	4.7	16.3
from Cows	5.3	5.5	15.0	20.4	27.7
Beasts[c]		1.0	10.0	13.3	7.1
Hogs	3.0	2.0	2.6	0.9	0.9
Furze[d]	2.9	0.8			
Hay and straw[e]				5.7	6.7
Misc.	3.2	2.8	3.8	3.8	9.0

(b) Expenditure (from Michaelmas)

	1840s	1850s	1860s	1870s	1880s
Labour	35.4	23.1	22.3	17.5	22.8
Rent and rates	33.5	19.0	17.2	13.4	13.3
Tradesmen's bills	10.7	15.8	26.6	33.0	26.0
Livestock	4.5	5.2	7.5	10.6	12.6
Corn, seed, and oilcake[f]	7.1	20.0	15.4	12.8	13.7
Manure	0.9	11.7	7.1	6.2	3.0
Misc.	7.9	5.2	3.9	6.5	8.6

[a] Not noted from 1865.
[b] Not noted between 1865 and 1878.
[c] Not noted until 1859.
[d] Furze sales noted until 1860.
[e] Hay and straw noted from 1874.
[f] Oilcake included in category from 1859.

Source West Sussex RO, Add. MSS 22769–73.

and meat production increased to meet the increase in consumption in local towns, and further afield via the rail network.

Conclusion

English farmers preferred tried and tested methods to new and experimental ideas, and yet, as our evidence confirms, change was taking place.[132] Agriculture moved, albeit slowly, from a low input/low output position in the early eighteenth century, to a higher input/higher output agricultural economy in the mid-nineteenth century, and this change came about primarily through the highways and byways of experimentation and innovation, whether of fertilizers, crops, and rotations, or animal breeding and feeding. The distance travelled was such that whereas the trials and tribulations for agriculture in the 1690s had been met by near-famine conditions, and those of the 1790s by enclosure (itself a form of innovation), in the 1880s and beyond the more entrepreneurial farmers could adapt quite successfully to altered conditions by changing their interests to suit the prevailing circumstances. Although deficient in some areas, the farm records show these changes. As such they give us confidence in their reliability as we move in the following chapters to look at how farming practice was reflected in production and output.

---

[132] Macdonald, 'The Diffusion of Knowledge', 30–9; N. Goddard, 'Information and Innovation in Early-Victorian Farming Systems', in B. A. Holderness and M. Turner (eds.), *Land, Labour and Agriculture, 1700–1920* (London: Hambledon, 1991), 165–90; id., 'The Development and Influence of Agricultural Periodicals and Newspapers, 1780–1880', *AgHR* 31 (1983), 116–31; id., *Harvests of Change* (London: Quiller Press, 1988).

CHAPTER 4

The Wheat Question

Medieval historians have always had available a substantial database of direct or near-direct measures of land productivity. The great seigneurial and ecclesiastical estates have left a legacy of land productivity measures that are the envy of modern historians.[1] It was not for another 600 years, until the national agricultural statistical database was conceived and put into place from the 1860s onwards, that such a rich collection of direct land productivity data again became available. Even so, by various routes, details of which will unfold in this and the next two chapters, we are able to piece together a long-run history of crop yields and animal output from the late seventeenth century to the 1880s. We begin with wheat.

Wheat, as the principal cash crop by the eighteenth century, and increasingly as the principal bread grain,[2] has always been regarded as the most sensitive indicator of agricultural productivity. Inevitably, therefore, it is the crop for which the long-run trend has been researched and analysed the most assiduously. The data currently available are given in Table 4.1, which combines contemporary estimates such as those of Arthur Young with figures available in the modern literature. The table brings together several

---

[1] Amongst a large literature perhaps the most obvious to see is J. Z. Titow, *Winchester Wheat Yields: A Study in Medieval Agricultural Productivity* (Cambridge: Cambridge University Press, 1972). But in terms of advancing the debate about medieval productivity see K. Biddick, 'Agrarian Productivity on the Estates of the Bishopric of Winchester in the Early Thirteenth Century', B. M. S. Campbell, 'Land, Labour, Livestock, and Productivity Trends in English Seignorial Agriculture, 1208–1450', and C. Thornton, 'The Determinants of Land Productivity on the Bishop of Winchester's Demesne of Rimpton, 1208 to 1403', all in Campbell and Overton (1991), 95–123, 144–82, 183–210.

[2] See E. J. T. Collins, 'Dietary Change and Cereal Consumption in Britain in the Nineteenth Century', *AgHR* 23 (1975), 97–115; C. Petersen, *Bread and the British Economy c.1770–1870* (Aldershot: Scolar Press, 1995); W. Ashley, *The Bread of our Forefathers* (Oxford: Oxford University Press, 1928).

Table 4.1 English wheat yields, c.1675–1914 (bushels per acre)

(a)

Date	Bushels/acre	Specific location (where known)	Source
c.1680	15.4	Norfolk/Suffolk	Overton
c.1675–99	14.7	Lincs.	Overton
c.1675–99	19.0	Herts.	Glennie
c.1675–99	21.8	Oxon.	Allen
c.1680–1709	15.9	Norfolk	Overton
c.1680s	16.9	Hants.	Glennie
c.1685–90	17.7	Norfolk/Suffolk	Overton
c.1690s	13.8	Hants.	Glennie
c.1700–25	16.5	Lincs.	Overton
c.1700–25	21.5	Oxon.	Allen
c.1710	14.8–15.6	Norfolk/Suffolk	Overton
c.1720	18.5	Norfolk/Suffolk	Overton
c.1725	19.2–19.5	Norfolk/Suffolk	Overton
c.1725	18.1	Lincs.	Overton
c.1730	20.2	Norfolk/Suffolk	Overton
c.1735	18.7	Norfolk/Suffolk	Overton
c.1750	15.0		Charles Smith
c.1770	23.8		Arthur Young
1794	16.8		Harvest inquiries
1795	15.6		Harvest inquiries
1800	21.9		Harvest inquiries
1801	22.6		1801 crop returns

(b)

	Bushels/acre			
	(a)	(b)	(c)	(d)
1815	37.0	25.7		
1816	25.3	17.6		

Table 4.1 *Continued*

	Bushels/acre			
	(a)	(b)	(c)	(d)
1817	33.4	23.2		
1818	32.6	22.6		
1819	27.7	19.2		
1820	37.3	25.9		
1821	30.9	21.5		
1822	30.9	21.5		
1823	25.8	17.9		
1824	29.0	20.1		
1825	34.9	24.2		
1826	35.3	24.5		
1827	34.7	24.1		
1828	27.1	18.8		
1829	27.7	19.2		
1830	33.6	23.3		
1831	30.3	21.0		
1832	35.7	24.8		
1833	34.3	23.8		
1834	41.5	28.8		
1835	32.8	22.8		
1836	36.7	25.5		
1837	34.0	23.6		
1838	33.1	23.0		
1839	31.2	21.7		
1840	45.1	31.3		
1841	39.7	27.6		
1842	54.8	38.1		
1843	50.0	34.7		
1844	54.5	37.8		
1845	44.9	31.2		

Table 4.1 *Continued*

	Bushels/acre			
	(a)	(b)	(c)	(d)
1846	45.7	31.7		
1847	46.4	32.2		
1848	44.1	30.6		
1849	57.0	39.6		
1850	41.8	29.0		
1851	48.9	34.0		
1852	45.0	31.3	23.3	
1853	37.9	26.3	21.3	
1854	57.3	39.8	35.4	
1855	46.3	32.2	27.9	
1856	52.7	36.6	27.5	
1857	57.3	39.8	33.7	
1858	57.9	40.2	32.0	
1859	55.1	38.3	26.5	
1860			22.5	
1861			25.7	
1862			29.8	
1863			39.4	
1864			35.9	
1865			31.1	
1866			25.5	
1867			21.4	
1868			34.7	
1869			27.5	
1870			30.5	
1871			24.4	
1872			24.4	
1873			22.9	
1874			29.8	

Table 4.1 *Continued*

	Bushels/acre			
	(a)	(b)	(c)	(d)
1875			23.3	
1876			25.4	
1877			27.0	
1878			30.5	
1879			15.8	
1880			24.9	
1881			24.4	
1882			26.0	
1883			28.5	
1884			29.9	
1885			30.8	31.5
1886			29.8	26.9
1887			28.9	32.3
1888			27.4	28.2
1889			30.0	29.9
1890			32.0	30.8
1891			30.0	31.3
1892				26.2
1893				25.8
1894				30.7
1895				26.2
1896				33.9
1897				29.0
1898				34.8
1899				32.8
1900				28.4
1901				30.8
1902				32.8
1903				30.1

Table 4.1 *Continued*

	Bushels/acre			
	(a)	(b)	(c)	(d)
1904				26.5
1905				32.7
1906				33.6
1907				34.0
1908				32.2
1909				33.6
1910				30.2
1911				32.6
1912				28.7
1913				31.3
1914				32.4

Sources 1689–1730: based on simplified versions of M. Overton, 'Estimating Yields from Probate Inventories: An Example from East Anglia, 1585–1735', *JEH* 39 (1979), 363–78; id., 'The Determinants of Crop Yields in Early Modern England', in Campbell and Overton (1991), 284–322; id., 'Re-estimating Crop Yields from Probate Inventories', *JEH* 50 (1990), 931–5; R. C. Allen, 'Inferring Yields from Probate Inventories', *JEH* 48 (1988), 117–25; P. Glennie, 'Measuring Crop Yields in Early Modern England', in Campbell and Overton (1991), 255–83.
1750–1801: based on M. E. Turner, 'Agricultural Productivity in England in the Eighteenth Century: Evidence from Crop Yields', *EcHR* 35 (1982), 503.
Column (a) taken from: M. J. R. Healy and E. L. Jones, 'Wheat Yields in England, 1815–59', *JRSS* 125 (1962), 578.
Column (b) ibid., but reduced in the ratio 50:72 on the suggestion of the authors, 576, because the original estimates were based on a wheat acreage that included paths, headlands, hedges, etc.
Column (c) taken from J. B. Lawes and J. H. Gilbert, 'Home Produce, Imports, Consumption, and Price of Wheat, over Forty Harvest-Years, 1852–53 to 1891–92', *JRASE* 3rd ser. 4 (1893), 133, rounded to the first decimal place.
Column (d) taken from *AHEW* vii. 1788.

series, which provide the context for our new estimates (Table 4.4). We shall discuss the existing data, our own estimates, and the effect of our findings on an understanding of the long-run trend.

Existing Estimates of Wheat Yields

Table 4.1 brings together the existing estimates of wheat yields over the period *c.*1680–*c.*1914. The data are clearly skewed with only a handful of date-specific contemporary estimates for the years between the 1730s and 1815, the key period about which we need to know more if we are to understand the agricultural revolution. For the period c.1680 to c.1735, the figures come from a method of translating the monetary valuations in probate inventories into a direct measure of land productivity through crop yields. The method was pioneered by Mark Overton,[3] and variations were proposed by R. C. Allen[4] and Paul Glennie.[5] Allen's revision of the method had the effect of raising the estimates of early modern crop yields, but Glennie's had the effect of lowering them again, although not to the level the original method had suggested.

Table 4.2 gathers together the estimates of wheat yields derived from probate inventories. It is based on the boundary dates employed by Overton, Allen, and Glennie. The variations in date arise partly because of the random survival of probate inventories, partly from the problems of sample selection which inevitably follow, and partly from questions of how the estimates of yield should be used. On the last point Glennie has argued that there is an important difference between identifying general yield trends over relatively long periods, with which Overton was primarily concerned, and identifying more or less absolute levels of crop yields, which was Allen's main interest.[6] However, the way these yields have been calculated is more or less the same regardless of the purpose for doing so. The yields are inferred by comparing valuations per acre of growing grain with valuations of stored grain per bushel after the same harvest.[7]

Table 4.2 also summarizes the outcome of inferring crop yields from probate inventories. Vertically it employs those short- and sometimes long-run windows of time that the original researchers adopted. Horizontally the table reflects the variations on the method, and the findings from the different counties in which it has been employed. The only columns that can be properly compared are the three for Hampshire, because they are drawn from the same sample. Column 4 is taken from Glennie's analysis of

---

[3] M. Overton, 'Estimating Yields from Probate Inventories: An Example from East Anglia, 1585–1735', *JEH* 39 (1979), 363–78; id., 'The Determinants of Crop Yields in Early Modern England', in Campbell and Overton (1991), 284–322; id., 'Re-estimating Crop Yields from Probate Inventories', *JEH* 50 (1990), 931–5.

[4] R. C. Allen, 'Inferring Yields from Probate Inventories', *JEH* 48 (1988), 117–25.

[5] P. Glennie, 'Measuring Crop Yields in Early Modern England', in Campbell and Overton (1991), 255–83.

[6] Ibid. 256.

[7] This is based on ibid. 258, esp. 9, which indicates some subtleties in the process of estimation. Nevertheless this was the essence of the method employed.

Table 4.2 English wheat yields based on probate inventories, 1673–1749 (bushels per acre)

	Overton (Norfolk/ Suffolk)	Overton (Norfolk/ Suffolk)	Overton (Lincs.)	Glennie (Hants.)	Glennie (Hants.)	Glennie (Hants.)	Glennie (Herts.)	Allen (Oxon.)
	1	2	3	4	5	6	7	8
1673–82	15.4							
1675–99			14.7				19.0	21.8
1680s				11.0	15.8	16.9		
1690s				8.6	10.5	13.8		
1680–1709		15.9						
1685–90	17.7							
1700–24			16.5					
1700–27								21.5
1700–39		18.5						
1700–49			18.1					
1708–11	14.8							
1709–16	15.3							
1713–16	15.6							
1710–39		19.2						
1718–31	19.5							
1721–25	19.5							
1725–49			18.7					
1727–32	20.2							

Sources As for Table 4.1, but specifically, Overton, 'Re-estimating Crop Yields', 934–5; Overton 'The Determinants of Crop Yields', 302–3; Glennie, 'Measuring Crop Yields', 273, 278; Allen, 'Inferring Yields', 123.

Hampshire probate inventories and employs Overton's original method. Column 6 is Glennie's demonstration of the difference that ensues when the same probates are analysed using Allen's adjustment of the technique. Finally, column 5 shows the effect of carrying out the same exercise but now using Glennie's further revisions of the technique. These figures therefore put into perspective the dramatic differences in outcome according to which method of estimation is employed. While the basic methodology may be plausible, if it produces these contrasting estimates for *c.*1700 there are serious implications for using them to construct a long-run history of English

Table 4.3 Estimates of wheat yields drawn from Hampshire inventories (in bushels per acre and index numbers)

	Original Overton (bushels/acre)	Revised Allen (bushels/acre)	Revised Glennie (bushels/acre)
1600–9	8.0 (100.0)	13.1 (100.0)	11.1 (100.0)
1619–28	9.1 (113.8)	14.5 (110.7)	12.2 (109.9)
1639–46	9.1 (113.8)	15.1 (115.3)	12.2 (109.9)
1659–69	10.2 (127.5)	15.4 (117.6)	14.1 (127.0)
1683–9	11.0 (137.5)	16.9 (129.0)	15.8 (142.3)
1690–9	8.6 (107.5)	13.8 (105.3)	10.5 (94.6)

Note The three estimates have been drawn from the same sample of Hampshire probate inventories used in Table 4.2. Index numbers in brackets; 1600–9=100.
Source Glennie, 'Measuring Crop Yields', 273.

crop yields and land productivity, particularly for identifying major turning points.[8]

Arguably, as long as the *trend* in yields from the probate inventory material alone is being analysed it hardly matters which method is used. This is demonstrated for Hampshire in Table 4.3.[9] From trough-to-peak and peak-to-trough the Allen method produces the least variation and the Glennie revision the greatest variation, but in spite of these varying amplitudes the direction of the trend is the same in all three cases. Interpretations of the seventeenth-century trends, therefore, would differ in only fairly minor ways. However, in the *longer-run trend* from the medieval period to the eighteenth century it makes a significant difference which absolute level of estimated yield is adopted in the late seventeenth century. Adopting Allen's method suggests that the medieval high point in yields was re-attained by or in the seventeenth century, but adopting Glennie's method suggests that it was delayed until the eighteenth century. The choice of method materially influences the recognition of turning points, and this is obviously of great importance when we are trying to identify the agricultural revolution.

---

[8] G. Clark, 'Yields Per Acre in English Agriculture 1250–1860', *EcHR* 44 (1991), 445–60, esp. 447–8.
[9] Glennie, 'Measuring Crop Yields', 273.

Clearly there are problems with the method of inferring crop yields from probate inventories, but the body of estimates for the quarter-century or so before 1700 suggest yields of just over 15 bushels per acre in Norfolk and Suffolk, just under 15 bushels per acre in Lincolnshire, and something between 12 and 16 bushels per acre in Hampshire. Much higher yields seem to have been achieved in Hertfordshire and Oxfordshire, ranging between 19 and 22 bushels per acre.

Detailed probate inventories tend to run out by about 1730, and figures in Table 4.1 for the rest of the eighteenth century rely on contemporary estimates. As with most of Arthur Young's figures, his 23.8 bushels per acre around 1770 is likely to be optimistic,[10] but no further national data-sets are available until the harvest inquiries of the 1790s, and the appropriate parts of the 1801 crop returns. These have been used by Turner to translate point-specific measures into county, regional, or national estimates of crop yields.[11]

After the Napoleonic wars there are no consistent data-sets available until the reports drawn up by assistant tithe commissioners in conjunction with tithe commutation in the wake of the 1836 Tithe Commutation Act. These reports do not cover the whole country, and they relate only to the tithable portion of the produce. They were estimates of what might be produced *on an average*, and the assistant tithe commissioner simply estimated the quantities involved. To these problems can be added the likely bias arising from the fact that many of the parishes commuting tithes after 1836 were still unenclosed, and such parishes were not likely to be noted either for progressive or innovative farming. Overall, estimates based on the tithe database may present a more pessimistic view of agricultural output than would be the case with a true random sample, and for this reason they are not included in Table 4.1.[12]

---

[10] M. E. Turner, 'Agricultural Productivity in England in the Eighteenth Century: Evidence from Crop Yields', *EcHR* 35 (1982), 503.

[11] Ibid. 489–510; M. Overton, 'Agricultural Productivity in Eighteenth-Century England: Some Further Speculations', *EcHR* 37 (1984), 244–51; M. E. Turner, 'Agricultural Productivity in Eighteenth-Century England: Further Strains of Speculation', *EcHR* 37 (1984), 252–7.

[12] R. J. P. Kain and H. C. Prince, *The Tithe Surveys of England and Wales* (Cambridge: Cambridge University Press, 1984), *passim*; R. J. P. Kain, *An Atlas and Index of the Tithe Files of Mid-Nineteenth-Century England and Wales* (Cambridge: Cambridge University Press, 1986), *passim*; A. D. M. Phillips, 'Agriculture and Land Use, Soils and the Nottinghamshire Tithe Surveys, circa 1840', *East Midland Geographer*, 6 (1976), 288; M. Sill, 'Using the Tithe Files: A County Durham Study', *Local Historian*, 17 (1986), 205–11; J. V. Beckett and J. E. Heath (eds.), *Derbyshire Tithe Files 1836–50* (Chesterfield, Derbyshire Record Society 22, 1995), pp. xxxii–xxxiii. Kain's estimated averages for *c.*1836 of 33 and 34 bushels per acre respectively for barley and oats also seem remarkably low, and confirm our view that tithe file evidence is not a good guide to the state of cultivation generally: *An Atlas*, 460.

Although the tithe files do not help us, fortunately for the period from 1815 we have wheat yields compiled by Healy and Jones, the summary estimates of the nineteenth-century Rothamsted experiments compiled by Lawes and Gilbert for the period 1852–84, and finally the 'official' yields which become available from 1885. These figures are all included in Table 4.1. The estimates compiled by Healy and Jones were made by agents representing Liverpool corn merchants. They covered what were known as the midland and southern circuits. The former included the area bounded by Liverpool, Warrington, Coventry, Chipping Norton, Hereford, and Chester; and the latter the area bounded by Liverpool to Bradford, York, Norwich, Gravesend, Tonbridge, Swindon, and Worcester, and touching many towns in between. The mean yields are presented in two forms: as they were originally estimated (column *a*); and after deflation in the ratio 50 : 72 as Healy and Jones suggested (column *b*).[13] The Lawes and Gilbert figures (column *c*) are their estimates of the United Kingdom wheat yield for the fifty years up to 1891–2, partly based on their experiments at Rothamsted. Subsequently J. R. Bellerby used these data in his estimates of United Kingdom agricultural output and income to measure wheat output up to 1883. In addition, he applied them to the easily measurable ratio between the yield of wheat and the other grains after 1883 to estimate the yields of those other grains before 1884 and hence their outputs.[14] Although the annual agricultural returns began in 1866, 'official' crop yields were not systematically collected or estimated until 1885 (column *d*). Even then they were 'estimates' of yield taken before and after the harvest, but checked for reliability by testing weights during threshing. It was the job of crop reporters to furnish these yield estimates. Each reporter was responsible for about 80,000 acres covering perhaps 1,200 holdings ranged over forty parishes. The national estimates that were produced were then derived by a system of weighting the many sample estimates.[15]

---

[13] M. J. R. Healy and E. L. Jones, 'Wheat Yields in England, 1815–59', *JRSS* 125 (1962), 574–9, with a map on 575. A. R. Wilkes has suggested that even these deflated estimates exert an upward bias in the yield per acre in view of contemporary estimates of prices, population, and the likely extent of the wheat acreage: 'Adjustments in Arable Farming after the Napoleonic Wars', *AgHR* 28 (1980), 93.

[14] J. B. Lawes and J. H. Gilbert, 'Home Produce, Imports, Consumption, and Price of Wheat, over Forty Harvest-Years, 1852–53 to 1891–92', *JRASE* 3rd ser. 4 (1893), 77–133; RHC, Bellerby MSS, D84/8/17, *passim*. See also Chapter 5 below.

[15] On this history see J. A. Venn, *The Foundations of Agricultural Economics* (Cambridge: Cambridge University Press, 1933), 435; On the publication of these annual crop yields see BPP, *Annual Agricultural Statistics*, from 1884. See also MAFF, *A Century of Agricultural Statistics: Great Britain 1866–1966* (London: HMSO, 1968), table 55; and J. Caird, *English Agriculture in 1850–51* (2nd edn.

We have not included in Table 4.1 any data from the *Agrarian History*. Volume v lists twenty-one examples of wheat yields covering just nineteen years in the long-run period 1671–1754. Fifteen of these examples are from a single farm at Arreton on the Isle of Wight.[16] These data are included (and extended to 1778) in the new series we present in Table 4.5. Chartres took the wheat yield in 1695 to be 16 bushels per acre and suggested it had risen to 18 bushels by 1750. His estimates were compiled before the probate inventory data became widely available, but his conclusions are not entirely out of line with the data in Table 4.2.[17] A. H. John, in volume vi of the *Agrarian History*, summarized wheat and other grain yields in the period 1750–1850 by presenting a table based on standard sources: Arthur Young's testimony in *c*.1770; the Board of Agriculture *General Views* for the period 1794–1817 (as summarized by J. R. McCulloch); James Caird's estimates for 1850; and the *Mark Lane Express* estimates of 1861, all of which are reproduced from Craigie's review of 1883. The Healy and Jones estimates are also reproduced.[18] All these sources are used in this chapter. Holderness, in estimating agricultural output, settled on wheat yields of 18 bushels per acre (gross and 16 net of seed) in *c*.1750, 21.5 (19.5) in *c*.1800, 22.5 (20.5) in *c*.1810, and 28 (26) in *c*.1850. These were based on his own review of the standard sources.[19]

What does this disparate evidence about wheat yields tell us about the course of land productivity? Though the variations on the basic method of inferring yields from probate inventories makes establishing a sound base in 1700 problematic, it is a starting point from which to establish a framework into which any new estimates can be fitted. On the basis of Table 4.1 we can suggest that from a base of about 16 bushels per acre wheat yields rose to 20 or 21 bushels per acre a century later. At times from *c*.1800 to *c*.1850 they hardly rose above this figure (and indeed the tithe evidence in *c*.1840 suggests a wheat yield of only about 22 bushels per acre, thus confirming our view that data from this source needs to be handled carefully), although more generally they appeared to be closer to 28 bushels. For the second half of the nineteenth century the evidence shows no great, or even small, or, more importantly, no systematic land productivity change. The Lawes and

---

London: Longman, 1852; Repr. London: Frank Cass, 1968), 474 for an estimate for the year 1850. For estimates for 1861, 1870, 1879, and 1882, and annual averages for 1876–82 and 1863–82: P. G. Craigie, 'Statistics of Agricultural Production', *JRSS* 46 (1883), 1–58, esp. 40–2.

[16] *AHEW* v (2). 882–3.   [17] Ibid. 443–4.   [18] *AHEW* vi. 1048–52.
[19] Ibid. 134–47.

Gilbert wheat yield estimates for the United Kingdom suggest an average of nearly 28 bushels per acre for the long period 1852–3 to 1891–2. This varied from just under 29 bushels in the 1850s, to nearly 30 bushels in the late 1880s, with a low of 25–7 bushels from the late 1860s to the mid-1880s. Yet over this period the United Kingdom acreage under wheat declined from 4 to 2.4 million acres. It might have been expected that the desertion of the least productive wheat lands would occur first, thus leading to an improvement in the overall unit land productivity. But land was only one of the factor inputs; the improvement in yield in the late nineteenth century was accompanied by a significant decrease in the agricultural labour force and only modest compensating increases in fertilizers and other land improvement inputs.[20]

New Estimates of Wheat Yields

Wheat yield data from farm records are given in Table 4.4. The earliest records which report wheat yields in the new series come from farms in Hitchin and Welbury in Hertfordshire, and from farms in Essex and Kent. They date from 1720 to 1722. More broadly in the 1720s wheat yields are collated from farms in various locations in Hertfordshire, Essex, Kent, Hampshire, and East Sussex. Invariably they occur only in discontinuous runs of years, though occasionally for some farms it is possible to construct quite long annual series. Consequently Table 4.4 inevitably includes a changing sample of observations, with an inconsistent geography. At its peak the series is based on over 100 wheat yield observations. In view of the data in Table 4.1, and of our general desire to locate material for the period between the probate inventories and the Healy and Jones index, as well as the availability of 'official' annual national and county estimates from the mid-1880s, we concentrated on extracting data from the 1720s to the mid-nineteenth century. For this reason the database and the series tail off towards the end of the nineteenth century.

The number of observations varies across time, but in view of our comments on farm records in Chapter 2 it is no surprise that the material

[20] Lawes and Gilbert, 'Home Produce', 132–3.

Table 4.4 Wheat yields—summary statistics

Period	No. of observations	Mean	Median	Standard deviation	Skewness
1720s	32	19.99	18.56	5.80	0.97
1730s	25	21.14	20.07	3.85	1.05
1740s	40	21.75	21.84	4.41	−0.07
1750s	28	22.42	21.70	6.84	0.01
1760s	47	21.82	21.43	6.38	0.28
1770s	57	19.68	20.00	4.76	0.12
1780s	41	18.88	17.16	6.04	0.57
1790s	83	18.97	18.50	4.74	0.55
1800s	91	20.98	21.95	5.17	−0.42
1810s	76	21.17	20.69	6.06	0.60
1820s	86	23.60	23.00	6.40	0.21
1830s	110	26.67	25.49	6.35	0.12
1840s	113	30.60	31.05	6.76	0.01
1850s	134	27.47	28.00	7.45	−0.05
1860s	112	28.57	27.85	8.46	0.69
1870s	90	28.92	28.00	7.08	0.37
1880s	54	26.47	25.93	6.32	−0.10
1890s	32	27.05	25.50	8.47	0.68
1900s	37	28.08	28.88	8.16	−0.27
1910s	12	33.34	33.11	5.58	0.17
	1,300	24.86	24.00	7.60	0.45

increases in both quantity and quality with time. By the end of the eighteenth century there are observations from seven or eight different counties annually for most years. In 1799 there are three farm observations from Essex, two each from Suffolk and West Sussex, and one each from Hampshire, Herefordshire, Norfolk, and Somerset. In 1813 there are eight observations from seven separate counties; in 1815 nine from six counties; in 1829 ten from eight counties; and in each of 1820, 1822, 1823, and 1829 there are nine observations from various combinations of seven counties.

The most intensively and extensively covered years are 1830, 1832, 1848, and the period 1851–8 inclusive in each of which years there are fourteen or fifteen separate observations from between seven and ten counties. Over the long run from 1720 to 1914 there is at least one observation for each year. In total there are 1,300 separate observations from more than 100 farms or holdings (where a holding may be the combination of more than one farm under the same tenure or management) located in twenty-seven of the thirty-nine English counties.

Table 4.4 summarizes the decennial averages of yields regardless of location and without any adjustments or weightings. The standard measures of central tendency more or less coincide. The standard deviations vary from decade to decade, but tend to increase over time. Even so, the standard deviations are quite small relative to the arithmetic means, indicating a tolerably small dispersion of the observations. However, because there are so many influences that control land productivity, especially environmental ones, any exercise in summary statistics is bound to produce some dispersion about the mean. Even so, the statistical summary suggests that we can be reasonably confident that these measures of central tendency are indicative of the trend of yields over time.

Figure 4.1 lends support to this contention. The figure is a boxplot of the quartile distribution of the observations on the same decade basis, showing also the median score (the black bar) in relation to the quartiles. The box itself contains two quartiles of the distribution, or 50 per cent of all observations. The outer limits of each plot show the maximum and minimum observations as long as those observations fall within plus and minus 1.5 times the inter-quartile range. Thus what we obtain is a graphic representation of data concentration, data range, and, broadly speaking, central tendency. Where observations are greater than plus or minus 1.5 times the inter-quartile range they are identified as outliers. For a database of 1,300 observations there are very few outliers, and most of them are on the high rather than low side of the decade distributions. The plot is also a measure of skewness. For the observations of the 1720s there is a skew, or long tail of observations, above the median. The 1730s shows a similar right-skewness, and indeed given that it contains an outlier which is over twice the size of the median score this does indicate an asymmetric distribution. Thereafter the distribution is more or less symmetrical until the 1780s when there is right-skewness once more. At the opposite end of the distribution, the 1890s is again right-skewed, and the 1900s relatively left-skewed. We conclude from this that the

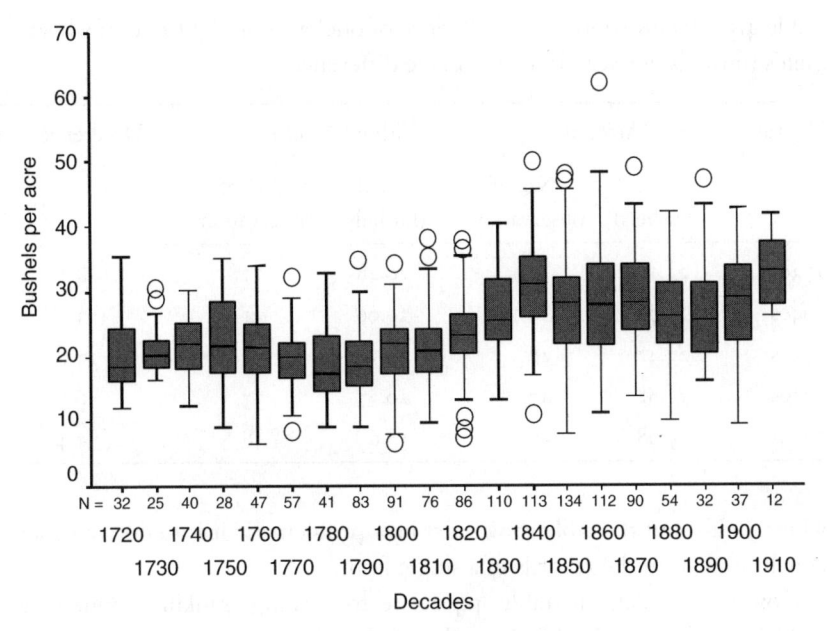

Fig. 4.1 Boxplot of English wheat yields

distribution and therefore the trend are influenced more by relatively higher-than-average yield measurements than by relatively lower ones.

We can demonstrate one influence of this on the trend. We noted earlier that the wheat yields reported in volume v of the *Agrarian History* for the period 1732–50 were predominantly from a single farm at Arreton on the Isle of Wight. The wheat yields from this farm were high relative to what has generally been considered to be the norm for wheat yields in the period. Bowden suggested that 'the probability is that land at Arreton was inherently more fertile than land in most other parts of the country, while the 1730s and 1740s were exceptional decades for good harvests'.[21] In other words, if the data do not fit the anticipated trend, try to explain it away. To test whether this was a satisfactory way of proceeding we went back to the farm records for Arreton and extracted the wheat yields to 1778. Table 4.5 demonstrates the influence of the Arreton yields on the overall trend. The inclusion or exclusion of Arreton data is clearly important, especially in the 1750s. Inevitably it will influence some of the conclusions we draw about

[21] *AHEW* v (2). 9.

Table 4.5 Demonstrating the influence of one farm on the trend of wheat yields (bushels per acre and percentage differences)

Decade	With Arreton Yield (bushels)	No. of observations	Without Arreton Yield (bushels)	No. of observations	Difference (%)
1730s	21.14	25	20.28	17	−4.1
1740s	21.75	40	20.99	31	−3.5
1750s	22.42	28	19.5	18	−13.0
1760s	21.82	47	20.37	37	−6.7
1770s	19.68	57	18.6	48	−5.4

wheat yields, but the table shows just how problematic it is to derive general conclusions from the records of a single farm.

How do the data in Table 4.4 relate to existing thinking about wheat yields? Estimates derived from probate inventories suggest an average wheat yield in the first quarter of the eighteenth century in Norfolk and Suffolk of 19–20 bushels per acre, of 18 bushels or so in Lincolnshire, probably on or over 19 in Hertfordshire, and between 21 and 22 in Oxfordshire. The new series suggests an average for the 1720s of around 19 bushels per acre rising to 21 or 22 by the mid-eighteenth century. Direct county comparisons of evidence drawn from probate inventories and farm records are not possible, except in the case of Hertfordshire where some early farm accounts have survived. The series in Table 4.4 opens with yields from an estate that straddles the parishes of Hitchin and Wellbury, Hertfordshire. In 1720 the wheat yields from the two sides of this estate were 14 and 18 bushels respectively (rounded to the nearest bushel). In 1721 they were 32 and 19; in 1722 26 and 16; in 1723 30 and 24; in 1724 25 and 13; and in 1725 25 and 22. These figures demonstrate a number of points. They show the annual variability in the harvest under the influence of exogenous environmental factors. They also suggest that Hertfordshire yields may have been higher than elsewhere, when set against the overall mean of 20 bushels per acre for the 1720s (from 32 observations, Table 4.4). That may, of course, be why Glennie's figures for the county derived from probate inventories were relatively high (Table 4.2). If we exclude Hertfordshire from the 1720s sample the number of observations reduces to 18 and the mean falls to 18.5 bushels. Clearly the precise geography involved in the sample at any given date is an important determinant of overall mean yields. In this case it suggests that Hertfordshire may have enjoyed higher yields than other wheat-growing counties, and it is further evidence of why relying on a single farm (as in the use of Arreton in volume v of the *Agrarian History*) may be an unreliable way of proceeding.

Arthur Young reported a large number of wheat yields from the many counties he visited on his tours of the 1760s and 1770s.[22] There are eight counties for which there is both an observation by Young and a measurement from the farm records. Of those eight, there is one observation for each of Huntingdonshire, Staffordshire, and Suffolk, but for the other five counties there are multiple observations; three for Lancashire, seven for Sussex, ten each for Berkshire and Lincolnshire, and twenty-five for Yorkshire (specifically the West Riding). The average wheat yield derived from the farm records is lower than the figures reported by Young in all of these five county comparisons. These differences ranged from 5 per cent in Lancashire and the West Riding, to 6 per cent in Lincolnshire, and 24 and 25 per cent differences in Berkshire and Sussex respectively. We conclude from this that our sample does not exaggerate wheat yields relative to Young's observations; indeed, given his well-known optimism, on which we commented above, it is only to be expected that our figures would come out below his. Nor do our data appear to exaggerate the true situation when we compare them with Holderness's wheat yield estimates for the period 1770–1800. He found that the modal wheat yield was 24 bushels per acre, whereas our mean wheat yield for the same period is just over 19 bushels per acre (which when all of our individual observations are rounded to the nearest bushel produces a modal score of 20 bushels per acre). All we know of the origin of Holderness's estimates is that they were not derived from the calculations of contemporaries, or from the computations of later historians.[23]

At 19 bushels per acre in the 1790s rising to 21 bushels per acre in the 1800s, the new wheat yield estimates are similar to the yields that have been

---

[22] They were summarized by Caird, *English Agriculture*, 474. See also R. C. Allen and C. ÓGráda, 'On the Road again with Arthur Young: English, Irish and French Agriculture during the Industrial Revolution', *JEH* 48 (1988), 93–116.

[23] B. A. Holderness, 'Productivity Trends in English Agriculture, 1600–1850: Observations and Preliminary Results', unpublished paper (Edinburgh: International Economic History Conference, 1978), and quoted in Turner, 'Agricultural Productivity in England', 490. Sadly Jim Holderness died before we were able to find out the sources of his data.

collated from the government inquiries of the 1790s, and from the 1801 crop returns.[24] The well-known history of the 1790s is clearly represented in our data with dramatically lower yields in the dearth years 1794 and 1795 than in the preceding years, followed by other poor years in 1799 and 1800, as we would expect; 1801 was a decidedly better year. The mean crop yield in the 1790s, based on 83 observations, is 18.97 bushels per acre, but when those famously problematic years are removed the mean crop yield increases to 20.1 bushels per acre, now based on 55 observations, and all other measures of central tendency, dispersion, and skewness improve. The whole series is constructed from a sample of observations which is self-selected by survival. We have no intention of selecting further from it to prove a particular point or interpretation, but the 1790s was in many respects an aberration and in this particular case we think it appropriate to identify the precise construction of the series to show how our sample fits with the known history of the period.

A further comparison of the period around 1800 can be made with fifteen of the individual county estimates produced for the Board of Agriculture's *General Views* of the 1790s and 1800s.[25] In only four cases are the new estimates higher than the equivalent Board of Agriculture averages. In five of the fifteen counties there are fewer than five observations per county, but in Table 4.6 we compare the average county estimates for the remaining ten counties. These are reported as the simple arithmetic means. For Berkshire and Norfolk, the new estimates are 22 and 15 per cent higher than the estimates from the Board of Agriculture, but for nearly all the other counties the new estimates are lower. For Herefordshire the new estimate is 24 per cent lower, for Essex, Lincolnshire, Suffolk, and Sussex the new estimates are over 10 per cent lower, and for Somerset the new estimate is 7 per cent lower. The two sets of estimates are more or less the same for Hampshire and Yorkshire. As with the comparison with Young, the new estimates seem not to exaggerate wheat yields in the period.

A rise in wheat yields took place in the nineteenth century, reaching a maximum of 30–1 bushels per acre in the 1840s. This rise essentially preceded the traditional period of High Farming. Yields then settled down to a plateau at around 27 or 28 bushels per acre for the rest of the century, before

---

[24] Turner, 'Agricultural Productivity in England', esp. 501.
[25] As listed in J. R. McCulloch, *A Statistical Account of the British Empire*, i (London: The Society for the Diffusion of Useful Knowledge, 1837), 482.

Table 4.6 Comparison of new wheat estimates with Board of Agriculture estimates of 1790–1809 (bushels per acre)

	New estimates (no. of observations)	Board of Agriculture
Berkshire	24.4 (9)	20
Essex	21.0 (17)	24
Hampshire	20.6 (35)	20
Herefordshire	15.1 (14)	20
Lincolnshire	20.9 (5)	24
Norfolk	22.9 (21)	20
Somerset	18.6 (7)	20
Suffolk	17.2 (12)	20
Sussex	19.4 (39)	24
Yorkshire	21.9 (5)	22

Sources New estimates reported in this chapter, and the Board of Agriculture estimates taken from the summary in J. R. McCulloch, *A Statistical Account of the British Empire*, i (London: The Society for the Diffusion of Useful Knowledge, 1837), 482.

rising modestly again in the early twentieth century, in the period of recovery that followed the great agricultural depression of the last quarter of the nineteenth century. This compares very well with the accepted trend in the growth of wheat yields that historians already recognize. For example, the long-run average wheat yields in the United Kingdom from the early 1850s to the early 1890s have previously been estimated at nearly 28 bushels per acre.[26] Our long-run average in the last quarter of the nineteenth century and into the first decade or so of the twentieth century agrees closely with the annual statistics on crop yields given in Table 4.1. Once into the twentieth century the official figures suggest that more often than not wheat yields exceeded 30 bushels per acre.[27] Our data, although less full by this date for reasons we gave earlier, show this upward trend, and also compare quite well with the more point-specific estimates of Craigie, as demonstrated in Table 4.7. Although the comparison is not identical, the new estimates are of a similar order of magnitude. Even the figure for 1879, which was the

[26] Lawes and Gilbert, 'Home Produce', 132–3.   [27] *AHEW* vii. 309–10.

Table 4.7 A comparison of the new wheat estimates with P. G. Craigie (bushels per acre)

Date	New estimates	Craigie's estimates
c.1850	28.9	26.3
1861	26.4	29.1
1863–82	29.3	27.9
1868	33.0	28.0
1870	31.3	29.6ss
c.1870	29.7	29.5
1876–82	27.5	24.6
1879	25.7	19.8
1882	27.7	26.0

Note The figures relate to the precise years or periods stated except for c.1850 and c.1870, which are calculated as the five-year averages centred on those years.
Sources The new estimates reported in this chapter and P. G. Craigie, 'Statistics of Agricultural Production', *JRSS* 46 (1883), 21.

most calamitous harvest of the century (in the sense of plunging below the trend of which it formed a part), was also the worst year in Craigie's sample trend.[28]

The long-run nineteenth-century trend of wheat yields is summarized in Table 4.8, and compared with estimates familiar from the standard literature. The comparison is achieved by splicing together the Healy and Jones wheat yields for the years to 1850, with the Lawes and Gilbert yields from 1850 to 1891, and the official estimates from 1884 onwards (Table 4.1). The new series starts slightly below the Healy and Jones figures for the 1810s, rises above them in the 1820s and 1830s, but is a good deal lower in the 1840s. In terms of the rate of growth, the Healy and Jones series increases from the 1810s to the 1840s by 54 per cent while the new series increases by 48 per cent. The new series compares well with the Lawes and Gilbert data, particularly in showing the same kind of plateau in wheat yields during much of the second half of the nineteenth century. In contrast, the official

---

[28] Craigie, 'Statistics', 21. Craigie reported estimates given to him rather than collecting his own data.

Table 4.8 A comparison of the new wheat estimates with those from the mid- and late nineteenth century

	New estimates	Existing estimates
1810s	21.2	21.7[a]
1820s	23.6	21.8
1830s	26.7	23.8
1840s	30.6	33.5
1850s	27.5	28.5[b]
1860s	28.6	29.4
1870s	28.9	25.4
1880s	26.5	28.1[c]
1890s	27.1	30.2
1900s	28.1	31.5
1910s	33.3	31.0

[a] Based on the five years 1815–19 from Healy and Jones.
[b] Based on Lawes and Gilbert, 1852–9. The equivalent figure from Healy and Jones is 34.8.
[c] Based on Lawes and Gilbert. The equivalent figure for the second half of the 1880s from the official returns is 29.8.
Sources From 1810s to 1840s, Healy and Jones, 'Wheat Yields', 578; from 1850s to 1880s, Lawes and Gilbert, 'Home Produce', 133; from 1890s onwards, BPP, *Annual Agricultural Statistics*.

annual returns from the 1880s were always higher than the new series by 10 per cent or more, with the exception of the short period 1910–14.

Long-Run Changes: Wheat Yields and the Agricultural Revolution

The course of change in long-run wheat yields has been regarded as the benchmark in identifying significant turning points in agricultural change and productivity. The pioneer work of Fussell, and Deane and Cole, suggested that whatever the processes of agricultural change in the eighteenth century and the resulting influences on productivity, it was unlikely that yields improved by more than 10 per cent.[29] This may be too cautious. A comparison of data in the probate inventories with material collected during the harvest crises of the 1790s suggests that wheat yields improved by rather more than 10 per cent during the eighteenth century.[30] When taken in conjunction with Charles Smith's estimates of *c.*1750 and Arthur Young's estimates and observation from the 1760s and 1770s, these findings tend to place the agricultural revolution earlier, rather than later, probably in the period before 1770 favoured by Allen, Clark, and, in the 1960s, Kerridge.[31]

Our new series questions these findings. Table 4.4 shows that wheat yields advanced by less than 10 per cent during the eighteenth century, and may have actually declined in the last quarter of the century, or at best stood still. Change was concentrated before *c.*1770, with a significant slowing down sometime after 1780 and lasting into the nineteenth century.[32]

In the absence of data on acreage it is difficult to say very much about the trend of agricultural output, but an alternative approach allows us to make certain inferences. For example, the well-known proposition that in the decades prior to the 1780s there was an extension of pastoral activity, aided and abetted by the enclosure of the heavy clays of the Midlands, may explain the observable increase in yields in the mid-century decades. The line of reasoning suggests that the terms of trade between pastoral and arable activity meant that those lands which were best suited to one or other method remained in, or converted to, that activity.[33] This resulted in an apparent increase in average crop yields. It also follows that where there was a

---

[29] G. E. Fussell, 'Population and Wheat Production in the Eighteenth Century', *History Teachers' Miscellany*, 7 (1929), 111; P. Deane and W. A. Cole, *British Economic Growth 1688–1959* (2nd edn. Cambridge: Cambridge University Press, 1969), 62–8; Turner, 'Agricultural Productivity in England', esp. 490, 499.

[30] Based on an assessment of Overton, 'Estimating Yields', *passim*, and Turner, 'Agricultural Productivity in England', *passim*.

[31] R. C. Allen, 'The Two English Agricultural Revolutions, 1450–1850', in Campbell and Overton (1991), 236–54; id., *Enclosure and the Yeoman: The Agricultural Development of the South Midlands 1450–1850* (Oxford: Oxford University Press, 1992), esp. part 1; E. Kerridge, *The Agricultural Revolution* (London: Allen & Unwin, 1967); G. Clark, 'Agriculture and the Industrial Revolution, 1700–1850', in J. Mokyr (ed.), *The British Industrial Revolution: An Economic Perspective* (Oxford: Westview Press, 1993). The 'timing' of the agricultural revolution is discussed in more detail in Chapter 1, above.

[32] The new figures more or less agree with Holderness, 'Productivity Trends' and Turner, 'Agricultural Productivity in England', 490, 503–5, and are at times in line with but at other times not in line with N. F. R. Crafts's assessment of the course of agricultural change in *British Economic Growth during the Industrial Revolution* (Oxford: Oxford University Press, 1985), 38–44. Part of the discordance is related to the use of different turning points by individual commentators.

[33] M. E. Turner, *English Parliamentary Enclosure: Its Historical Geography and Economic History* (Folkestone: Dawson, 1980), 135–51.

conversion to pastoral farming, the resulting increase in animals meant an increase in the production of natural fertilizers. This suggests a mechanism for increasing yields on the lands that remained in arable production. Conversely, from c.1780 and especially during the French wars, the terms of trade in some locations moved back in favour of arable production, influenced and reinforced by trade disruptions and bad harvests, and their combined effects on food supplies. Enclosure activity during the French wars points to an increase in arable activity. It was also a period when marginal land was enclosed, especially in the first decade of the nineteenth century. To this extent relatively inferior land came under the plough, and this may help to explain why average crop yields fell.

The course of change in the nineteenth century is a different story. From c.1800 until the 1870s wheat yields increased by 30 to 40 per cent. If there was an agricultural revolution at all, and if wheat yields are a sensitive parameter for identifying the existence and the outcome of that revolution, then surely it can be located in these decades. The coincidence with the period of High Farming in the third quarter of the century is initially appealing, but this diminishes on closer inspection of the new data. Instead it raises a number of important questions about the course of agricultural change and its traditional interpretation.

In the post-French war depression wheat yields reached an unprecedented height. At a time when farms were being taken in hand because tenants were having trouble paying their rents, yields began a slow but unmistakable rise towards a mid-century plateau.[34] It is usually argued that hard on the heels of this post-war period inputs of non-farm manure, such as imported natural substances and industrial and artificial fertilizers, were being heaped on the land in ever-increasing quantities. Yet what effect did this application of fertilizer have? The data suggest that there was no obvious response in wheat yields. Instead, average yields seemed to reach something of a plateau. They fluctuated between a low of 26 bushels per acre in the 1830s and 1880s with a high of 30 bushels per acre in the 1840s, but around a mean which

---

[34] Supporting evidence for the difficulties faced by tenants can be found in TBA *passim*; Wilkes, 'Adjustments in Arable Farming'; and also littered throughout the pages of *Report from the Select Committee Appointed to Inquire into the Present State of Agriculture*, BPP [612], V (1833); *First Report from the Select Committee on the State of Agriculture: with Minutes of Evidence and Appendix*, BPP [79], VIII, part 1 (1836); *Second Report from the Select Committee on the State of Agriculture*, BPP [189], VIII, part 1 (1836); *Third Report from the Select Committee on the State of Agriculture: with Minutes of Evidence, Appendix and Index*, BPP [465], VIII, part 2 (1836); *Report from the Select Committee of the House of Lords on the State of Agriculture in England and Wales*, BPP [464], V (1837).

Table 4.9 Estimated arable acreage and fertilizer inputs

	Arable acres (000 acres)	Fertilizer 000 tons	lbs/acre
1820s	11,143	39	7.84
1836 tithes	15,093	73	10.83
1851 Caird's England	13,667	263	43.10
1854	15,262	383	56.21
1872	18,136	781	96.46

Notes Acreages for the 1820s, the tithes (excluding seeds in the western counties), Caird, the partial agricultural census of 1854 (but grossed up), and the official returns for 1872, all taken from Kain; fertilizers based on net volume retained for 1825–31, 1837–42, 1851–3, 1854–8, and 1872–6, and including bones, superphosphates, nitrates, guano, both home and imported, but adjusted for net home consumption, plus small amounts of unenumerated manure imports, taken from Thompson.

Sources R. J. P. Kain, *An Atlas and Index of the Tithe Files of Mid-Nineteenth-Century England and Wales* (Cambridge: Cambridge University Press, 1986), 459; F. M. L. Thompson, 'The Second Agricultural Revolution', *EcHR* 21 (1968), 76–7.

remained more or less constant at 27–8 bushels per acre. Possibly historians have exaggerated the impact of home-produced and imported non-natural manure-based additives. Table 4.9 is a crude demonstration of this doubt. It takes Thompson's estimates of home and overseas fertilizer inputs for selected cross-sections, and relates those inputs to the available arable area.[35] The rate of growth of the application of these fertilizers was exponential, but at 96 pounds per arable acre we question the overall impact. Even if we modify the estimates in terms only of wheat lands the overall impact is still modest, rising to 0.2 tons per acre in the 1870s or a little under 5 hundredweight.

What precisely was the value of external inputs if all they did was to allow English farmers to maintain a stable wheat yield? Possibly they were cost saving since artificial fertilizers or external additives were more concentrated relative to traditional muck and lime. Whether this made them relatively labour saving remains an open question. What is not in doubt is the failure of English agriculture to break through the seemingly upper limit of

---

[35] F. M. L. Thompson, 'The Second Agricultural Revolution, 1815–80', *EcHR* 21 (1968), 62–77.

30 bushels of wheat per acre. The much vaunted advantages of imported and other off-farm manure and additives, especially during High Farming, seems to have been more apparent than real. We temper this conclusion by pointing out that it presupposes that contemporary farmers were not content with raising yields to an upper limit of close to 30 bushels per acre. The conclusion also neglects other ways of measuring productivity improvements, such as seeding rates.[36]

A slump in wheat yields shows up in the 1880s, as we would expect to be the case during the darkest decade of the agricultural depression. Several years of bad weather injured arable and at times livestock farmers alike. Crop yields recovered in parallel with the modest general recovery in agriculture prior to the Great War, although there was no corresponding recovery in the acreage of land devoted to wheat. On the contrary, the wheat acreage tumbled throughout the period. In the late 1860s the English wheat acreage stood at 3.4 million acres, but by 1890 it had fallen to 2.2 million acres, and to 1.7 million acres by 1900. At its lowest point there were just 1.3 million acres in wheat production in 1904. In the decade prior to the Great War it averaged close to 1.7 million acres.[37] This fluctuation in the acreage under wheat is not replicated by a corresponding fluctuation in yields, which remained fairly flat at something close to 28 bushels per acre. Wheat output declined, 'officially', from close to 70 million bushels in the late 1880s to a low point of 35 million bushels in 1904. More generally in the years adjacent to 1904 output varied from 50 to 55 million bushels, settling a little above 52 million bushels in the decade prior to 1914.[38]

It is perhaps surprising that wheat yields did not fluctuate given the changing acreage involved. On the one hand, we would have expected the relatively inferior wheat-growing lands to be converted first into other uses, with the result that average yields would rise. On the other hand, these were difficult times for arable producers as the costs of production remained high relative to agricultural incomes. Conjecturally, arable producers may have maintained unit acre yields by reducing inputs such as artificial manure. Alternatively, perhaps the best wheat land did not stay in wheat production, but instead went into higher-value production such as milk, or market gardening.

---

[36] These are discussed in Chapter 5. [37] *AHEW* vii. 1770–1.
[38] These are figures based on the official annual acreage and estimated yields of the time, as reported ibid. 1789. The official yield estimates are generally higher than our own.

Table 4.10 English land utilization classification

Land classification	Site characteristics	Soil characteristics
I.		
1A First Class Land	level or gently sloping, favourable aspect not too elevated	deep, easily worked, fertile loams, silts, mild peats
2A General Purpose Farmland	similar to 1A	well-drained soils, good depth, workable for much of the year
2AG General Purpose Farmland	similar to 1A with higher rainfall	well-drained soils, good depth, workable for much of the year
3G First Class	similar to 1A with higher water table	good soils but liable to flood, heavier less tractable soils
4G Good, Heavy Land	similar to 3G	better clays, heavy loams, good depth, high fertility
II.		
5A Downland/Lightland	moderate elevation, relatively gentle slopes	chalk, limestone, and sandy soils, shallow, light, often stony
5G Downland/Lightland	moderate elevation, relatively gentle slopes	chalk, limestone, and sandy soils, shallow light outcrops of rock, difficult to plough
6AG General Purpose Farmland	defective due to relief, some steep slopes	heavy, shallow, or stony soils
III.		
7G Poor Quality Heavy Land	often low lying, in need of extensive drainage	heavy, intractable clays
8H Mountain or Moorland	higher altitude, extreme elevation or ruggedness	thin, poor soils
9H Lightland	subject to drought	'hungry' soils, coarse sands and gravels, overdrained
10 Poorest Land	unsuited to agriculture without major outlay on reclamation	salt marshes, sand dunes, etc.

Notes I = good-quality land (highly productive when under good management); II = medium-quality land (land of only medium productivity even under good management); III = poor-quality land (land of low productivity).

Source L. D. Stamp, *The Land of Britain: Its Use and Misuse* (London: Longman, 1948), 363–6, 482–3, just over a million acres, or 3.4% of predominantly urban land omitted.

Land use	Area (acres)	Percentage of English land
arable, often intensive	1,958,300	6.1
eastern ordinary arable crops	6,998,800	21.8
western ley farming	2,436,400	7.6
grasslands, fatting pastures, best dairy	1,156,400	3.6
grassland pastures	4,716,300	14.7
arable, barley–turnip–sheep land	2,296,300	7.2
downland or fescue pastures	220,300	0.7
crops, long-term ley in western England	6,877,100	21.5
generally in grass	591,100	1.8
natural vegetation	2,782,600	8.7
heathlands, wastes	805,700	2.5
	100,200	0.3

Environmental Influences on Wheat Yields

The analysis thus far has tended to assume that a kind of homogeneity existed in English farming. In what follows we relax this assumption and explore the underlying detail in the long-run trend in English wheat yields by looking at questions of land quality, county/regional distinctiveness, and the basic details of farming systems. We have little direct indication of land quality on the farms in this study, but we know where they were located. Consequently we can use the base map of English land utilization classification constructed in the 1930s to say something in broad terms about the relationship between output and land quality.[39] From this data we can locate farms broadly within different land qualities.

All of the farms in the sample were located on what the land utilization survey considered was either good-quality land, which was highly productive when under good management, or medium-quality land of only medium productivity even under equally good management (see Table 4.10). Land that was classified in a category called 1A was first-class land on level or gently sloping aspects, with deep, easily worked fertile loams, silts, and mild peats. It included the drained fens that had become good-quality land by the 1930s. Although generally arable and intensively cultivated, by the 1930s 1A occupied only 6.1 per cent of English land area. This was the best of the good-quality land. The best of the medium-quality land was defined as category 5A, downland or lightland, of moderate elevation with relatively gentle slopes, perhaps on chalk or other limestone, and also sandy soils which were shallow and often stony. It was mainly under arable in the 1930s, or under a combination of barley, turnips, and sheep. Table 4.10 completes the categorization.

We cannot correct for the fact that some of the data for some of the farms represented conditions that pertained 200 years before the land utilization survey of the 1930s was conducted. But it was Stamp's view that there was 'a considerable, often remarkable, stability of land use on the best land and on the poorest land'. Most change and fluctuation was associated with land of intermediate quality.[40] This is not the same as saying that there had not been

---

[39] Numerous publications by L. D. Stamp explain this classification. The most useful are 'Fertility, Productivity, and Classification of Land in Britain', *Geographical Journal*, 96 (1940), 389–412; and *The Land of Britain: Its Use and Misuse* (London: Longman, 1948), 362–81, and especially the summary table on 366. See also J. Sheail, 'Elements of Sustainable Agriculture: The UK Experience', *AgHR* 43 (1995), 178–92, esp. 187–90.
[40] Stamp, *The Land of Britain*, 359. See also Sheail, 'Elements', 189.

adjustments in soil quality. We recognize that the actions of geology, God, and man have altered soil condition, especially man, and perhaps especially with respect to field drainage or, more radically, fen drainage. Yet unless there is evidence to the contrary we have assumed that relative land quality did not alter fundamentally over the two centuries prior to the 1930s. On individual farms this assumption will be tempered by the special actions of farmers, but in general the assumption maintains that relative land quality in the 1930s was broadly the same as in the 1850s, 1780s, and 1720s, and intervening decades.

When the data are replotted they produce few surprises. Figures 4.2–4.4 show the long-run trend in wheat yields adjusted for some of these *broad* differences in land quality. Fig. 4.2 shows the trend for good-quality land according to the land utilization classification. It varied from first-class land to general-purpose farmland. In the main this includes land that was generally better suited for arable purposes, and includes farms in most of those counties not located broadly in the west of England. But it also includes west of England farms that can be classified as first-class and general-purpose farmland, but which ordinarily were regarded as better suited to grass. These form a distinctive clutch of farms in the counties of Hereford, Worcester, Gloucester, Somerset, and Devon. Despite the yearly variations, and even the large dispersion within years, the upward trend in the first three-quarters of the nineteenth century is unmistakable and clearer than the general trend that we have already identified. There are 905 observations of wheat yields on these good-quality land types.

An equally clear pattern emerges from Fig. 4.3, the yields from farms on medium-quality soils. This sub-set is based on 395 observations of wheat yields. These medium-quality soils were described in the 1930s land utilization survey as downland or general-purpose farmland, better suited to arable or genuine downland farming. The latter included fescue pastures of the type found on chalk and other limestone soils, and also found on shallow sandy soils. It therefore included land that historically was considered marginal for arable production and better suited to sheep walks.[41] As such, it was susceptible to over-fertilization and also to soil erosion on the slopes once the original grass surface had been removed for arable use. This category of land also included mixed crop farming and the long-term leys of the type

---

[41] About which William Cobbett reported in his *Rural Rides*, i (Everyman edn. London: J. M. Dent & Sons, 1912), 114.

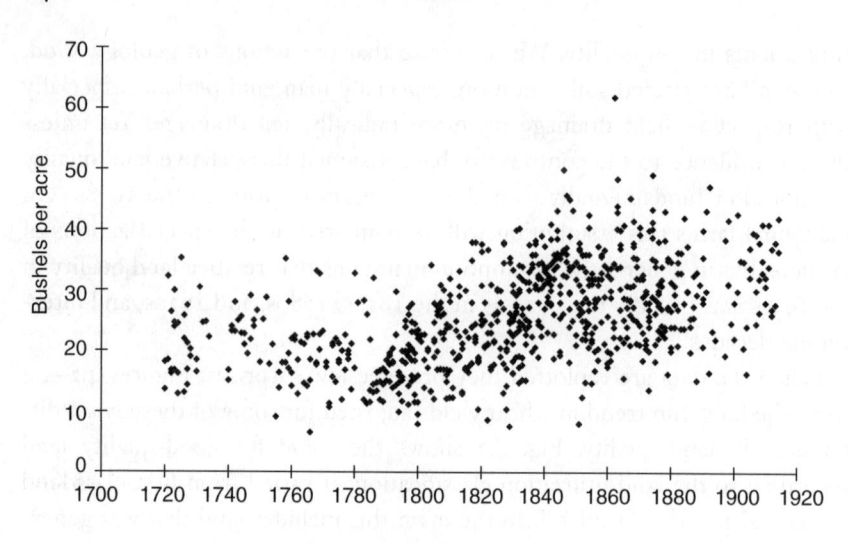

Fig. 4.2 Wheat yields on good-quality soils

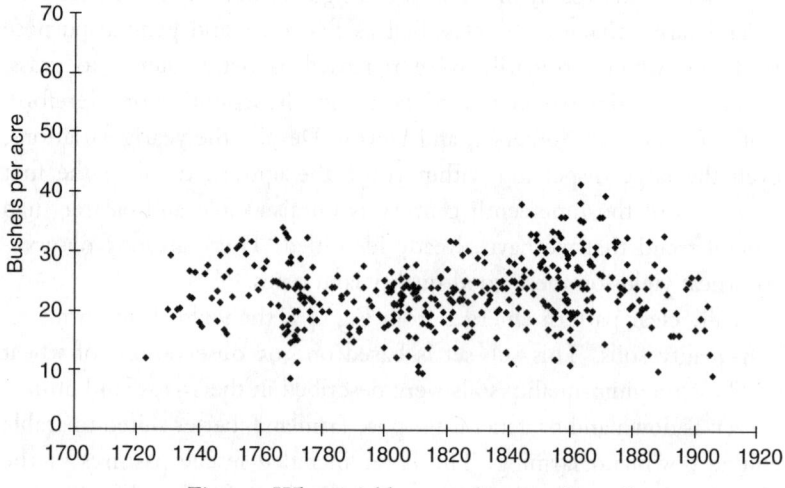

Fig. 4.3 Wheat yields on medium-quality soils

found in western England. Altogether these medium-quality soils included farms in Berkshire, Cambridgeshire, Hampshire, Wiltshire, Lincolnshire, the Isle of Wight, and East Sussex, along with Norfolk, Essex, and Bedfordshire, and finally Gloucestershire, and a few from the North and West Ridings of Yorkshire. There was very little if any general variation in the long-term trends. Wheat yields remained on or around 20 bushels per acre throughout the period, though with an unmistakable, if modest, rise in

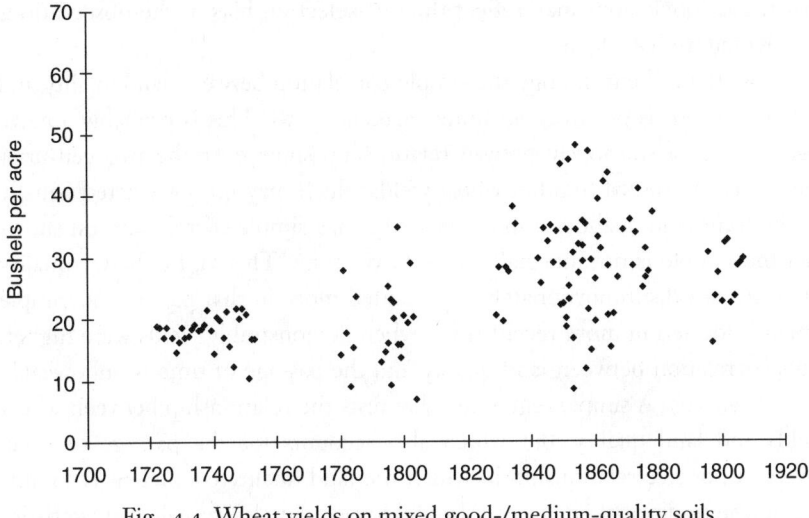

Fig. 4.4 Wheat yields on mixed good-/medium-quality soils

average yields in the mid-nineteenth century, a time of increasingly successful sheep/corn and shed cattle/corn mixed farming.

Finally, Fig. 4.4 shows the long-run trend of wheat yields from farms with, broadly, mixed-quality soils. That is, they were a cross between good and medium quality and included general-purpose farmland and downland or lightland (2A and 5A in the land use classification in Table 4.10). These were found in Lincoln, Norfolk, Suffolk, and Kent, with just a few farms also in Dorset. The resulting trend is based on 129 observations, and despite the wide annual dispersion, for the four broad periods from which they arise there is an unmistakable upward trend from $c.1800$ to the 1860s, from about 20 bushels per acre to more like 30 bushels per acre.

Further refinement of the database may more narrowly identify the major regions or soil types that underlie the main trend, but they are unlikely to alter the main trend. What we may say is that, contrary to prior expectations, the good-quality land did not necessarily always produce the highest unit acre yields. If anything, before 1800, the medium-quality soils produced the best results, if only marginally. The good-quality soils produced wheat yields which were more often than not on or below 20 bushels per acre, while on the medium-quality soils they tended to be on or above 20 bushels per acre. What this partial disaggregation shows, perhaps, is the greater changes that occurred after $c.1800$. The differential character of those changes according to broad soil type becomes clearer, although we accept that these conclusions may reflect the self-selection bias in the observations over which we have little control.

Across the full chronology the simple correlation between land quality and unit acre yields is positive, but quite weak, at 0.158. This is probably a naive test showing a spuriously precise relationship since over the two centuries there was an upward trend in wheat yields which may not have acted equally on different land qualities. On the other hand, a simple correlation test shows that the sample is not internally biased over time. That is, the better-quality lands are not disproportionately represented more in that part of the sample which is located in more recent times when demonstrably yields were higher. This correlation between land quality and the passage of time is only weakly positive at 0.07. A simple regression that tests the relationship between wheat yields and land quality, but which also accounts for the passage of time, reveals a positive relationship in which the land quality coefficient is significant at the traditional 99 per cent level. However, it also reveals that, statistically, not very much is being explained about the variation in yields.

Ideally, a model which tests more fully for variations in land quality is required, or a model which includes variables which themselves contribute to land quality, such as topography, aspect (southward facing slopes, etc.), drainage, rainfall, temperature, altitude, and so on.[42] This is possible only on the crudest county, or, at best, sub-county, regional basis.[43] However, since there was clearly a difference between the broad trend of yields in the eighteenth century compared with the nineteenth, a difference which is observable across the whole sample, we can offer duplicate tests for the two periods. Ordinarily we would suspect technological changes in agriculture to be at work, but in this case the explanation is likely to be more mundane. There was the conjunction, internally, of poor harvests in the period 1795–1800. These poor harvests produced low yields with which to terminate the eighteenth-century trend, but conversely a low starting point with which to begin the nineteenth-century trend. This was probably exacerbated by the extension of the area under arable land in the French wars when soils that under other conditions would have been better suited to pasture were ploughed. Even if the first reaction to a plough up was the release of nutrients, this was unsustainable without regular inputs of manure. Some of this

---

[42] For which see L. Brunt, 'Nature or Nurture? Explaining English Wheat Yields in the Agricultural Revolution', *University of Oxford, Discussion Paper in Economic and Social History*, 19 (1997).

[43] For example, the land utilization surveys of the 1930s often gave long-term local area rainfall and other climatic data.

explanation is based on firmer ground than mere conjecture. The state of the harvests and the extension of the arable were facts.

Carrying out the same correlation tests separately for the two century-based chronologies begins to isolate the effects of any bias in the distribution of the sample farms on varying quality of land. For the 353 observations of the eighteenth century the simple correlation coefficient between yield and land quality is moderately strongly negative at −0.25. That is, the better the land the worse was the yield! It might be thought that this result shows up the distorting effects of the bad harvests of the 1790s because the inclusion of those years means that the eighteenth-century long-run trend ends with a downturn of yields. In fact this seems not to be the case: the effect of removing the poor harvest years of 1794, 1795, and 1799 hardly changes the result at all (it moderates the coefficient to −0.231). There may be a second bias, this time in terms of sample distortion over time (essentially whether the distribution of good and bad land in the sample is not evenly spread through time). Whether this is strong enough to have such an adverse affect on the result is problematic. When tested (the simple correlation between the passage of time and the quality of soil), this bias is weakly negative at −0.019. That is, there is a very small bias explained by a slightly disproportionate distribution of poorer land quality observations in the eighteenth century. Perhaps a more likely explanation of what we see in the eighteenth century is that there is very little sign of the agricultural revolution. We would expect it to show up in the trend of wheat yields, but it is just not there. In contrast, for the 947 observations in the nineteenth century the converse was the case, a marginally stronger but positive correlation between land quality and yields (correlation coefficient of 0.27) and a slightly stronger and positive bias in the occurrence of observations from places with better land quality (0.082).

CHAPTER 5

Barley and Oats

We have singled out wheat for particular attention because it is so frequently taken as a proxy for the state of the harvest more generally. Our findings begin to suggest a significant dating for the agricultural revolution in the first four decades or so of the nineteenth century, but to conclude that we have therefore solved the puzzle which has occupied generations of agricultural historians would be premature. Wheat was a proxy for the agricultural economy, not the agricultural economy in its entirety. Indeed, the 1801 crop returns show that in some counties barley and oats were just as important as wheat within the grain harvest.[1] Barley was still used as a bread grain in some areas, and under some agricultural conditions, although its main use by the eighteenth century was in the brewing industry for malt, and as livestock feed. Oats were grown extensively for fodder and, in some parts of northern England, as a bread and cereal grain.

Unfortunately, until the late nineteenth century we cannot be sure of the national distribution of wheat, barley, and oats, let alone the regional pattern. Ashley's restatement of Gregory King's assertion that 38 per cent of the nation's bread in *c*.1696 came from wheat was not much more than a guess.[2] Charles Smith's detailed distribution of wheat, barley, oats, and rye in 1766 was based on a number of assumptions about the size of the population, and human and animal consumption of grain and its derivatives.[3] The 1801 crop

---

[1] M. E. Turner, 'Arable in England and Wales: Estimates from the 1801 Crop Returns', *Journal of Historical Geography*, 7 (1981), 291–302, esp. 296.
[2] W. Ashley, *The Bread of our Forefathers* (Oxford: Oxford University Press, 1928), 8.
[3] Charles Smith, *Three Tracts on the Corn-Trade* (2nd edn. London, 1766), 161–2, 166, in which the ratio of wheat to barley to oats to rye is 1:1.14 : 1.05:0.26. But see M. Overton's critique of Smith in 'Agricultural Productivity in Eighteenth-Century England: Some Further Speculations', *EcHR* 37 (1984), 248. In spite of the very substantial criticism of the Smith estimates they still persist as the

returns may not be particularly dependable when used to assess the main parameters of agricultural production, but for this period they are the nearest we can come to something approaching a reliable census of land use if only because they have a rather firmer basis than the estimates of contemporaries.[4]

Once we have detailed figures in the 1860s the position is clearer. In the second half of that decade the wheat acreage varied between 3.1 and 3.4 million acres, and barley between 1.8 and 1.9 million acres. More or less 1.5 million acres were under oats. The proportions changed thereafter, notably in the agricultural depression, but a rough and ready estimate of barley and oats occupying together about the same acreage as wheat cannot be too far from the truth. Of course, acreage is only one measure of the relationship between the different crops. Yield estimates are not available until 1885 when the official production was 74 million bushels of wheat, 67.5 million bushels of barley, and 67 million bushels of oats. Since bushels are unreliable as a comparative measure, the most accurate common denominator is their value: in 1880 wheat was £13.7 million, barley £10.9 million, and oats £4.8 million.[5] Whichever way we cut the figures, barley and oats together represented about half the corn harvest.

Barley

Figure 5.1 and Table 5.1 show the trend in barley yields. An upward trend can be discerned from a little under 30 bushels per acre in the second quarter of the eighteenth century to 35 bushels per acre in the third quarter of the nineteenth century. Contemporary estimates would lead us to expect that the acreage under barley contracted from about 1700 to 1830. Holderness has speculated 'that inferior ground was abandoned early'.[6] If so, it is not entirely demonstrated in the trend in average decennial yields. These stayed fairly flat around 29–30 bushels to the 1770s, dipped significantly in the 1780s and 1790s, before recovering in the following two decades. From the 1820s yields

---

basis not only for estimating various aspects of grain output and bread consumption in the mid-18th century, but also as the basis for a number of subsequent estimates. In this context see C. Petersen, *Bread and the British Economy c.1770–1870* (Aldershot: Scolar Press, 1995), 136, 185–214.

[4] Turner, 'Arable in England and Wales'. See also M. E. Turner, 'Counting Sheep: Waking up to New Estimates of Livestock Numbers in England, *c.*1800', *AgHR* 46 (1998), 142–61, for a critical review of contemporary (*c.*1800) estimates of livestock numbers.

[5] *AHEW* vii. 1770, 1784–7, 1789, 1906.   [6] *AHEW* vi. 106.

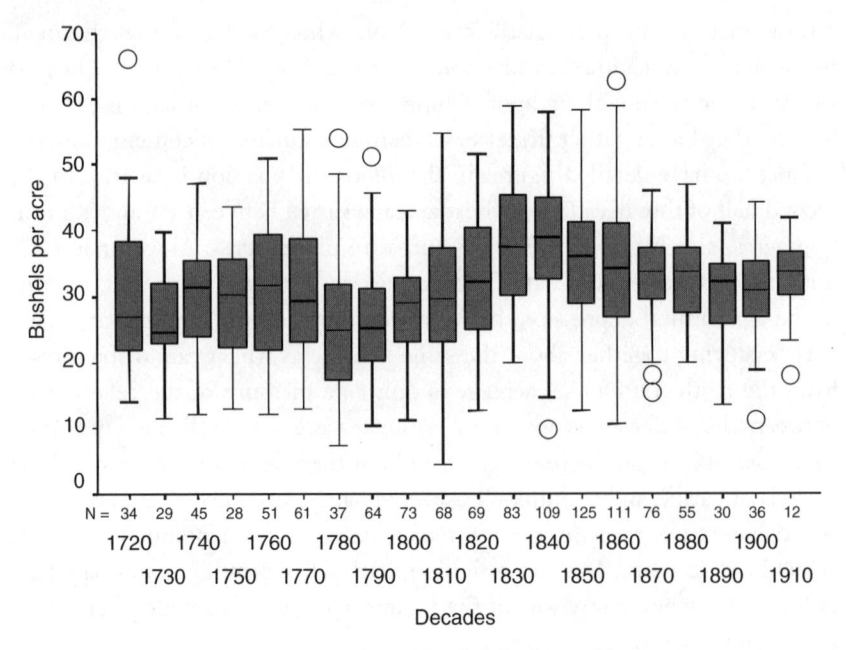

Fig. 5.1 Boxplot of English barley yields

were higher than at any previous time, reaching remarkable levels in the 1830s and 1840s. While they fell back slightly from this peak, only in the depths of the late nineteenth-century depression did they return to the levels of the eighteenth century and even then to the higher rather than lower yields. While the parallel with wheat is not absolute, the decade-on-decade increase in yields during the first half of the nineteenth century suggests that we are looking at a trend which is by no means dissimilar. As with wheat this looks like the key period for tracing a real upward shift in yields.

Unlike the equivalent trend for wheat, the annual deviations are more pronounced. This is hardly surprising because barley is susceptible to variable weather, especially to wet conditions during seedtime and harvest. It also has a sensitivity to soil fertility. Consequently there is a greater tendency for the annual quality to fluctuate. One contemporary thought that there was 'no grain perhaps more affected by soil and cultivation than barley', and according to J. C. Loudon:

in Britain, barley is a tender grain, and easily hurt in any of the stages of its growth, particularly at seed time: a heavy storm of rain will then almost ruin a crop on the

Table 5.1 Barley yields—summary statistics

Period	No. of observations	Mean	Median	Standard deviation	Skewness
1720s	34	30.2	26.9	11.3	1.09
1730s	29	26.1	24.5	7.2	−0.13
1740s	45	29.7	31.2	8.6	0.01
1750s	28	29.0	30.3	8.1	−0.23
1760s	51	30.7	31.7	10.9	−0.17
1770s	61	30.6	29.2	9.0	0.48
1780s	37	25.5	24.9	11.2	0.75
1790s	64	26.0	25.1	9.1	0.56
1800s	73	28.3	29.0	6.7	−0.12
1810s	68	30.2	29.5	10.2	0.05
1820s	69	32.8	32.3	10.2	−0.02
1830s	83	37.7	37.5	9.8	0.09
1840s	109	38.1	39.0	9.1	−0.59
1850s	125	34.9	36.0	9.6	−0.27
1860s	111	34.9	34.4	10.4	0.29
1870s	76	33.2	33.6	6.0	−0.56
1880s	55	33.2	33.8	6.7	0.11
1890s	30	30.6	32.3	6.8	−0.93
1900s	36	30.4	30.8	7.1	−0.50
1910–14	12	32.5	33.8	6.7	−0.87
	1,196	32.3	32.3	9.7	0.10

best prepared land; and in all the after processes greater pains and attention are required to ensure success than in the case of other grains.[7]

Not surprisingly, we find a wide dispersion about the mean for the period 1795–1804, from 10 bushels at the lower end to 51 bushels at the upper end.

---

[7] *The Library of Agricultural and Horticultural Knowledge* (3rd edn. Lewes: Baxter, 1834), 36; J. C. Loudon, *An Encyclopaedia of Agriculture*, iii (London, 1831), 822. And on the influences of soil, climate, weather, tillage and sowing, and seed variety see J. M. Wilson (ed.), *The Rural Cyclopedia; or, A General Dictionary of Agriculture*, 4 vols. (Edinburgh, 1847–9), i. 345, 348.

While this gives cause for concern when it comes to settling on average scores, such large fluctuations were also often quoted by contemporaries. To this extent we are confirming prior expectations.[8]

Barley also had different end uses, which may have affected yields. Whereas wheat was almost entirely grown for bread, the majority of barley—and an increasing majority until the late nineteenth century—was for malting. Gregory King reckoned that malting barley constituted about 70 per cent of all barley in the 1690s. Although our usual strictures about the reliability of contemporary guesswork apply, if this is taken as a vaguely accurate base point the figure is likely to have risen by the mid-eighteenth century.[9] By about 1900 it was estimated that two-thirds of the entire barley crop was malted, and the proportion may even have been higher in the 1870s when more than 90 per cent of output was sold off farms.[10] Assuming the figure is unlikely to have fallen, King probably started from an optimistic base, but through the eighteenth and nineteenth centuries it seems likely that something in the region of 60–70 per cent of barley was for malting. Wilson argues that the best malting barley was grown in the drier areas of south and east England, and that 'the demand for malting barley was a principal prop of light-land arable farming in the east of England and was responsible for its improvement'. From its base in Bury St Edmunds in Suffolk the Greene King brewery boasted accessibility to 'Britain's best malting barley on its doorstep'.[11] On the strength of such remarks, perhaps we can establish a basis for analysing our barley yields regionally.

Where do we draw the precise boundary? If we isolate the areas of low rainfall and light land, we might separate out Lincolnshire, East Anglia,

---

[8] Holderness suggested that the spread of barley yields in c.1800 was of the order of 14 to 45 bushels per acre: *AHEW* vi. 142.

[9] Ibid. 130.

[10] The figures for proportion sold off farm were 93% in 1873, 64% in 1894, and 68% in 1911. Part at least of the decline must have been because farmers required the barley for feed as their grain acreage diminished and their animal stock increased: F. M. L. Thompson, 'An Anatomy of English Agriculture, 1870–1914', in B. A. Holderness and M. E. Turner (eds.), *Land, Labour and Agriculture, 1700–1920* (London: Hambledon, 1991), 234.

[11] T. R. Gourvish and R. G. Wilson, *The British Brewing Industry, 1830–1980* (Cambridge: Cambridge University Press, 1994), 75, 183, 121. On the advantages of the east of England, the strength of metropolitan demand for barley and malt, and the comparative cost advantage of transportation coastwise in the 17th and 18th centuries see *AHEW* v (2). 22–3, 87–8. See also Joan Thirsk's comments on the importance of barley to East Anglian farmers in the 17th century in the light of the public debate regarding the corn bounty, in *AHEW* v (2). 331–2. Chartres has estimated that 41% of all barley exports in 1749–50 came from East Anglia and a massive 84% of all malt exports: *AHEW* v (2). 453.

Essex, Cambridgeshire, and Hertfordshire.[12] In doing so we observe a consistent difference in barley yields between these east of England counties and the yields produced elsewhere. Except in the 1880s barley yields in the east were consistently higher than in other counties, and usually more than 10 per cent higher. In the 1780s they were 44 per cent higher. The separate trends over time are different. In the east the barley yields varied between 31 bushels (1770s) and 41 bushels (1830s), but the long-run trend was fairly flat. The yields elsewhere fluctuated rhythmically from 22–3 bushels to 27 bushels per acre from the 1720s to the 1770s, reining back to just 22 bushels in the 1790s, before rising steadily to 35 bushels in the 1830s and 1840s. There then followed the most clear-cut trend with a steady fall to 27 bushels by the 1900s.[13]

This attempt to explain the broad regional differences in barley yields falls foul of the particular attributes of the crop. The concentration of malting barley production in the east of England should, according to the science and art of producing the crop, result in lower than average yields. This is the opposite of what we have found. More than most crops perhaps, the quality as well as, or perhaps even more than, the quantity of the crop is important. Nitrogen is essential for high yields. It produces a high protein content. For feeding barley—whether intended for human consumption in the form of bread or soup, or as corn feed for animals—a high protein level is beneficial. By contrast, malt production relies on a high maltose content, which is the sugar involved in fermentation. Consequently good malt barley needs to be high in starch content and low in protein. For malting barley it is the quality of the crop rather than its quantity that is critical, and high-yielding barley is usually incompatible with a good malting sample. In other words the best malting barley did not necessarily produce the highest yields. Top-quality malting barley requires light soils, ideally loam derived from a calcareous

---

[12] This rather wide geography belies what Christine Clark calls the 'crucial relationship between soil and barley cultivation', in C. Clark, *The British Malting Industry since 1830* (London: Hambledon Press, 1998), 10–23, and esp. 16. In our analysis we are forced for lack of detail to adapt this relationship and go for the more sweeping geographical approach. Brown and Beecham referring to barley cultivation in Hertfordshire talk about the 'healthy demand from Ware's maltsters': *AHEW* vi. 281. Elsewhere Brown comments that the largest maltsters were to be found in Hertfordshire; that London, easily the biggest market for malt, had historically acquired its sources from the surrounding corn-growing counties, mainly Hertfordshire and Bedfordshire; and that Ware malt was the premium malt: *AHEW* vi. 509, 516–17.

[13] We have excluded the above average barley yields from the same farm in Arreton on the Isle of Wight that also produced high wheat yields. The inclusion of these yields for the 1730s–1770s raises the yields in those decades and therefore accentuates the decennial rhythm already identified.

subsoil, hot summers, and timely rainfall. The east of England is well suited for this purpose. On other soils barley tended to have too much nitrogen, and therefore in general was higher yielding, and consequently better suited as a feed.[14] Although the differences in yields between the eastern counties and elsewhere ran counter to our general expectations based on agricultural texts, within the eastern counties the story perhaps falls more into line. The barley grown around Bury St Edmunds may have been of the very best for malting purposes, as Wilson suggests, but generally the yields in Suffolk in our sample, covering the 1790s, 1830s–1860s, and 1900s, were consistently lower than in the other neighbouring eastern counties.

Farmers were able to adjust between quality and quantity. In a typical Norfolk rotation when barley followed the well-manured root crop, there was often too much nitrogen for a good malting barley, and the crop was more inclined to lodge—it would fall over.[15] On the Wessex Downs barley traditionally followed wheat and this was found to produce a superior crop to that grown after the root course. On such occasions the straw was stronger and the grain was of a better quality. Even in its native county the Norfolk rotation eventually came under fire. When writing of 'recent improvements', C. S. Read noted in 1858 that the 'quality of wheat-stubble barley is invariably finer than that of the turniplands'. He added that after sheep had consumed roots and cake on the land it was 'in too rich a condition to be safe for barley'. Although the practice of taking two successive corn crops was considered bad farming by the experts, farmers discovered through experience what scientists later found to be sound practice.[16]

Oats

Figure 5.2 and Table 5.2 give our data for oats. There was considerable annual variation in the yield of oats, but the long-run trends are clear. From a base of 20–5 bushels per acre in the 1720s and 1730s yields rose significantly

---

[14] D. H. Robinson (ed.), *Fream's Elements of Agriculture* (14th edn. London: John Murray, 1962), 229–36; Gourvish and Wilson, *British Brewing*, 186.

[15] In a wider context, the general advantages in terms of output arising from the Norfolk four-course rotation, both the mix of grass and crops and the different kinds of crops, are modelled in R. S. Shiel, 'Improving Soil Productivity in the Pre-fertiliser Era', in Campbell and Overton (1991), esp. 70–7.

[16] C. S. Read, 'Recent Improvements in Norfolk Farming', *JRASE* 1st ser. 19 (1858), generally 284–5, quotations on 285.

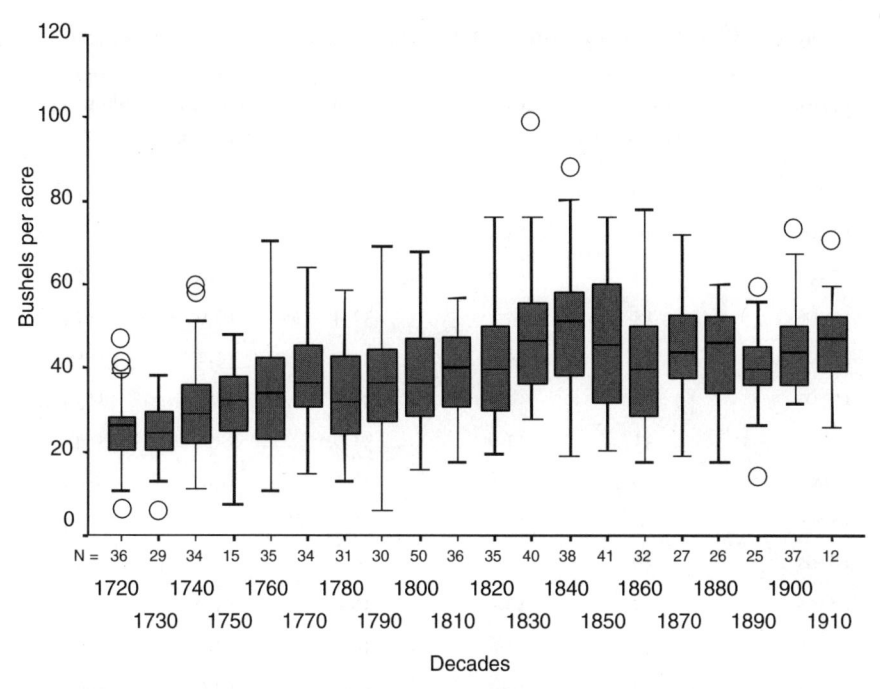

Fig. 5.2 Boxplot of English oats yields

from the 1730s until the 1770s. As with barley, there was then something of a fallback, but it was less pronounced. And, again as with barley, the first half of the nineteenth century saw a considerable increase in yields to reach nearly 50 bushels per acre in the 1840s. The figure then fell back slightly, but not to eighteenth-century levels. It stayed above 40 bushels an acre, a figure which had not been achieved before 1800. The general trend is similar to the one that Holderness constructed from standard contemporary sources, which showed a range of oats yields from 32 to 38 bushels per acre in *c*.1800, and a median of about 37 bushels. Holderness's conclusion that 'the productivity of oats apparently increased with good cultivation more than that of any other grain' is something we cannot test, although our evidence must surely point in that direction.[17]

Although the cultivation of oats is usually considered to be best suited to the north and west of England where summers are cooler and moist, the output of oats could be much higher in the south than in the north.

---

[17] *AHEW* vi. 141.

Table 5.2 Oats yields—summary statistics

Period	No. of observations	Mean	Median	Standard deviation	Skewness
1720s	36	25.4	26.1	8.7	0.32
1730s	30	24.5	23.9	10.1	0.21
1740s	34	30.8	28.9	11.4	0.85
1750s	15	30.3	32.4	10.9	−0.74
1760s	35	35.8	34.0	15.2	0.59
1770s	34	38.0	36.5	11.2	0.34
1780s	31	33.9	31.9	12.2	0.44
1790s	30	36.8	36.7	14.2	−0.04
1800s	50	37.8	36.3	12.2	0.34
1810s	36	38.9	40.4	10.7	−0.27
1820s	35	40.5	39.9	13.0	0.44
1830s	40	48.6	46.5	15.7	1.07
1840s	38	49.8	51.1	16.6	0.07
1850s	41	45.2	45.6	16.3	0.13
1860s	32	40.7	39.8	15.4	0.70
1870s	27	45.7	44.0	14.6	0.21
1880s	26	43.4	46.2	11.3	−0.54
1890s	25	40.5	39.8	9.7	−0.52
1900s	37	45.2	44.0	10.4	0.76
1910–14	12	46.8	47.1	12.0	0.26
	644	39.1	38.5	14.6	0.44

Oats were grown very successfully with high yields on the Fens.[18] However, as with barley, for some farmers the quality of the crop was as important as the quantity harvested, and seed varieties also had an important impact on yields. In the north, where oats remained significant for human consumption, and where livestock often relied on the straw throughout the winter,

---

[18] J. Haxton, 'Essay on the Cultivation of Oats', *JRASE* 1st ser. 12 (1851), 109; J. C. Morton (ed.), *Cyclopedia of Agriculture* (London, 1855), 409.

the nutritional value of the crop was critical. In his prize essay of 1851 John Haxton observed of oats that:

In the southern part of the island, where this grain is principally used for feeding horses and fattening stock, the main object hitherto seems to have been to obtain as much bulk of straw and as many bushels per acre as possible, without much regard to the quality of either; and hence we find the coarser varieties—such as the 'tartarian' and the Red sorts—principally cultivated . . . In Scotland, and the north of England, the quality of both straw and grain is a material point, as the former constitutes the principal fodder of live stock from Martinmas to Whitsuntide, while the latter made into meal is ... the main article of food of the Border peasantry.[19]

In such circumstances a key factor affecting the yield of oats was the importance a farmer placed on the quality of the straw. Traditionally, straw was the principal winter feed.[20] William Marshall noted that in the southern counties and the London basin, in unfavourable winters, straw was fed for up to five months from mid-November until mid-April.[21] This was particularly important prior to the introduction of rotations using root crops and clover leys, and even after mixed rotations were adopted straw was often used to bulk out the feed.[22] When the production of straw was important to the farmer, it was possible to select varieties of oats that yielded high quantities of straw relative to grain. Common oats gave long straw while Tartarian produced poor straw best suited to litter.[23]

In northern England, where oats were a principal corn crop, plants used for livestock feed tended to be harvested young before the grain had fully ripened. When this occurred, the transfer of both protein and soluble carbohydrates from the straw into the seed which takes place during the ripening process was avoided. This increased the feed value of the straw while lowering the yield of the grain.[24] Until the last twenty years of the nineteenth

---

[19] Haxton, 'On the Cultivation of Oats', 121.

[20] G. E. Fussell (ed.), *Robert Loder's Farm Accounts, 1610–1620* (London: Royal Historical Society, Camden Society 3rd ser. 53, 1936), 32–3; William Ellis, *The Modern Husbandman*, iv (London: T. Osborne, M. Cooper, 1744), 95–6; Robert Brown, *The Compleat Farmer* (London: J. Coote, 1759), 26.

[21] W. Marshall, *Minutes, Experiments, Observations, and General Remarks, on Agriculture, in the Southern Counties, a New Edition. To which is Prefixed a Sketch of the Vale of London, and an Outline of its Rural Economy*, i (London: G. Nicol, 1799), 389–90.

[22] W. Marshall, *The Rural Economy of Norfolk*, ii (London, 1787), 100; S. Jonas, 'On the Farming of Cambridgeshire', *JRASE* 1st ser. 7 (1846), 69–70.

[23] J. F. Burke, *British Husbandry: Exhibiting the Farming Practices in Various Parts of the United Kingdom*, ii (Library of Useful Knowledge, Farmer's Series: London: Baldwin & Craddock, 1837), 180, 184–5; J. R. Walton, 'Varietal Innovation and the Competitiveness of the British Cereals Sector, 1760–1930', *AgHR* 47 (1999), 40–2.

[24] T. B. Wood, *Animal Nutrition* (London: W. B. Clive, 1924), 48–9.

century oats were grown in some regions as much for the resulting straw as for the grain.[25]

Oats were often grown in inferior soils or on land as it was broken up from pasture. They were also grown as a prelude to land going out of arable and into long-term ley. These different circumstances affected yield. Where pasture was broken up, the land had a high concentration of nutrients but any pests and weeds present would compete with the oats on a significant scale. Where land was about to go into ley, it would be exhausted and yields would necessarily be low.[26] When oats (or for that matter barley) were undersown with clover or vetches, the additional nitrogen fixed by these plants increased the yield. However, if the nitrogen content were too high the crop would lodge, making harvesting more difficult and lowering the yield.

Table 5.3 brings together data for the second half of the nineteenth century from the agricultural statistics and contemporary estimates. Leaving aside specific years or short periods, the table shows that there was no obvious improvement in long-run unit acre productivity for any of the grains. Only in one respect would we modify this conclusion. In Ireland wheat yields improved markedly, and barley and oats yields moderately. However, this might be an unfair comparison since in Ireland there was a decline in grain acreages which was more dramatic than in the rest of the UK.

In addition, wheat, barley, and oats did not necessarily stand or fall together in terms of the annual harvest. Under most circumstances, the market for barley and oats was different from wheat. Barley and oats were not the principal grain crops in England as a whole,[27] but during years of dearth, such as occurred 1795–1801, they were mixed in with wheat and other products to produce bread, and thereby they eked out the diminished supply of wheat flour. In 1795 the yield of both crops was very high, and considerable quantities were diverted to supplement the wheat crop.[28] In years when the wheat harvest was poor, it was not always the case that barley and oats suffered a similar experience.[29]

---

[25] Robinson, *Fream's Elements*, 243.   [26] Burke, *British Husbandry*, 2, 181.

[27] E. J. T. Collins, 'Dietary Change and Cereal Consumption in Britain in the Nineteenth Century', *AgHR* 23 (1975), 97–115.

[28] PRO Privy Council Minutes, PC 4/6, witness reports of Oct. 1795, *passim*.

[29] M. E. Turner, 'Agricultural Productivity in England in the Eighteenth Century: Evidence from Crop Yields', *EcHR*, 35 (1982), 493; id., 'Corn Crises in Britain in the Age of Malthus', in M. E. Turner (ed.), *Malthus and his Time* (London: Macmillan, 1986), 118–19.

Table 5.3 United Kingdom grain yields, c.1850–c.1914

(a) Wheat (bushels per acre)

England		UK		Ireland	
Year	Quantity	Year	Quantity	Year	Quantity
c.1850	26.3	1852/3–1859/60	28.75	1850s	13.3
1861	29.1				
1863–82	27.9	1860/1–1867/8	28.875	1860s	11.7
1868	28.0				
1870	29.6	1868/9–1875/6	27.125	1870s	13.4
c.1870	29.5				
1876–82	24.6	1876/7–1883/4	25.25	1880s	14.7
1879	19.8				
1882	26.0	1884/5–1891/2	29.875	1890s	16.6
		1852/3–1891/2	27.875[a]	1900s	19.2

[a] Full period average.

(b) Barley and oats (bushels per acre)

England			Ireland		
	Barley	Oats		Barley	Oats
1861	37.7	46.8	1850s	13.4	16.7
1863–82	34.3	43.3	1860s	12.0	15.1
Pre-1878	35.2	47.5	1870s	13.0	16.0
1876–82	30.1	39.0	1880s	13.5	15.5
1879	25.2	40.7	1890s	15.1	17.2
1882	32.4	44.7	1900s	16.9	18.6

Table 5.3 *Continued*

(c) Wheat, barley, and oats (cwt per acre)

	England and Wales		
	Wheat	Barley	Oats
1885–9	16.3	15.6	13.6
1890–4	15.9	15.9	13.8
1895–9	17.2	15.9	13.8
1900–4	16.4	15.1	14.2
1905–9	18.3	16.5	14.8
1910–14	17.2	15.3	13.5

Sources English wheat, barley, and oats from P. G. Craigie, 'Statistics of Agricultural Production', *JRSS* 46 (1883), 21, 40–2. But note, calculating separately averages from Craigie's tables on 40–2 differs marginally from his report of the same on 21.

UK wheat from J. B. Lawes and J. H. Gilbert, 'Home Produce, Imports, Consumption, and Price of Wheat over Forty Harvest-Years, 1852–53 to 1891–92', *JRASE* 3rd ser. 4 (1893), 132–3, though there are two versions, the one based on 60 and the other on 61 lb per bushel. The one quoted here is 60 lb per bushel.

Irish wheat, barley, and oats from BPP, *Annual (Irish) Agricultural Statistics*.

England and Wales wheat, barley, and oats from MAFF, *A Century of Agricultural Statistics: Great Britain 1866–1966* (London: HMSO, 1968), 108–12.

Other Crop Yields

Table 5.4 presents summary statistics we have been able to compile from the farm records for a number of crops. As is clear, we have far more observations for wheat, barley, and oats than for other crops, and this means we have to be rather more tentative in our conclusions (Figs 5.3 and 5.4). The data are summarized on the basis of twenty-five-year cross-sectional aggregates.

Data for rye suggest a trend not dissimilar to that of the other corn crops, with a significant increase in yields in the course of the nineteenth century. Other crops grown on farms, notably the main pulse crops, were generally consumed as animal fodder. The unit yield of beans at up to 30–1 bushels per acre was just about as high in the first half of the eighteenth century as at any time during the nearly 200 years for which we have data. A slump in yields in the third quarter of the eighteenth century gave way to a steady rise

Table 5.4 Crop yields—summary statistics

	Wheat		Barley		Oats		Rye		Hay	
	Mean bushels/ acre	No. of observations	Mean bushels/ acre	No. of observations	Mean bushels/ acre	No. of observations	Mean bushels/ acre	No. of observations	Mean cwt/ acre	No. of observations
1700/24	19.9	18	31.2	16	22.1	13				
1725/49	21.3	79	28.5	92	27.7	87			39/41	5
1750/74	21.1	110	30.3	117	34.5	71			33/34	82
1775/99	19.2	146	26.7	124	36.5	74	22.3	2	23,6/24	22
1800/24	21.6	209	30.0	171	38.8	100	23.8	6	31/32	3
1825/49	27.8	267	36.9	231	47.1	99	29.7	10	55/57	9
1850/74	28.3	291	34.7	275	43.6	87	26.4	13	21	4
1875/99	27.1	131	32.4	122	42.8	64	37.0	25	33	8
1900/14	29.4	49	30.9	48	45.6	49	31.0	12		
Totals		1,300		1,196		644		68		133

Table 5.4 Continued

	Beans		Peas		Peas and Beans		Vetches		Ryegrass seed	
	Mean bushels/ acre	No. of observations	Mean bushels/ acre	No. of observations	Mean bushels/ acre	No. of observations	Mean bushels/ acre	No. of observations	Mean bushels/ acre	No. of observations
1700/24	31.2	8	16.9	11						
1725/49	29.7	21	19.4	42			14.9	8		
1750/74	20.0	15	19.2	15	20.1	6	11.5	6		
1775/99	23.6	17	20.4	34	20.0	1	13.3	12		
1800/24	25.4	36	20.6	62	20.9	2	14.5	29	9.1	2
1825/49	31.4	64	28.7	45	29.0	16	23.1	3	22.5	9
1850/74	28.6	45	28.6	32	37.0	1	20.0	3	19.7	18
1875/99	30.2	7	24.5	7						
1900/14	25.8	9	30.9	14	29.7	10	25.5	4		
Totals		222		262		36		65		29

Note The original yields were measured in a variety of different ways. They have all been converted to bushels per acre in this table except hay, which remains in cwt per acre. On occasions the hay yields were also measured in loads, and these have been converted into cwt by means of two different conversion factors, hence the range of measurements listed for that crop on some occasions.

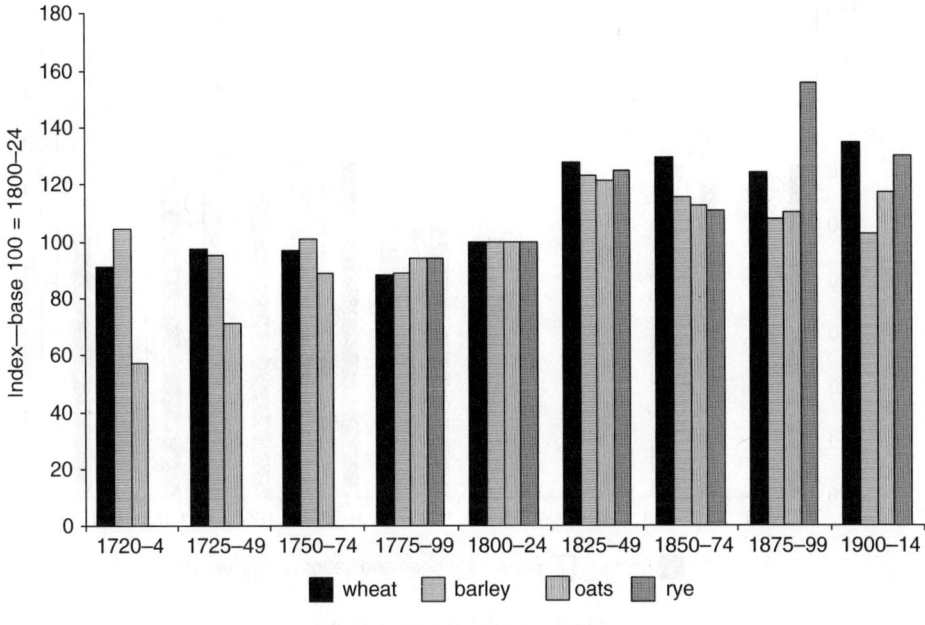

Fig. 5.3 Index of grain yields

over the next seventy-five years, and for the second half of the nineteenth century they remained more or less level at 28–31 bushels per acre. This trend mirrors some of those we identified in the grains, if not in degree then certainly with the emphasis on the increase in the nineteenth century rather than in the eighteenth century. The equivalent trend for peas almost perfectly mirrors the main trend in wheat yields—a fairly stable, plateau-like shape in the eighteenth century giving way to a new and higher plateau in the nineteenth century. In the case of peas, this new plateau settled down about 45 per cent higher than the eighteenth-century levels. For comparison, in the case of wheat it was about 33 per cent higher.

We also list the trend in yields for peas and beans together, and finally for vetches and ryegrass, in order to complete the record, but the samples are so small that we are not inclined to draw conclusions.

Finally we include estimates for the yield of hay, in terms of hundredweight per acre. These were reported variously in tons, hundredweight, and loads. Following Primrose McConnell we have converted the loads into

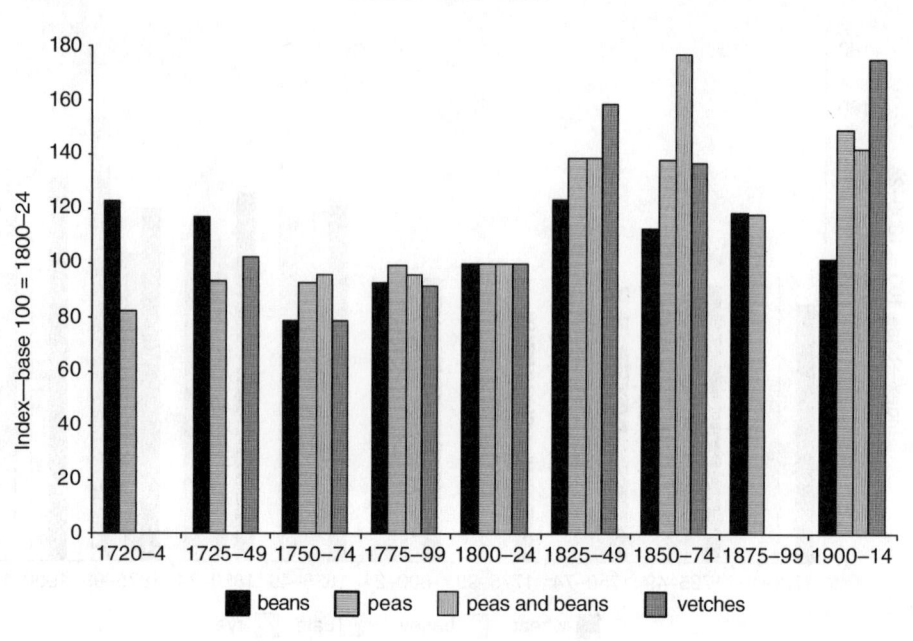

Fig. 5.4 Index of pulses yields

hundredweight using two conversion factors, of 18 hundredweight, or 19 hundredweight and 32 pounds, per load.[30] Certainly the depressed trend in eighteenth-century yields identified in a number of other crops seems also to be present in the case of hay.

Seeding Rates

Crop yields are not the only indicator of agricultural or productivity change. Table 5.5 gives a sample of direct measurements in the form of the seeding rate for wheat, barley, and oats. Seeding rates are an indicator of productivity change. Allen has argued that the well-attested grain yield levels in

---

[30] P. McConnell, *Note-Book of Agricultural Facts and Figures for Farmers and Farm Students* (10th edn. London: Crosby, Lockwood & Son, 1922), 28, 40, the two estimates being the estimated weight for old and new hay. The two different weights for hay account for the range of mean yields reported in Table 5.4.

Table 5.5 Seeding rates, crop yields, and yield rates

Crop and sample size	Period	Seeding rate (bushels/acre)	Yield (bushels/acre)	Yield rate (yield/unit seed) (bushels)
Wheat 88	1720s	3.2	20.0	6.3
	1740s	2.5	21.8	8.7
	1750s	2.4	22.4	9.3
	1760s	2.1	21.8	10.4
	1780s	3.1	18.9	6.1
	1790s	2.5	19.0	7.6
	1800s	2.7	21.0	7.8
	1810s	2.6	21.2	8.1
	1820s	2.7	23.6	8.7
	1840s	1.9	30.6	16.1
	1850s	1.6	27.5	17.2
	1870s	2.4	28.9	12.1
	1890s	2.8	27.1	9.7
	1900s	2.3	28.1	12.2
Barley 60	1720s	7.0	30.2	4.3
	1740s	4.6	29.7	6.5
	1750s	4.3	29.0	6.7
	1760s	4.2	30.7	7.3
	1780s	5.2	25.5	4.9
	1790s	4.7	26.0	5.5
	1800s	3.5	28.3	8.1
	1810s	5.2	30.2	5.8
	1840s	3.1	38.1	12.3
	1850s	2.7	34.9	12.9
	1890s	3.3	30.6	9.3
	1900s	2.7	30.4	11.2
Oats 43	1720s	6.7	25.4	3.8
	1740s	5.7	30.8	5.4

Table 5.5 *Continued*

Crop and sample size	Period	Seeding rate (bushels/acre)	Yield (bushels/acre)	Yield rate (yield/unit seed) (bushels)
Oats 43	1750s	5.1	30.3	5.9
	1760s	5.4	35.8	6.6
	1780s	6.3	33.9	5.4
	1790s	6.3	36.8	5.8
	1800s	4.1	37.8	9.2
	1810s	7.5	38.9	5.2
	1820s	5.0	40.5	8.1
	1840s	3.3	49.8	15.1
	1890s	4.0	40.5	10.1
	1900s	4.0	45.2	11.3

Note The seed rates are based on simple averages of the examples available; the yields based on decennial averages as indicated.

medieval Norfolk were the result of a high sowing rate. By the late eighteenth century Norfolk farmers had apparently reduced the sowing rate, but managed to maintain the yield to seed ratio. In any assessment of productivity change these adjustments must be taken into account.[31]

There are very few sowing or seeding rates recorded in the *Agrarian History*. For the period 1649–87 the wheat rate was between 2.25 to 3.37 bushels per acre for farms in Surrey, Hertfordshire, and Wiltshire, with a single quotation from Shropshire in 1745 of 2 bushels per acre. In estimating profitability in the early eighteenth century Bowden assumed sowing rates of 2.25 bushels per acre for wheat, 3.75 for barley, and 4.25 for oats.[32]

---

[31] R. C. Allen, 'Tracking the Agricultural Revolution in England', *EcHR* 52 (1999), 222. See especially B. M. S. Campbell, 'Land, Labour, Livestock, and Productivity Trends in English Seignorial Agriculture, 1208–1450', in Campbell and Overton (1991), 144–82, esp. 161, 163 for a demonstration of the relative stability of yield ratios on the Winchester, Westminster, and Norfolk estates over the period 1225–1453. Others have argued that high sowing rates coupled with high germination rates but fixed nutrient levels in soils could increase the competition for those nutrients and lead to low yields. See L. Brunt, 'Nature or Nurture? Explaining English Wheat Yields in the Agricultural Revolution', *University of Oxford Discussion Paper in Economic and Social History*, 19 (1997), 15.

[32] *AHEW* v (2). 87.

Table 5.6 Seed and yield rates at Hurstpierpoint, East Sussex

	Crop	Seeding rate (bushels/acre)	Yield (bushels/acre)	Yield rate (bushels)
1727–8	wheat	3.5	20.2	5.8
1727–9	barley	7.0	23.5	3.4
1727–9	oats	6.9	25.7	3.7
1727–9	peas	4.0	10.5	2.6
1728–9	peas[a]	4.0	17.1	4.3
1727–9	beans[b]	3.3	18.3	5.6
1743–7	wheat	2.6	24.7	9.6
1743–50	barley	5.5	36.3	6.6
1744–8	oats	5.1	48.2	9.5
1743–8	peas	3.1	33.6	11.0
1746	beans[c]	2.0	38.7	19.3

[a] Excluding 1727 when the yield was only 4.2 bushels per acre.
[b] Based on a sample of only 5.6 acres.
[c] Based on sample of only 1.25 acres.
Source East Sussex RO, DAN 2199/2201.

The data in Table 5.5 would suggest that his figures are too low, especially for barley and oats. Table 5.6 is a summary of the stated seeding rates and crop yields from a farm at Hurstpierpoint in East Sussex for two periods in the late 1720s and the mid-1740s. The table includes the calculated yield rate. For all of the crops listed the average seed rate between these two periods was significantly reduced, although this may have been affected by the land area involved, which was much larger for the 1720s than the 1740s. Even so, it is in line with the national figures in Table 5.5. At Hurstpierpoint the resulting effect on the yield rate was an increase of over 100 per cent for four of the listed crops, with a high of 157 per cent for beans, although the latter figure was based on too small a sample to be reliable. For the fifth crop, wheat, the improvement was 67 per cent. The result in terms of yield rate again follows the national picture. The only equivalent measure available in print for roughly the same period is the crop–seed ratios associated with a Northamptonshire farm for 1737–8 in which the wheat ratio is 6.1, barley 4.1, and beans 3.5. The rates or ratios for the two grains are not

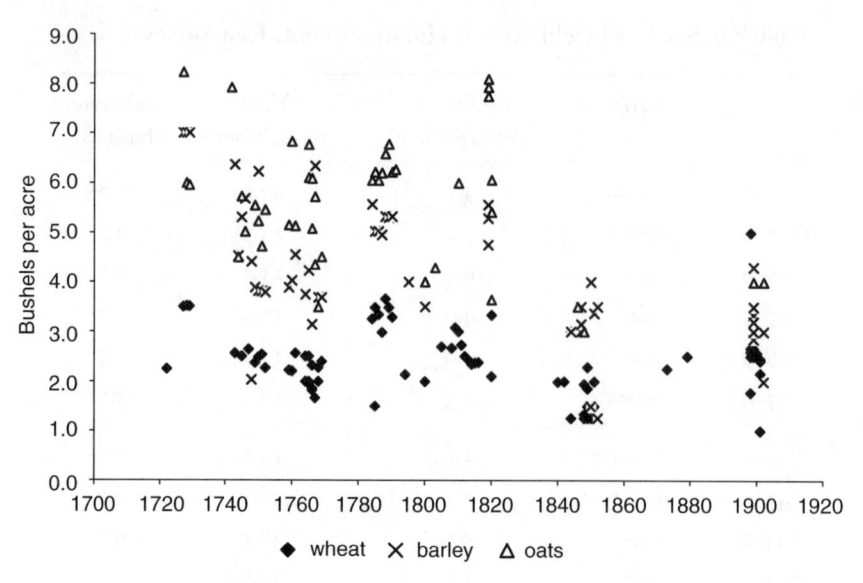

Fig. 5.5 Seeding rates of the grain crops

significantly dissimilar to the Hurstpierpoint ratios for the 1720s, though they are quite some way short of the equivalent measure in the 1740s.[33]

We may be on more secure ground when discussing seeding rates for the late eighteenth and the nineteenth centuries. The choices that farmers faced when it came to the density of sowing seeds can be neatly listed. They included the time of year, the changing technology (dibbling or drilling), the changing attitudes to soil dressing in preparation for the seed, and a general appreciation of the productivity effects of dense or sparse sowing rates. Yet when these considerations are taken into account the common rate of sowing at 4 bushels per acre before about 1800, for wheat at least, gave way to a sowing rate of 2 to 3 bushels per acre thereafter. The general impression to be gathered from the literature is that the medieval yield rate of bushels of wheat harvested per bushel sown improved to 6 or 7 in the seventeenth century and about 8 in the mid-eighteenth century. On the basis of the Board of Agriculture reports the modal yield rate for the 1790s–1800s for wheat was 9.[34] The average yield rates that Slicher van Bath meticulously extracted, calculated or inferred, listed, and averaged out, he principally drew from Arthur Young's works. They suggest a wheat yield rate in c.1770 of 9.5 rising

[33] *AHEW* v (2). 881.   [34] *AHEW* vi. 142–3.

marginally to 9.8 in the late 1790s. The rate for barley was 9.7 in both periods, and 8.5 for oats. The addition of William Marshall's evidence in the 1790s slightly raises the rates for barley and oats.[35] By 1846 (based on McCulloch's England), the equivalent yield rate we estimate was 7 for all of the crops. This seems an unlikely and large fall in the yield rates, and suspiciously all too constant across all of the crops. The precise rate derives from the difference between McCulloch's estimate of total output and his estimate of net output (the difference giving the total seed sown), as a ratio of his total output estimates. As with so much of McCulloch's work there is rather too much informed guesswork and rather too little direct measurement or observation, and on this occasion we suspect that he has assumed a yield rate in order to derive his seeding rate and hence his net output.[36]

The data in Table 5.5 and Fig. 5.5 strongly suggest that there was a long-run downturn in the seeding rates from the second quarter of the eighteenth century to the middle of the nineteenth. If this trend is correct and if the yields we have already measured are representative the estimates summarized in Table 5.5 seem reasonable. They put a slightly different gloss on the trend of productivity based only on yields, but they also confirm the overall impact. For wheat, based on yields alone, we detected little or no productivity improvement in the eighteenth century. What we can now say is that there was greater efficiency, and therefore a productivity improvement in the use of seeds. The seeding rate declined, the yield was maintained, but the yield rate increased. All three trends look to have been maintained into the nineteenth century, and they established a significant efficiency gain in the yield rate. These trends, and this efficiency rate, also occurred with the other grains. In the problem decades of the 1780s and 1790s when wheat yields dipped the crisis was apparently met by an increase in seeding rates.

We can only speculate on the changing farming mechanisms which induced these results. Logically we would propose that the rate of change out of broadcast sowing and towards successively dibbling by hand and the

---

[35] Our estimates based on B. H. Slicher van Bath, 'Yield Ratios, 810–1820', *AAG Bijdragen*, 10 (1963), 53–6, 134–7, 172–4. By 1808 these yield rates in Sussex at least, on the basis of Young's *General View* of the county, had fallen to 7/8 for wheat and 8 for each of barley and oats, as quoted in Slicher van Bath, 'Yield Ratios', 64–5, 144, 180–1. See also B. H. Slicher van Bath, *The Agrarian History of Western Europe A.D. 500–1850* (London: Edward Arnold, 1963), 330–1, and also his 'Agriculture in the Vital Revolution', in E. E. Rich and C. H. Wilson (eds.), *The Cambridge Economic History of Europe*, v: *The Economic Organization of Early Modern Europe* (Cambridge: Cambridge University Press, 1977), esp. 79–82.

[36] *AHEW* vi. 1046. See Holderness's critique of McCulloch's estimates ibid. 175–6.

introduction of seed drills is an obvious explanation. Unfortunately, as we saw in Chapter 3, the farm records do not offer us conclusive evidence about the pace of adoption, particularly of seed drills. Brunt's conclusion that this was a critical innovation for raising yields remains to be tested empirically.[37]

Conclusion

After completing his research for the *Agrarian History*, Holderness regretted that 'one thing it has been impossible to do is construct a long-run annual series of grain yields linking the eighteenth and nineteenth centuries'.[38] The findings we have presented in Chapters 4 and 5 suggest that there are data in farm records which take us a long way towards bridging this 'impossible' gap. If we accept that changes in agricultural productivity are sensitively measured by the course of crops yields in general, and wheat in particular, then we must be closer to identifying the major turning points in agricultural history and in locating the agricultural revolution than we have been before. Yet, as we have also suggested, to take wheat as a simple guide underplays the importance of other grains, notably barley and oats. It also overlooks entirely the pastoral side of English farming. We proceed in Chapter 6 to ask if this also passed through an agricultural revolution?

[37] Brunt, 'Nature or Nurture?', 21.   [38] *AHEW*, vi. 135.

CHAPTER 6

Livestock

Measuring an agricultural revolution purely in terms of the corn harvest provides us with a sensitive indicator of national well-being because it can be linked directly to the staple food supply. However, it is not sufficient in itself. Much of the agricultural economy was concerned with livestock, and the production of meat and related products including milk played an important role in feeding the nation. This role increased as population grew and, particularly in the course of the nineteenth century, as dietary demands expanded. Unfortunately, measuring livestock output is exceptionally complicated. In the late nineteenth century, in spite of the advent of the annual agricultural statistics, the regular collection of carcass weights, wool clips, milk yields, and so on was not undertaken. This difference between the treatment of crops and animals demonstrates the seasonal finality and therefore stock-taking of the crop harvest. Relatively speaking, in the process of animal production, the output from some animals could take years to be realized, so that even an annual animal census is less a final measure of output than the annual grain harvest. Add to this the internal trade in animals, and the movement from breeding to grazing to fattening grounds, and the measurement of output in the animal sector becomes, seemingly, almost impossible.

Consequently, as we turn in this chapter to the pastoral sector of the agricultural economy, our claims must inevitably be modest. Although it is possible to reconstruct estimates of livestock yields for the later nineteenth century,[1] for the period before c.1850 there are no systematic reports or estimates. Attempts by contemporaries and historians alike to count the British

[1] *AHEW* vii. 311–12.

sheep flock in the late eighteenth century show just how little we have to go on. Contemporary figures ranged between 12 million sheep in 1769, 29 million in 1770, and 12 million in 1771. At the extremes there are estimates of 10–12 million animals in 1774 and 54 million in 1782, and the most widely accepted estimate of all is John Luccock's 1809 estimate of 26 million sheep. More recent estimates suggest that this is too high, and a figure of 12 million sheep in *c*.1800 is more likely.[2] Even if—and it is a substantial 'if'—we can count the number of animals with some accuracy we are still a distance from measuring output. An estimate of gross beef output, for example, is the arithmetic product of the number of beef animals slaughtered and their average weights. This is a basic calculation before we consider precisely the different products contained in the animals—the offal, the bones, the hide, let alone the prime meat. And the traditional method of weighing and recording animals by so-called 'carcass' weights even excludes a number of these products, making the task of measuring animal output ever more complex. A full assessment would also include the supply of manure over the lifetime of a particular animal, and the list of 'products' can be extended further—wool, milk, tallow, hoofs, and so on. Even the basic elements in the initial calculation of the meat involved—the number of animals and their average carcass size—have evaded satisfactory estimation, let alone official collection.

These provisos are critical to an understanding of the data we present in this chapter. Evidence from the farm records offers a wealth of new observations on carcasses. We shall use these to point in the direction of what we see as the major trends in animal weights during the agricultural revolution, but we shall necessarily be circumspect in our claims.

Carcass Weights

Our basic data in this chapter take the form of carcass weights. The weight of an animal was determined by its breed, by its age and condition at slaughter, and according to the intended final use of the carcass. Farm records give

---

[2] See G. E. Fussell and C. Goodman, 'Eighteenth-Century Estimates of British Sheep and Wool Production', *Agricultural History*, 4 (1930), 131–51, esp. 132–3 for a list of 18th-century estimates. See also J. Luccock, *The Nature and Properties of Wool* (Leeds, 1805), esp. 338; id., *An Essay on Wool* (London, 1809), 38 and 339. Conversely see M. E. Turner, 'Counting Sheep: Waking up to New Estimates of Livestock Numbers in England, *c*.1800', *AgHR* 46 (1998), 142–61, for a critique of the plausibility of 18th- and early 19th-century sheep estimates. For cattle estimates see G. E. Fussell, 'The Size of English Cattle in the Eighteenth Century', *Agricultural History*, 3 (1929), 160–81.

a great deal of information on carcass weights, but our sample is insufficient to add anything to the general literature on the relative sizes of different animal breeds. We can say more about age and condition.

The *age* of an animal was obviously one of the determining factors in its final weight. However, the precise length of time between birth and slaughter was not fixed. It could be influenced by a whole range of factors including the availability of feed, both in total and at sensitive times of the year, the body composition of the animal, and the principal end product for which it was intended. With sheep, for example, the question arises of whether they were raised for their wool, their mutton, or their function as providers of manure. In other words a combination of environment, physiology, and the market determined age at slaughter. While the inhibiting influence of many of these factors was reduced significantly during the period,[3] farm records are inadequate for the precise identification of change. There is a wealth of scattered information in the *General Views* and other contemporary writing as to age at slaughter of cattle and sheep. None of it is systematic, but farmers generally indicated the degree of maturity of the animals through the use of age-specific nomenclature, although only rarely did they give detailed information in terms of weeks, months, or years.

The *condition* of the animal also affected its carcass weight. It would have been useful to us had farmers operated in the context of weights which isolated different categories of livestock in terms of their life cycle. For cattle these boundaries or weights might have defined them as breeding, store, or fat animals. Yet it is clear that only the most scientifically minded farmers even considered keeping records of the progress of the life cycle of their animals. Although feeding experiments produced occasional series of animal weights—from Arthur Young in the late eighteenth century to Lawes and Gilbert in the mid- to late nineteenth century—for most farmers animals were simply sold by the head and the price paid was based on their condition at the moment of sale.[4] It was only at the point of slaughter that weight, *per se*, became important. From the farm records it seems that farmers entered the weight of their animals at slaughter, but generally just for those

---

[3] C. S. Read, 'On the Farming of Oxfordshire', *JRASE* 1st ser. 15 (1854), 268; T. D. Acland, 'On the Farming of Somersetshire', *JRASE* 1st ser. 11 (1850), 678–9; L. G. L. Guilhaud de Lavergne, *The Rural Economy of England, Scotland and Ireland* (Edinburgh, 1855), 19–20, 27–8.

[4] A. Young, 'Experiments in Weighing Fatting Cattle Alive', *Annals of Agriculture*, 14 (1790), 140–63; J. B. Lawes and J. H. Gilbert, 'On the Composition of Oxen, Sheep, and Pigs, and of their Increase whilst Fattening', *JRASE* 1st ser. 21 (1860), 433–88.

in fat condition. Only a handful of farmers recorded animal weights from animal purchases.

We may have no difficulty understanding how breed, age, and condition affected carcass weights, but interpreting the evidence which is available to us leaves room for imprecision. For example, it may be unclear whether the figures recorded referred to live weight, carcass weight, or even, particularly in records generated by the butcher, the weight of a single quarter. This problem of interpretation is not unique to us. In his study of animal breeding, Russell assumed that the minimum weight of fat oxen required by Elizabethan naval purveyors, 6 hundredweight, referred to the carcass weight. When using a conversion factor of carcass to live weight of 0.6 (or 8 : 14), this produced a live weight of approximately 1,100 pounds. Trow-Smith, on the other hand, believed that the naval purveyors were referring to the live weight, and hence to animals of 672 pounds.[5] Fortunately, farm records generally provide sufficient information for us to be reasonably sure as to what is being measured. In some cases even the sale of individual parts of the animal was recorded. At Chicheley, Buckinghamshire, sales of sheep skins and rough fat in the 1780s were individually noted, while at Nymet Rowland, Devon, the farmer regularly recorded the sale of skins in the 1820s and 1830s.[6]

Live animals not yet ready for slaughter were usually sold by the head, and the weights recorded in farm records were generally for fat animals ready for killing, in other words carcass weights. The farmer received payment only for the edible parts of the animal, leaving the proceeds from the sale of the 'offal', such as the hide, tallow, entrails, feet, and head, to the butcher. As a result, when meat prices were recorded they were often quoted as so much per pound (or other unit of weight), 'sinking the offal'. The value of the offal went to the butcher rather than to the farmer, and was reckoned to be the 'fifth' quarter of the carcass.[7] In weight, the offal was not considered as part of the carcass, hence the fact that the carcass weight was less than the live animal weight. In general the edible carcass averaged

---

[5] N. Russell, *Like Engend'ring Like: Heredity and Animal Breeding in Early Modern England* (Cambridge: Cambridge University Press, 1986), 128; R. Trow-Smith, *A History of British Livestock Husbandry to 1700* (London: Routledge & Kegan Paul, 1957), 177.

[6] Buckinghamshire RO, D/C/4/30; Devon RO, 1107M/E1.

[7] J. F. Burke, *British Husbandry: Exhibiting the Farming Practices in Various Parts of the United Kingdom*, ii (Library of Useful Knowledge, Farmer's Series: London: Baldwin & Craddock, London, 1837), 395.

approximately 60 per cent of the animal for sheep and cattle and 80 per cent for pigs (where the head and feet of the pig were consumed), but in practice the precise amount of offal depended on the condition of the animal. In 1804, James Quantock of South Petherton in Somerset had his live animals weighed on a weighbridge at Illminster both before and after slaughter. His ox weighed 1,736 pounds live with a carcass weight of 1,026 pounds, a ratio of 100:59. His sheep, which he noted was 'remarkably fat', weighed 281 pounds fat, and had a carcass weight of 172 pounds, at a ratio of 100:61.[8]

Variations in the unit of measure further complicated the issue. Pigs were generally, though not always, weighed by the score of 20 pounds, but there was greater variation with the weights of sheep and cattle. In the London meat markets, and other areas of south-east England, a stone of 14 pounds was used when measuring live weight, while an 8-pound stone was used for carcass weight. In northern and eastern England the 14-pound stone was the norm, with local variations such as the 12-pound stone occasionally found in Herefordshire. Generally in the north-west, as well as in the west, and parts of the south, the score of 20 pounds was used.[9]

Important implications arise from these uncertainties regarding animal weights. A very fat beast would have a higher ratio of edible meat to offal while a lean animal would have less. The proportion of 'meat' from a given live weight could, and often did, vary by more than 15 per cent.[10] The quantity of fat was partly determined by the nature of the animal, but it also depended to a large extent on taste. In the late seventeenth century and again in the mid-Victorian period fat animals with large carcasses were fashionable. Such carcasses would have contained too much fat for the modern consumer.[11] These variations in the proportion of meat, tallow, and even hide had a considerable impact on output calculations. Since the farmer was generally paid just for the meat, farm records only occasionally provide information about the composition of the carcass. Coton Hall Farm, at Bridgnorth in Shropshire, is exceptional in this respect, since the accounts provide an insight into the relationship between fat, hide, and meat in the mid-eighteenth century (Table 6.1). The farmer noted the weight of the

---

[8] Devon RO, DD/MR110.
[9] J. C. Morton (ed.), *Cyclopedia of Agriculture*, ii (London, 1855), 1126. See also Appendix 2 below.
[10] See, for example, T. H. Horne, *The Complete Grazier* (12th edn. London, 1877), 178.
[11] Russell, *Like Engend'ring Like*, 128; F. Gerrard, *Meat Technology* (3rd edn. London: Leonard Hill, 1964), 106.

Table 6.1 The composition of cattle carcasses at Coton Hall Farm, Bridgnorth, Shropshire, 1744–1769

Animal	Percentage of the weight of the carcass meat of:	
	Hide	Fat
Beef	15.1	5.3
Buffalo	11.7	2.8
Bull	17.2	3.5
Bullock	15.8	7.7
Calf	16.0	
Cow	14.7	7.3
Heifer	16.1	5.9
Ox	13.6	8.6

Source RUL, SAL 5/1.

meat, the hide, and the fat of the carcass, although not the total weight of the carcass.[12]

Even if the farmer kept good records, how accurate were the data with which he was dealing? In 1887 the Markets and Fairs (Weighing of Cattle) Act required markets in England and Scotland to make provision for livestock weighbridges. In the absence of weighbridges before the 1860s, it is unclear how weights were determined.[13] Some farms, particularly those specializing in fatstock sales, may have had steelyards or other such machinery for weighing livestock. In the *Book of the Farm* Stephens described suitable steelyards that cost between £15 and £25 in 1848.[14] Burke thought that the weight of livestock could be

duly ascertained by means of the steel-yard, which no extensive grazier should ever be without, for by its constant use while the cattle are fattening, he can instantly ascertain the state and progress of the beasts, ... [but] ... when judged by eye, it is,

---

[12] RUL, SAL 5/1.
[13] R. Perren, *The Meat Trade in Britain 1840–1914* (London: Routledge & Kegan Paul, 1978).
[14] H. Stephens, *The Book of the Farm*, ii (2nd edn. Edinburgh, 1851), 145–6.

however, to an inexperienced person, a matter of much uncertainty, and even when conquered there is the still greater difficulty of estimating the difference between the *live and dead weights*.[15]

While some farmers may have weighed their animals and even gone through the calculations for reducing live weight into carcass weight, it is unlikely to have been the norm. Even in the 1890s when official returns were made for animals sold by liveweight, very few animals were, in fact, sold in this way.

In these circumstances it is inevitable that many weights recorded in accounts were based on estimation, but was the farmer a good estimator? Butchers purchased animals according to actual or, more often, estimated weight. They were experienced at judging the proportion of meat, tallow, hide, and other potential products of an animal, and they paid the farmer or the grazier accordingly. Arthur Young noted that the farmer was at a disadvantage when dealing with the butcher:

Report speaks of men so accurate in judging of the weight of a beast, as to pronounce within a few pounds; of necessity such skilful persons must be either salesmen or butchers, no farmer or grazier, without other experience than what his own business affords, can, I apprehend, pretend to rival such accuracy of eye.[16]

The same sentiment was repeated a century later by Albert Pell who noted that 'One class of experts, the butchers, are continually checking their judgement of live cattle by weighing the carcasses; they are therefore better guessers than those who feed for them.'[17]

To help the farmer, it became common from the early years of the nineteenth century to calculate the dead weight of an animal by physically measuring its length and girth. Burke recounted the method that was employed:

The girth is taken by passing a cord just behind the shoulder-blade, and under the fore legs: this gives the circumference; and the length is taken along the back from the foremost corner of the blade-bone of the shoulder, in a straight line to the hindmost point of the rump. ... or to that bone of the tail which plumbs the line with the hinder part of the bullock: ... The girth and length are then measured by the footrule, and this mode of measurement has been adopted by all the writers upon the subject, as being equally applicable to every kind of animal.[18]

---

[15] Burke, *British Husbandry*, 392. [16] Young, 'Experiments', 140.
[17] A. Pell, 'On Weighing Live-Stock', *JRASE* 2nd ser. 25 (1889), 447.
[18] Burke, *British Husbandry*, 393.

Once these measurements had been taken the farmer could consult one of many size-to-weight tables designed to help farmers arrive at an accurate weight equivalent.[19] They were generally constructed on the basis of carcass rather than live weight.[20] Trials that compared the calculations with actual dead weight suggest that the tables tended to underestimate the actual weight of the meat.[21] Although such 'weights' were more reliable than estimates made by eye, their use was limited and the information in farm accounts was derived from formal calculations rather than the actual weight which might have been obtained by putting each animal on scales.

Finally in assessing carcass weights, we have to take account of the feed, and its regularity and quality, available to individual animals. In a traditional three-course system involving arable with separate pasture and meadow, the density of livestock on a holding—or more widely in a community—was limited by the supply of feed in the winter and early spring, before a new flush of grass appeared. By this time the stubble had been grazed to exhaustion, but accumulated straw from the previous harvest or harvests was an important if limited feed supplement. Oats were stored for the horses and other beasts of burden. Even so, feed was often scarce. One solution to the problem was to send livestock out of the locale for winter-feeding, but with the animals went their manure and this obviously had an effect on sustaining the fertility of the arable. Animals could emerge from the winter in a poor condition and take time to recover, thus lengthening the time before they were ready for slaughter. Animals that could not be fed were slaughtered, possibly before they were fully ready for the market.

Despite the numerous problems we have identified with the estimation of animal weights it is still possible to construct long-run series. In what follows we turn to the actual end result of animal husbandry in terms of carcass weights found in farm records and how they help us to understand the contribution of the pastoral economy to the agricultural revolution.

---

[19] Burke, *British Husbandry*, 393–4; Stephens, *The Book of the Farm*, 43–4; Horne, *The Complete Grazier* (10th edn. London, 1851), 88; C. Hillyard, *Practical Farming and Grazing* (4th edn. London, 1844), 226–9; *The Library of Agricultural and Horticultural Knowledge* (3rd edn. Lewes: J. Baxter, 1834), 437–8; D. Low, *Elements of Practical Agriculture* (Edinburgh, 1834), 519–20; J. Christison, *New Tables for Computing the Weight of Cattle by Measurement* (Edinburgh: Stirling Kennedy & Co., 1839), 1–2.

[20] Stephens, *The Book of the Farm*, 143–4; Lord Somerville, 'The Farmer's, Grazier's and Butcher's Ready-Reckoner; a Short Table, by which the Weight of Stock, According to the Different Usages in England, can be Ascertained; and the Value of Stock of Any Size, with the Difference, at once Discovered', Letters and Papers on Agriculture, *Journal of the Bath and West of England Society*, 9 (1799), 289–90.

[21] Burke, *British Husbandry*, 394–5; Stephens, *The Book of the Farm*, 144–5.

Sheep

Table 6.2 is a reconstruction of sheep carcass weights gathered by Fussell and Goodman in 1930. It was compiled from figures produced by contemporary commentators ranging from D'Avenant and King towards the end of the seventeenth century, to Young, Marshall, and a number of less well-known eighteenth-century commentators. The table is divided into breeds on the basis of their fleece characteristics—whether long- or short-wool sheep. We have excluded breeds for which Fussell and Goodman admitted that they had few 'observations'. For the long-wool sheep these include the Romney breed of Kent (with weights in 1749 and 1771 ranging from 96 to 120 pounds), and the Cotswolds (with weights in 1789, 1792, and 1795 ranging from 80 to 190 pounds). The latter was an important breed used in the development of mixed characteristic cross-bred sheep varieties. The long-wool sheep in the summary also includes the Teeswater Devon breeds for which Fussell and Goodman quote only a single estimate of 120 pounds in 1786. Of the short-wool varieties we exclude the Western breed (with weights in 1786, 1791, 1792, and 1795 of between 56 and 184 pounds).[22]

Obviously it is difficult to have very much confidence in trends over time from what in reality are little more than a few scattered observations. Long-wool varieties were, in the main, heavier than the short wool, but there is no way of measuring trends over time with much accuracy. The average size across the breeds in general was said to be 32 pounds in Gregory King's day, rising to 70 or 80 pounds in the 1770s and 1780s and thereafter, but these trends are based on too few contemporary reports to be of use in identifying long-run productivity and short-term turning points. From evidence in the *Annals of Agriculture* in the 1780s and in first editions of the *General Views* from 1794, we can suggest an average carcass size for long-wool sheep of from 56 to 160 pounds, and of short-wool sheep from 24 to, improbably, 400 pounds. The disparities in these estimates may reflect a mixture of live-weight and carcass estimates. Therefore, if the final observation refers to a live animal, on the basis of the usual ratio of live to carcass weight of 14:8 this animal was probably nearer to 229 pounds. If we discard this figure altogether the range becomes the more reasonable or accessible 24 to 180 pounds.[23]

---

[22] Fussell and Goodman, 'Eighteenth-Century Estimates', 134–7. On breeds see also Luccock, *The Nature, passim*; id., *An Essay, passim*.
[23] Fussell and Goodman, 'Eighteenth-Century Estimates', 137.

Table 6.2 Eighteenth-century estimates of sheep carcass weights (lbs)

Date	Norfolk (long)	Lincoln (long)	Leicester (long)	Wiltshire (short)	Heath (short)	Southdown (short)	Hereford (short)
1737	55						
1749				28			
1770					48–56		
1778	48–72	80–120+					
1781	56–60						
1786	72	100	100		40–60	72	56
1787	40–60	60–100		104			
1789							64/72
1792			152+	172		104	
1795			79	82.5	40–80	84	
1800		100	130			125	60
1803			72–120		48–64	72	48

Source G. E. Fussell and C. Goodman, 'Eighteenth-Century Estimates of British Sheep and Wool Production', *Agricultural History*, 4 (1930), 134–7.

Neither sheep nor other animal livestock weights figure prominently in the various volumes of the *Agrarian History*.[24] Sales of seventy-five animals are recorded for the period 1694–1745, ranging in size from a top weight of 88 pounds in the 1740s at Ickworth in Suffolk, to a bottom weight of 36 pounds in the late 1730s at Frampton in Dorset. The average weight over the whole half-century ranged from 40 to 63 pounds, and these are not dissimilar weights to the ones we report below.[25] Also recorded is the sale of twenty wethers in the early 1740s from Stanton in Suffolk (average weight 48 pounds, and ranging from 40 to 53 pounds), eleven ewes in the early 1740s from the same Suffolk estate (average 44 pounds, ranging from 33 to 58 pounds), and just two lambs in 1737–8 and 1740–1 of 18 and 26 pounds respectively. None of these observations is duplicated below. Moore-Colyer has collected and collated the sheep weights for two periods, 1750–1800 and 1801–50, on the basis of type of breed, but his evidence is extracted almost wholly from standard reference sources.[26] For only two of the reported English breeds, the Leicester and the Morfe, or Shropshire Down breed, are there reported weights for both periods. In the first case they are for different ages of sheep (100–120 pounds for two- to three-year-old sheep in 1750–1800, and 80–100 pounds for twelve- to fifteen-month-old sheep in 1801–50, or alternatively 120–150 pounds for two-year-olds in 1801–50). The range of observations is small, and there is the ever-present ambiguity of whether these are carcass weight quotations for live or dead animals.

The need to settle on some average weight and give some sense of productivity change encouraged Holderness to suggest that the average weight increased from 56 pounds in *c.*1750 to 86 pounds in *c.*1850, an increase of 54 per cent.[27] The figure for 1850 is based on Robert Herbert's estimate published in the *JRASE* in 1859.[28] Changing weights alone are not sufficient to identify productivity change. Much of the size of animals was to do with consumer fashion. P. G. Craigie and R. H. Rew quoted contemporary estimates of

---

[24] *AHEW* v (1), *passim*, sheep flock sizes. [25] *AHEW* v (2). 884.

[26] *AHEW* vi. 327–31. For the first period Moore-Colyer has an average weight for the Lincoln breed of 100 pounds taken from Trow-Smith's *History of British Livestock*, with an alternative weight of 104 pounds taken from C. Vancouver, *General View of the Agriculture of the County of Cambridgeshire* (London, 1794). Neither of these weights is remotely similar to the ones we have collected. For the second period he has taken the sheep weights for English breeds almost entirely from J. Wilson, 'On the Various Breeds of Sheep in Great Britain, Especially with Reference to the Character and Value of their Wool', *JRASE* 1st ser. 16 (1855), 222–49.

[27] *AHEW* vi. 154–6.

[28] R. Herbert, 'Statistics of Live Stock and Dead Meat for Consumption in the Metropolis', *JRASE* 1st ser. 20 (1859), 475–6.

English sheep slaughter weights that show a modest productivity advance from 56–60 pounds in the early 1870s, to 69 pounds by the mid- to late 1870s. By the turn of the century these settled at an average of 65 pounds. These figures do not differ greatly from those we have extracted from the farm records.[29]

We can improve on these estimates by gathering together examples from many farm records, although we also face the same problems encountered by contemporary writers in that there was a large variety of breeds and an equally large regional variation. The entire database of sheep carcass weights amounts to 6,218 animals from 3,554 separate sales transactions. Only rarely did these transactions represent the sale of a single animal. For example, forty wethers were sold from Sandford in Devon in both October 1832 and February 1834. Twenty-eight ewes were sold from Sandford in 1832 and, in February that year, thirty-nine wethers. Thirty-nine shearhogs were sold off a farm at Eathorpe in Warwickshire in 1828.[30]

In what follows we adopt a pattern that we subsequently apply also to cattle and pigs. We have produced a simplified master table of the course of animal weights over the period 1700–1914 based on twenty-five-year averages (Table 6.3). Then, separately for each animal type, we have produced a pair of graphs. The first is a pictorial representation of all individual animal carcass weights at the point of sale, and the second reduces the data to annual averages. Put in another way, the first graph shows the annual dispersion of different sizes of animals, and the way this dispersion changed over time. The second graph captures the mean annual trend. If we take a sub-set of the entire sheep database and look at the pattern of ewe carcass weights we find that the average weight over time of the 861 ewes was 61 pounds (to the nearest pound). In the raw state, as in Fig. 6.1a, there was a pattern of rising weights from something on or near 50 pounds from 1720 to 1770, to something above 50 pounds thereafter, and to 65 pounds average weight in the last quarter of the nineteenth century. The average weight in the mid-nineteenth century stabilized at around 60 pounds, but in some years exceptionally large animals came to market, and this helps to explain the relatively high dispersion about the mean. This could also have arisen from the need

---

[29] P. G. Craigie, 'Statistics of Agricultural Production', *JRSS* 46 (1883), 25–6, 29; R. H. Rew, 'Production and Consumption of Meat and Milk: Second Report', *JRSS* 67 (1904), 368–84, esp. 371–8. See also id., 'Production and Consumption of Meat and Milk: Third Report', ibid. 385–412, and id., 'Observations on the Production and Consumption of Meat and Dairy Products', ibid. 413–27. See other estimates for the same period for GB and UK in *AHEW* vii. 311–12, and accompanying text.

[30] Devon RO, 1283M/LP1/1; RUL, WAR 2/2.

to sell an individual animal at a particular time of the year, perhaps to meet unexpected bills, although this is little more than speculation.

Just as much of a problem is the concentration of sales from specific locations, which determined the trend. In other words it is a trend based on a moving geographical sample. The beginning of the trend in the 1720s and 1730s is determined by the sales from a farm in Kent, and in the 1760s it is the product of ewe sales of an unnamed breed at Bridgnorth in Shropshire. In the 1800s the Southdown breed in Devon determined the trend. In the 1820s it is partly determined by the Down breed in Somerset, and unnamed breeds from Devon, Warwickshire, Lancashire, and Nottinghamshire. From 1830 to the mid-1850s an unnamed breed in Warwickshire determined the trend, though with influences from Devon and Nottinghamshire. Thereafter the downturn in average weights is the product mainly of two north of England breeds, the Cheviots in Northumberland, and in $c.1890$ the sale of an unnamed breed of ewes at Castle Eden in Durham. There was also some influence from the sale of Cornish ewes in the 1880s. It has proved impossible to standardize the database or indeed to standardize a sub-set from it to control for changing geography and its concomitant changing breeds. Figure 6.1b attempts to pick out the general trend, compromising the annual dispersion by taking the annual mean.

We can also isolate the sale of 487 lambs from 1723 to 1902. Figure 6.2a shows the trend of the lamb carcass weights over time. It declined from something like 30 pounds per animal in the 1720s to a little over 20 pounds by 1760, then rose to 30 pounds by 1810 or 1820. There was a steady rise thereafter until the end of the nineteenth century when the average carcass weight was widely dispersed about a mean of 50 pounds per animal. However, many of those sales took place in individual years, and involved animals from the same farm. Therefore Fig. 6.2b shows the mean annual carcass weights. This has the effect of reducing the number of observations to sixty-one, the number of years for which there are data. The trend becomes clearer, but the failure to control for a wide geographical mix remains even if it is less pronounced than for ewes. The lamb sales early in the 1720s are from the same farm in Kent as the equivalent early ewe sales, and in $c.1760$ the lamb sales are dominated by animals from Bridgnorth, Shropshire. Those in $c.1900$ arose from sales at Castle Eden. In intervening years the geography of the sample is much more mixed and includes sales from Worcestershire, Devon, Berkshire, Cheshire, Lancashire, Buckinghamshire, Essex, Warwickshire, Somerset, and Northumberland. The average weight in the third quarter of the nineteenth century of 34 pounds agrees closely with the average estimate of 33 pounds by

Table 6.3 Average weight of the different kinds of animals, c.1700–c.1914 (lbs)

	Cows and heifers		Beef cattle		Calves		Ewes		Lambs	
	Average wt.	Number	Average wt.	Number	Average wt.	Number	Average wt.	Number	Average wt.	Number
1700–24	371.3	15	530.6	88	79.0	12	49.1	22	29.4	24
1725–49	568.6	182	635.0	46	76.0	31	46.5	60	29.6	89
1750–74	692.0	2	650.6	117	95.0	17	45.7	15	22.2	70
1775–99	429.3	26	775.1	29	108.6	8	84.0	1	39.0	1
1800–24	590.6	142	1008.2	43	140.0	85	56.2	165	30.6	64
1825–49	539.6	61	655.8	51	142.0	340	66.4	407	31.6	28
1850–74	525.0	6			141.2	12[2]	59.5	159	34.3	9
1875–99					113.8	31	65.5	30	50.4	156
1900–14							58.5	2	52.2	46

Long-run averages and total samples

	Average wt.	Number	Average wt.	Number	Average wt.	Number	Average wt.	Number	Average wt.	Number
1700–1899	556.5	434	671.9	374	134.2	646	61.0	861	37.6	487

	Wethers		'Sheep'		Bacon pigs		Porkers		Hogs	
	Average wt.	Number	Average wt.	Number	Average wt.	Number	Average wt.	Number	Average wt.	Number
1700–24	53.1	7	55.5	15					192.5	2
1725–49	55.5	20	55.6	299	271.0	1	88.3	39	212.9	16
1750–74	53.6	7	62.4	630	246.0	59	91.0	89	284.2	122

1775–99	64.0	299	56.2	235	208.9	64	90.1	218	247.6	100
1800–24	53.6	44	75.6	716	221.4	32	90.8	119	417.4	185
1825–49	43.2	210	68.8	654	224.0	4	108.4	15	219.6	361
1850–74	60.6	55	78.1	495	317.4	54	71.0	2	191.0	1
1875–99	56.1	62	68.3	948	291.0	27	86.0	7		
1900–14	53.0	13	60.4	109						

Long-run averages and total samples

1700–1914	55.7	717	68.1	4,101	253.7	241	90.7	489	279.3	787

	Sows		'Pigs'	
	Average wt.	Number	Average wt.	Number
1700–24	205.0	4		
1725–49				
1750–74	259.9	9	199.0	21
1775–99	334.8	5	153.3	32
1800–24	251.3	15	217.3	134
1825–49	344.0	9	233.8	255
1850–74	339.9	5	234.8	159
1875–99			90.9	71

Long-run averages and total samples

1700–1899	285.1	47	210.8	672

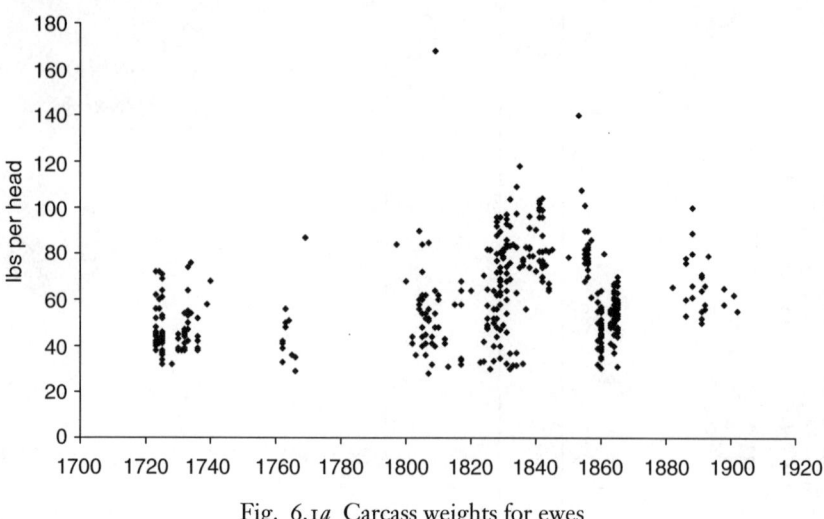

Fig. 6.1*a* Carcass weights for ewes

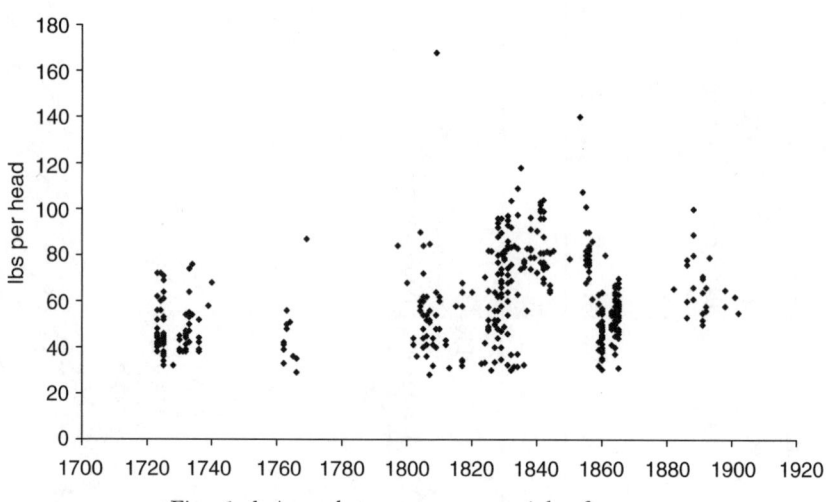

Fig. 6.1*b* Annual average carcass weights for ewes

the contemporary observer Robert Herbert,[31] though in our case this average weight is dominated by the sale of animals from a farm in Northumberland. However, it is in the middle of a clear long-term trend of lamb carcass weights across what is otherwise a wide spectrum of counties.

[31] Herbert, 'Statistics', 475–6.

LIVESTOCK 189

Fig. 6.2a Carcass weights for lambs

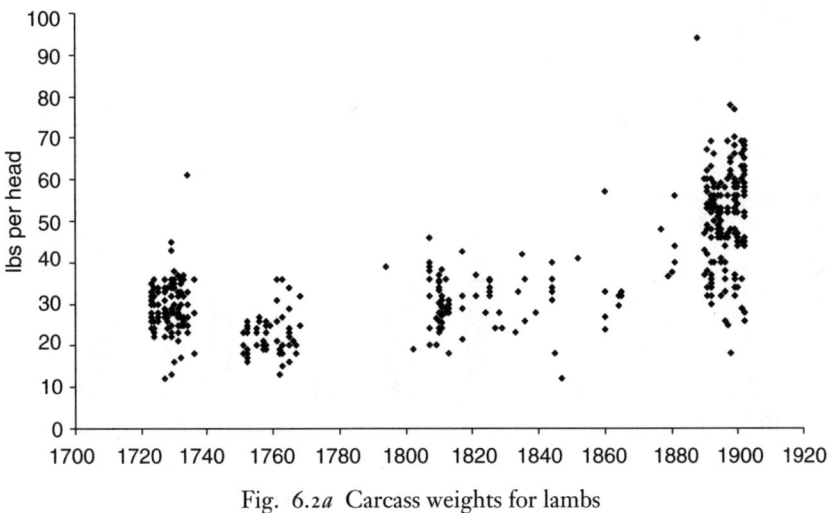

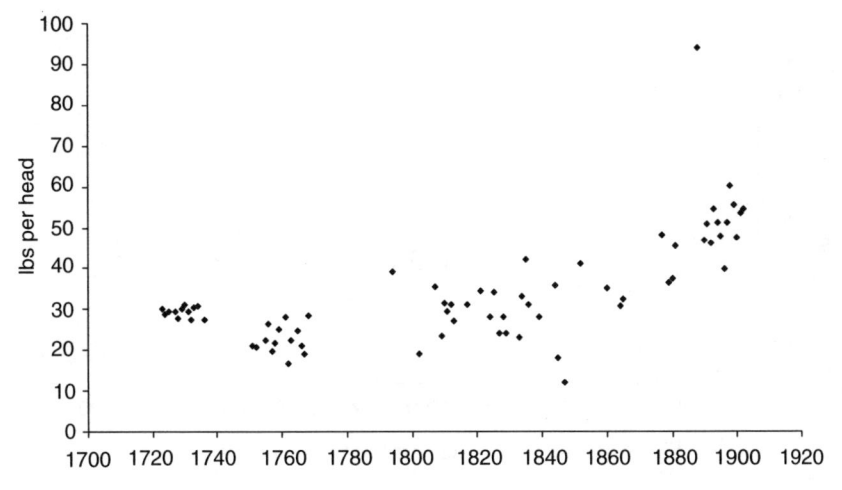

Fig. 6.2b Annual average carcass weights for lambs

The equivalent trend line for the sale of wethers is shown in Figs 6.3a and 6.3b. It arises from the sale of 717 animals. The trend is fairly flat over the period with an average carcass weight of 55 to 60 pounds, but as with some of the other trends there are moments when there is a concentrated geography—Kent in the 1720s, Shropshire in the 1760s, and Durham in the 1890s. In between these dates there was also geographical concentration—Buckinghamshire in the 1780s, Devon in the 1800s, Somerset in the 1820s,

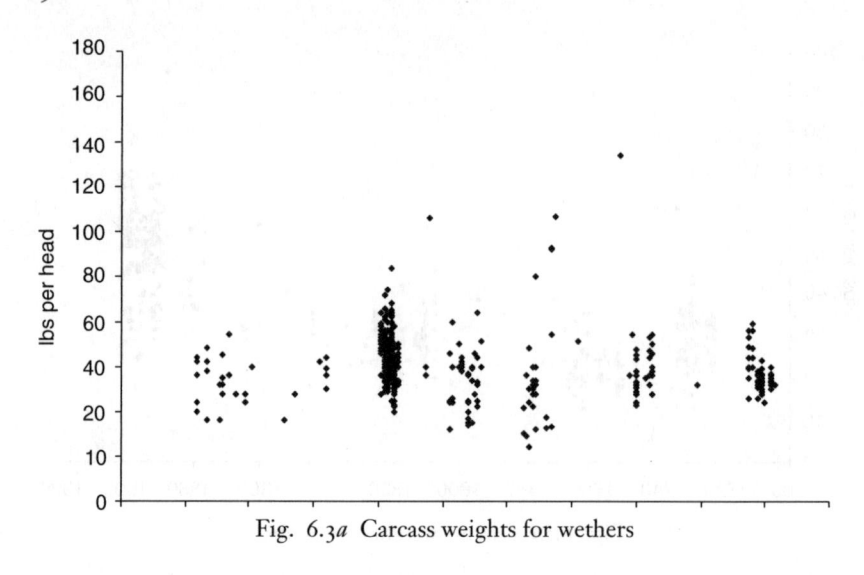

Fig. 6.3*a* Carcass weights for wethers

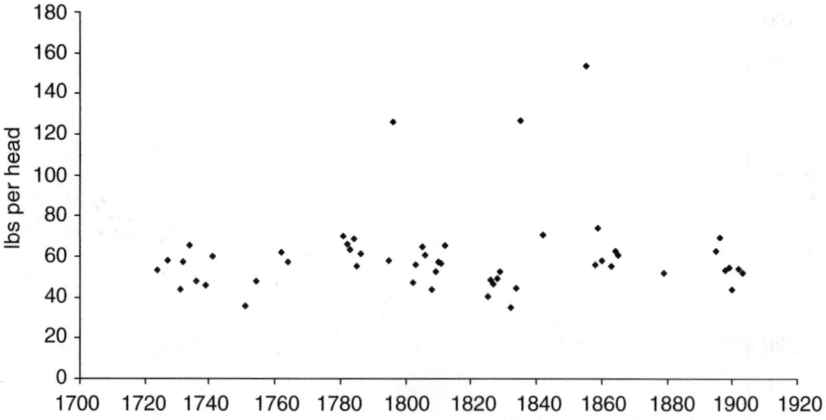

Fig. 6.3*b* Annual average carcass weights for wethers

and Northumberland *c.*1860. The incidence of geographical concentration makes it difficult to analyse beyond the fact that apart from some outstandingly large animals the average weight of wethers hardly changed at all over a very long time.

There are too few observations of other descriptive types of sheep such as ballards, threaves, hoggets, and shearhogs, to constitute a sample. These terms refer either to the age or sex of an animal. For example, tup and ewe lambs are respectively male and female animals between birth and weaning,

Fig. 6.4a Carcass weights for 'sheep'

Fig. 6.4b Annual average carcass weights for 'sheep'

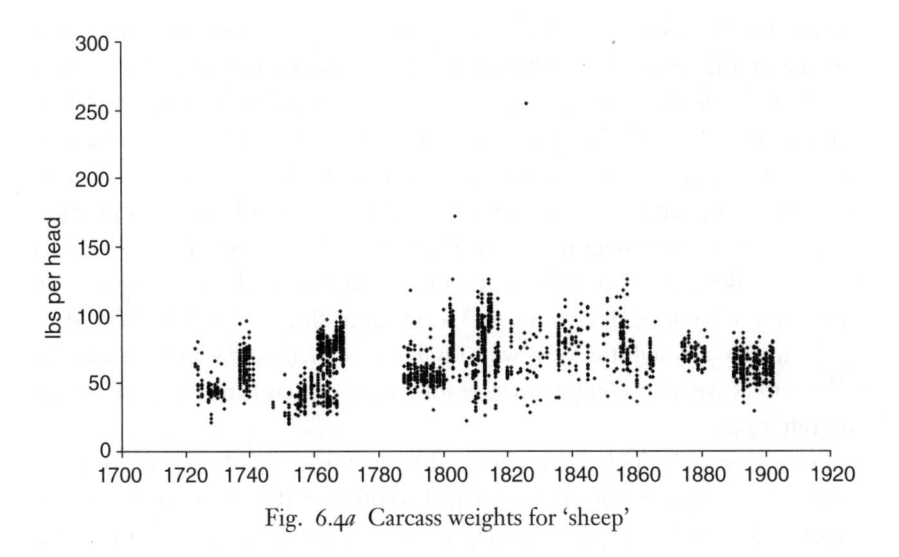

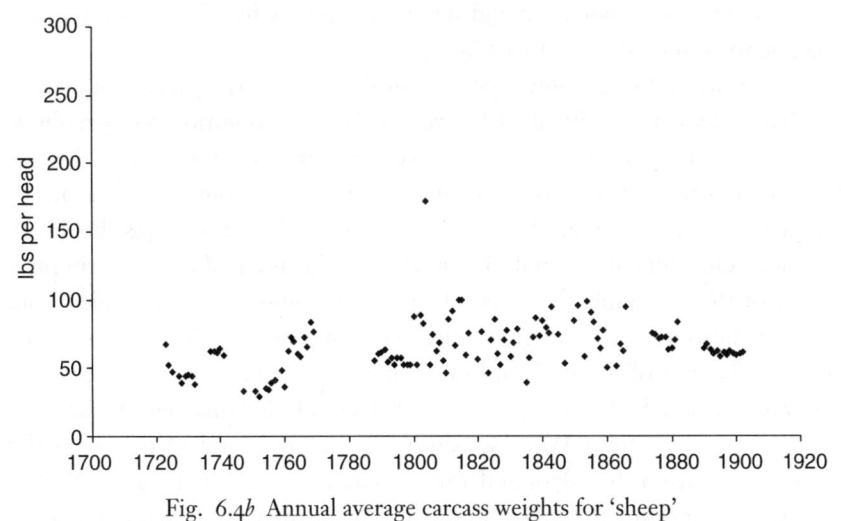

whereas a hogget is a male (tup hogg) or female (ewe hogg) animal between weaning and first shearing, unless it has already been shorn as a lamb. A shearling or shearhog is a male animal after first shearing but before the second shearing, and a threave is the equivalent female. But there still remains the largest sub-set of all, namely those animals generically described simply as 'sheep'. The dataset contains 4,101 'sheep'. These sales are plotted in Figs 6.4a and 6.4b. They include sheep for locations in Essex, Warwickshire,

Devon, and Somerset when for twelve years twenty or more sheep were sent for sale at the same time. This included the sale of fifty-four sheep from Sandford in Devon in 1835, fifty-two from the same farm in 1832, and fifty from a farm in South Petherton in Somerset in 1800. On other occasions there were relatively large numbers of sheep sent for sale in any single year from the same farm. In 1737, forty-four sheep were sold off a farm from Chicheley in Buckinghamshire from February to December. This seemed to have been the pattern of sales on this farm because in the next three years there were forty-nine, forty-one, and forty-seven sheep sales. The sales were made steadily through the year, with up to five animals in each of the months of June to November going to market (or sold elsewhere) and the peak month was July 1738.

The clearest chronological pattern that emerges regarding sheep carcass weights is an asymmetric convex-shaped trend over the two centuries or so. There was a rise from approximately 50 pounds per carcass at the beginning to 75 pounds by about 1810 and nearly 80 pounds by 1870, followed by a decline to 60 pounds per carcass by 1900.

Even if we feel reasonably confident of the chronology, we cannot be sure we have adequately controlled for geography and condition. Geography is perhaps the most difficult effect to overcome in discerning trend. In Fig. 6.4*b* there is what appears to be a seesaw trend, rising from the 1720s (40–50 pounds) to 1740 (60 pounds) and then falling back to the 1750s (below 40 pounds). Our difficulty is in deciding whether this is a real trend, or simply a result of the geography involved. The figure begins with sheep sales from Kent, followed by Buckinghamshire, and then from Shropshire by the 1750s. The sale of small Shropshire animals continues into the 1760s, but the average weight for sheep sales in that decade is bolstered by the sales of larger animals from Worcestershire. At the other end of the trend the observations from the 1870s and 1880s come from a farm at Audley End in Essex, and those for the 1890s are from Castle Eden in Durham. The convex shape of the long-run trend is sustained in the second and third quarter of the nineteenth century from observations mainly from Devon, Worcester, and Warwickshire. The likely distortion from this changing geography is almost impossible to calculate, but we can do no more than collate the evidence available to us in farm records.

End use of the animal also raises difficulties because sheep were not raised only for their meat. Until 1750 sheep were mainly regarded as a source of wool, with a secondary role as a source of manure, and only finally for the meat they provided.[32] The emphasis on these three roles varied through time and across space. On the Wiltshire Downs prior to the 1790s half a million sheep were kept principally for their manure-bearing qualities. In the 1790s this emphasis altered, and farmers changed to breeding the animals with the new object of improving carcasses, both of ewes and lambs, and particularly improving the quality of lambs rather than their quantity.[33] Competition from cotton and indeed imported wool reduced the wool element in the composite use of the sheep and this had an effect on the numbers that were required primarily to supply the woollen industry.[34] Changes in living standards and diet enhanced the value of the meat.

The age at which the animals were slaughtered varied with their end use, and as the nineteenth century advanced the average size may have stabilized or even decreased. This might suggest a slowdown or reversal of productivity, but as long as there was an increase in animal throughput the productivity of sheep production as a whole may actually have increased. This is the classic accounting problem of stocks and flows, which unfortunately the data available to us cannot separate. As we saw in Chapter 3 the development of new breeds did not automatically result in an increase in carcass weights but it did allow animals to be fattened more quickly so that they might be with the butcher in two rather than four years. Rew put it succinctly in 1904 when he contended that 'it is probably true to say that animals, being sold at an earlier age, are not now fattened to so great a size as formerly'. This is a clear enough warning about using slaughter weights as an indicator of productivity, but Rew also commented on 'the progressive improvement of farm stock [that] has tended to the more general production of animals which, in butchers' phraseology, "die well", and has possibly increased to some extent the proportion of the "carcase" to the live weight'.[35] In short, this suggests more meat, but less tallow, hide, skin, bone, and so on.

---

[32] *AHEW* vi. 314.

[33] T. Davis, Sr., *General View of the Agriculture of the County of Wiltshire* (London, 1794), 20–1. See also J. Donaldson, *General View of the Agriculture of the County of Northampton* (London, 1794), 59–60. And in c.1840, on the production of meat from sheep juxtaposed with using cattle to make manure see P. Pusey, 'On the Progress of Agricultural Knowledge during the Last Four Years', *JRASE* 1st ser. 3 (1842), 204–5, and twenty years later during the High Farming period on whether to use cattle to produce manure to apply to grain crops or to produce meat for the market see P. H. Frere, 'On the Feeding of Stock', *JRASE* 1st ser. 21 (1860), 233–4. See also Hillyard, *Practical Farming*, 207–8.

[34] See the discussion and estimates of sheep numbers in Turner, 'Counting Sheep', especially 152–9.

[35] Rew, 'Observations', 416–17.

Cattle

The business of putting sheep carcass weights into a long-run context is a salutary exercise in how much we still do not know, and may never know, about some of the basic parameters of English agricultural history, and much the same is true of cattle. The cattle population at any given date remains something of a mystery. Deane and Cole have argued that the size of the cattle population showed little variation over the course of the eighteenth century and the first half of the nineteenth century. Their view is derived from a reading of Gregory King's estimate of 4.5 million animals at the end of the seventeenth century for England and Wales, and a calculation that by the 1830s there were about 5.2 million animals in the whole of the United Kingdom. They deduce from this that if there was a significant rise in meat production, as the evidence from the meat markets suggests, then this must have been due to an increase in average animal size, animal quality, or a more rapid turnover of stock.[36] Holderness has argued that King exaggerated the bovine population, and that cattle numbers in England and Wales varied between 2.8 and 3.5 million in the eighteenth century. He infers that the cattle population of England and Wales, which included store cattle brought from Scotland and Ireland and subsequently fattened in England, increased by only 0.5 million from 1750–1850, with the greatest concentration after 1800. This calculation is based on the reported throughput of the number of cattle sold at London's Smithfield market, and the throughput of store cattle from Ireland which were fattened in England and Wales in the early nineteenth century.[37]

Figures for carcass sizes are no more reliable. King estimated the size of bullocks killed at Smithfield in the late seventeenth century at 370 pounds. A century later Sir John Sinclair put the average size at 800 pounds. Deane and Cole believed that King underestimated carcass sizes, while Holderness thought that Sinclair overestimated them.[38] If both of these modern authorities are correct then it narrows the degree of productivity advance which the contemporary figures suggest. The same estimates were also used by Fussell in reviewing cattle weights in the eighteenth century.[39] He brought

---

[36] P. Deane and W. A. Cole, *British Economic Growth 1688–1959* (2nd edn. Cambridge: Cambridge University Press, 1969), 69.
[37] *AHEW* vi. 151–2.   [38] Deane and Cole, *British Economic Growth*, 69; *AHEW* vi. 153.
[39] Fussell, 'The Size of English Cattle', 160–81.

together many individual references to cattle weights reported by contemporaries, especially Arthur Young and the Board of Agriculture reporters. If nothing more, Fussell showed that there is no shortage of individual carcass weights, but this is not the same as finding agreement on how they changed over time. Certainly any search for an average animal size across different breeds is futile, although Fussell's work suggested that within individual breeds the average size did not change very much. This led him to conclude that the achievements of Robert Bakewell in breeding could not have been replicated widely, and that breeding was geared towards producing more meat and less bone rather than increasing the size of the animal. To this extent Fussell attributed the increase in meat that was sold at the well-known markets not to an increase in animal size (though marginally he allows that this happened), but to a combination of replacing draft oxen by horses, and slaughtering animals at an earlier age.[40] Once again this highlights the problem of distinguishing between stocks and flows.

Since Fussell's pioneer work, the database has been only marginally increased in size. Bowden found relatively little new information for the period 1640–1750. He noted the sale of three oxen between 1692 and 1729, weighing in at 2,037, 2,083, and 1,262 pounds respectively, the smallest animal coming from Surrey and the two larger from Yorkshire. He found data for a further thirty-eight animals sold between 1691 and 1747, variously described as steers, bullocks, Welsh beasts, and Scotts cattle. In the light of some contemporary estimates these animals varied quite narrowly in size, between 394 and 531 pounds. They came from Hertfordshire, Derbyshire, Huntingdonshire, Devon, and Suffolk. Such figures suggest that perhaps King did indeed underestimate the size of animals sent to market. Twenty-nine cows were sold from farms in Surrey, Sussex, and Shropshire between 1726 and 1749, varying in average size from 300 pounds to 622 pounds. They included two animals at 1,741 and 813 pounds respectively, and two Surrey heifers which in the late 1730s weighed in at 768 and 808 pounds.[41]

Late eighteenth-century estimates of the size of all manner of cattle vary between 450 pounds and 1,000 pounds. By the mid-nineteenth century there was a narrowing in this range from 660 to 700 pounds with important differences in size according to different cattle breeds, particularly between English and Welsh breeds.[42] Over the period 1750–1800 the size of English

---

[40] Ibid., 176, 179–81.   [41] *AHEW* v (2) 884–5.

[42] Fussell has commented on the small size of Welsh and Scots cattle, and also the cattle from E. Anglia, in Fussell, 'The Size of English Cattle', 167. The evidence we present below is entirely

cattle varied from shorthorn cows weighing in at 560 pounds minimum and 840 pounds maximum, and Devon cows at 640 pounds, and shorthorn oxen varying between 840 and 1,540 pounds. Devon oxen at five years of age weighed 880 pounds. Some Welsh breeds match these weights, but the average seems to have been somewhat smaller. For the period 1800–50 the weights have increased through most breeds by something close to 25 per cent. However, the evidence arises from so few observations, coming from such disparate geographies, that confidence in the generality of those original estimates and the inferred productivity increases is not high.[43]

Clearly there is much need for a reappraisal of cattle carcass weights, and as with sheep it is possible to look at a number of different animals—calves, cows, and so on—within the broad definition of 'cattle'. Our results, achieved by analysing the data in this way, are collated in Table 6.3 and Figs. 6.5–6.8. We start with calves in Figs 6.5a and 6.5b. These show average carcass weights based on the slaughter weights of 646 animals. In many years for which we have continuous data from individual farms it is clear that the sale of calves was a regular annual process. In a number of years there are outliers when relatively small animals were slaughtered.[44] A clear trend can be discerned if we allow for these outliers. There was a steep increase in calf carcass weights in the late eighteenth century, from 50 to 90 pounds in c.1760 to on, or a little over, 100 pounds by 1800. They reached a maximum size of about 170 pounds, and settled at an average size of about 160 pounds in the early 1820s. Equally clearly, the trend thereafter was downwards, though more gently. By 1890 calf carcass weights settled at about 110 pounds.

The eighteenth-century observations come mainly from farms in Shropshire and Gloucestershire. In the 1820s and 1830s they are from Buckinghamshire and Warwickshire, and in the 1840s and 1850s mainly from Warwickshire and West Sussex. For the 1860s they come from West Sussex, apart from the outliers in Northumberland, and by the late 1870s from Worcestershire. Animals from Worcestershire dominate the trend in

from English counties, and apart from the possibility of some Welsh store cattle being fed on English farms, they are entirely English cattle.

[43] *AHEW* vi. 345–6.

[44] In 1761 there was the sale of a single small animal in Shropshire. In 1796 there was the sale of another in Gloucestershire, and in 1820 one from Devon. In 1860 and 1864 there are similar sales of very small animals (relative to trend). In the former year it involved six very small Northumberland animals from the same farm, ranging from 32 to 91 pounds in weight, with only one animal which was 'on trend' at 150 pounds. In the latter year there was the sale of just one animal from the same Northumberland farm, which was reportedly only six months old and dying.

LIVESTOCK

Fig. 6.5a Carcass weights for calves

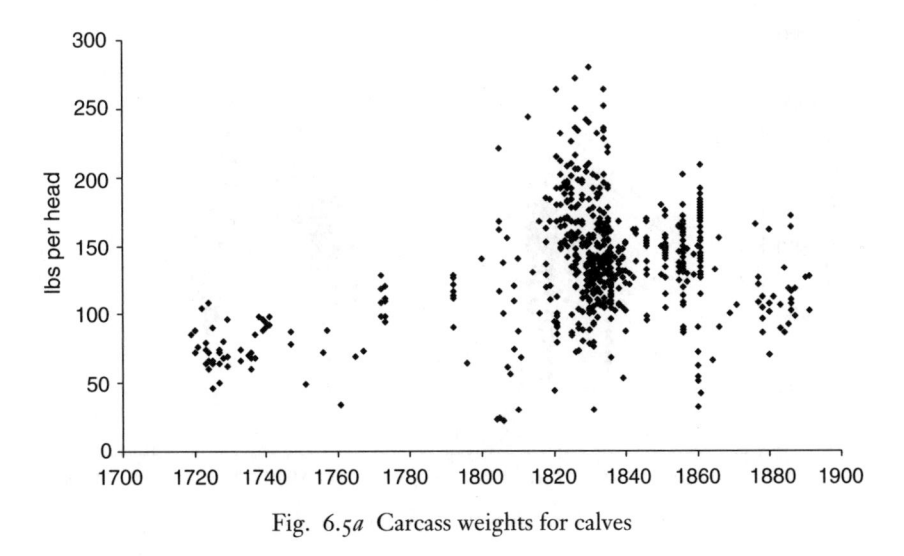

Fig. 6.5b Annual average carcass weights for calves

the early 1880s. By the late 1880s the trend arises from the sale of animals from West Sussex and Worcestershire. Recutting these data in order to gather together the counties in broad regional groupings does not alter the basic conclusion regarding trend that we have identified.

The database contains the sale of 434 female animals described variously as cows and heifers. Figures 6.6a and 6.6b plot the data from the earliest sales at East Sutton in Kent in 1727 to the last at Brancepeth in Durham in 1877.

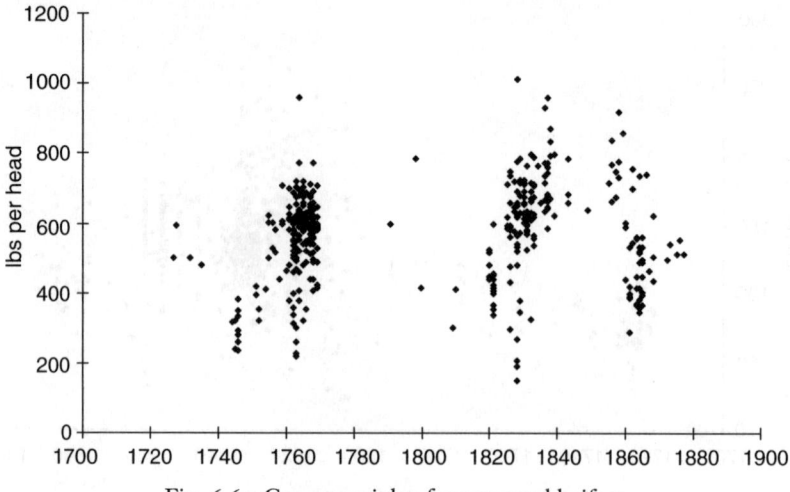

Fig. 6.6a Carcass weights for cows and heifers

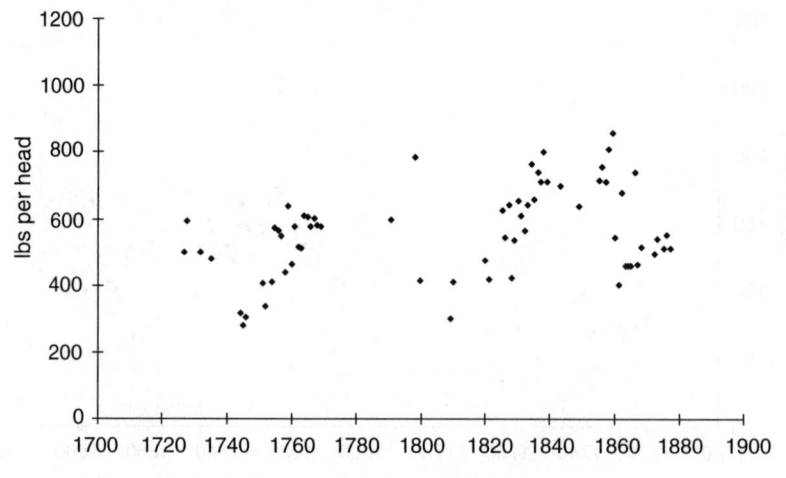

Fig. 6.6b Annual average carcass weights for cows and heifers

In the case of cows and heifers there is sufficient geographical separation in the sample to allow some marginal speculation on productivity shifts over the long period. Figure 6.7 differs from Fig. 6.6b in that it filters out those isolated years when only a small number of sales took place. Therefore the sale of isolated animals from a Kent farm in the 1720s, from a Devon farm in the mid-1790s, and in four years of the 1830s have been removed. So have the isolated sales from a Northumberland farm in the period 1814–18, odd years on a Somerset farm 1818–60, two years for a farm in Buckinghamshire,

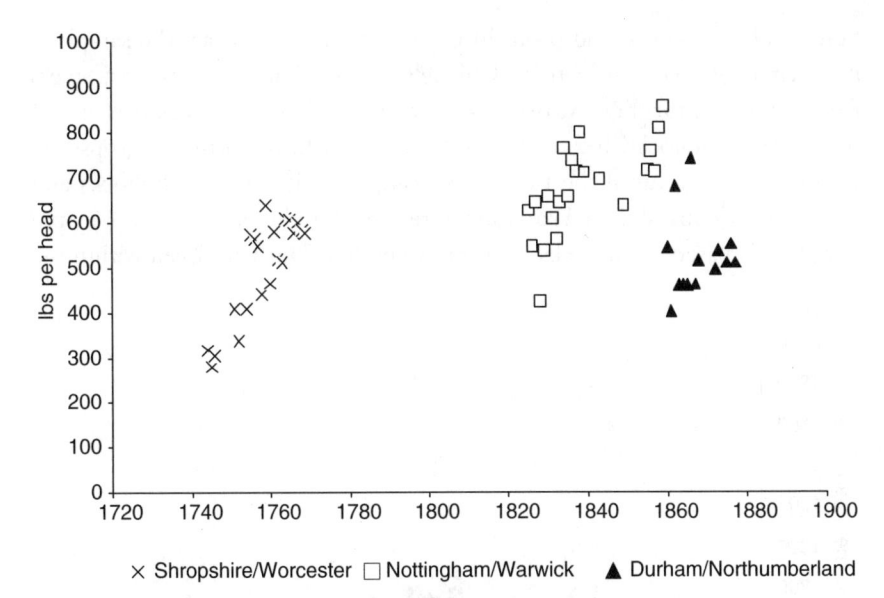

Fig. 6.7 Annual average carcass weights for cows and heifers: regional examples

and odd isolated sales in Kent, Wiltshire, and Worcestershire. What remains is the sale of 385 animals giving three distinct series. Taken together they have an internal chronological and geographical consistency. The remarkable doubling of carcass weights in the period 1744–69 arises from the sale of 192 animals from two farms, one at Bridgnorth in Shropshire and the other at Ombersley in Worcestershire. When these two farms are isolated, they almost mirror one another in displaying this clear trend. The rather more scattered, but nevertheless clear, upward trend in carcass weights from 1825 to 1859 arises from the sale of 136 animals from a farm at Eathorpe in Warwickshire. The aberrant year in this trend, 1828, arises from the sale of twenty animals, of which nine barely struggled to 200 pounds each, and five others averaged only 149 pounds each, against a trend for the period of 500–600 pounds per animal. Finally, the trend from 1860 to 1877 arises from the sale of 57 animals from two farms, Park End in Northumberland, and Brancepeth, county Durham. What was otherwise an upward trend was distorted by the sale of five animals. Each of them was on or over 700 pounds, and this was quite out of trend with adjacent sales from these farms. Four of these occurred in 1862, and the fifth was a single animal sold in 1866.

Of the remaining cattle in the database, 374 are described as adult, or relatively adult, male animals, and include a mixture of beef cattle, beasts, bulls, bullocks, steers, and oxen. In the first twenty-five years the sample is dominated by beef cattle from Chicheley in Buckinghamshire and steers from Kent, and for the next twenty-five years it is based on an assortment of beasts, bulls, bullocks, beef cattle, and oxen from Bridgnorth in Shropshire. From about 1790 until the turn of the century it is based on bullocks and oxen from Ombersley in Worcestershire, and for the last ten years in the sample it is based on bullocks from Brancepeth in Durham. Even within the

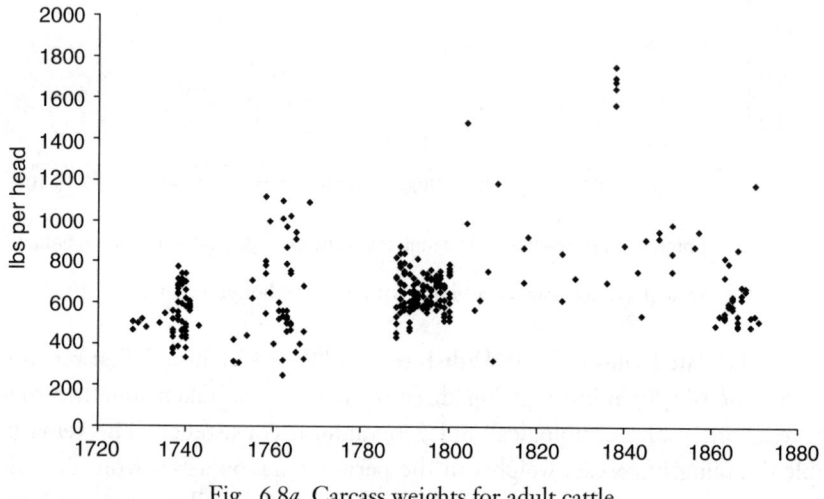

Fig. 6.8*a* Carcass weights for adult cattle

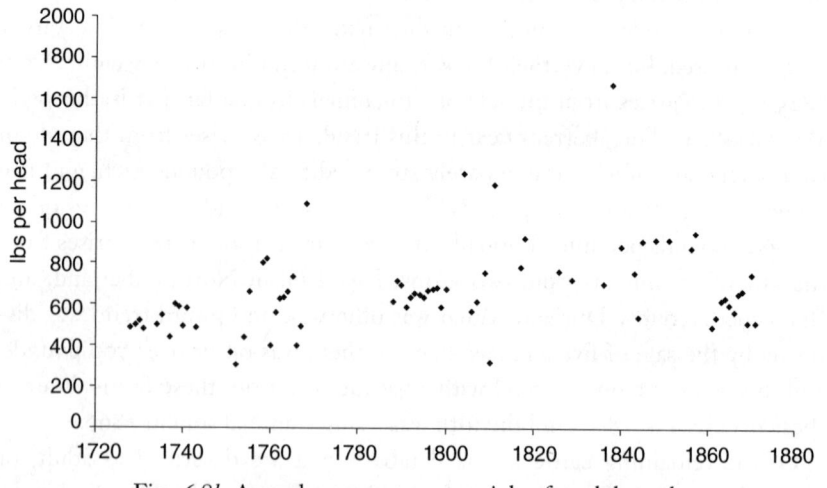

Fig. 6.8*b* Annual average carcass weights for adult cattle

same animal description, for the same farms, in closely associated years, there was a great variation in individual animal sizes. Therefore while it may seem obvious that bulls might have unduly influenced average carcass weights, because ordinarily we might assume they were the largest beasts, in reality, animals described as bullocks could be just as large. At times we might be picking up the variation in breeds which Fussell identified.[45] For example, in the 1760s in Shropshire, the bullocks, so called, which were sent to market averaged 557 pounds and ranged from 250 pounds to 1,096 pounds.

We have not tried to disaggregate this sample further, preferring to describe them as adult or relatively adult male cattle. The trend in changing carcass weights is shown in Figs 6.8*a* and 6.8*b*. The geographical variation in the sample, as well as the descriptive variation, might easily influence this trend. The different pattern of animal sizes in the north-east of England has been discernible in a number of animals in this analysis so far, for example for cows and heifers. For adult males this again appears to be the case. But for the presence of Durham bullocks just after 1860, there was otherwise an unmistakable increase in average carcass weights of adult beasts over the previous 120 years. That part of the sample that arises from the north-east of England comprises animals that were always below the size indicated by trends. At an average of around 600 pounds per head they were clearly smaller than the average size reported by the contemporary Robert Herbert.[46]

Pigs

Identifying patterns in the pig population—whether in terms of the size of the stock or of the average size of each animal—is just as complex. The numbers and sizes of pigs can fluctuate wildly. Porkers are usually smaller than bacon or ham pigs, while all pigs can be slaughtered early and in large numbers as a stop-gap measure at times when there is a shortfall in both human and animal fodder. Never was this more clearly demonstrated than during the Irish famine of the 1840s when the failure of the potato harvest had a direct effect on the food supply both of animals and humans.[47]

---

[45] Fussell, 'The Size of English Cattle'.   [46] Herbert, 'Statistics', 475–6.
[47] M. E. Turner, *After the Famine* (Cambridge: Cambridge University Press, 1996), 48–9, 54–6.

However, pigs can recover their numbers quickly because they have multiple litters and a relatively short gestation period. Unlike cattle and sheep stocks, for which it is imperative to keep relatively large breeding stocks, it is necessary to keep only a small pig breeding stock relative to the total population. A distinction also has to be drawn between the widely kept cottage pig (often an adjunct to the household economy whether rural or urban) and pigs kept on farms for market disposal.[48]

In 1695, a 'guess' by Gregory King put the swine and pig population for England and Wales at around 2 million.[49] Recent estimates suggest that the pig population in England c.1800 was about 1.8 million animals, rising only to a little over 2 million animals at the end of the nineteenth century.[50] Most agricultural historians have accepted that the pig population was around 2 million animals in the eighteenth and nineteenth centuries. By contrast, there is little consensus about the size of pigs at slaughter. Holderness suggested that in the eighteenth century 'the largest swine killed out at 1 cwt or more', with a plausible increase in size over the next 75–100 years of one-third to two-fifths.[51]

We have located 2,236 pig carcass weights from a number of different kinds of animals which, when separated, constitute large enough sub-set samples to allow individual trend analyses. These are bacon pigs, hogs, porkers, sows, and a category simply called pigs (and there is a further sub-set of 'other' pigs, including boars, though the sample is not large enough for separate treatment). These are collated in Table 6.3 and Figs. 6.9–6.12.

There are 241 bacon pigs in the sample. The trend of carcass weights of these pigs is shown in Figs 6.9a and 6.9b. For the decades either side of c.1760 a farm in Shropshire determines the trend. By the early 1780s a Buckinghamshire farm is dominant, and for the period 1788–1801 the trend depends on a farm in Worcestershire. In the early nineteenth century a second farm in Buckinghamshire is important, and for the third quarter of the nineteenth century a farm in Durham dominates the figures. Within these geographical sub-sets it appears there were chronological trends, with carcass weights falling, but with the average easily exceeding the estimate of 1 hundredweight.

---

[48] R. Malcolmson and S. Mastoris, *The English Pig: A History* (London: Hambledon Press, 1998) looks primarily at the incidence of domestic pig keeping, although chapter 4 includes comments on breeding. The authors quote a comment from *The Complete Farmer; or, General Dictionary of Agriculture and Husbandry* (5th edn. London: R. Baldwin, 1807), to the effect that 'almost every county or district is in possession of a particular kind', 71, but that the situation was changing rapidly, 75.
[49] Turner, 'Counting Sheep', 152.   [50] *AHEW* v (2). 445.   [51] *AHEW* vi. 154–5.

Fig. 6.9a Carcass weights for 'bacon' pigs

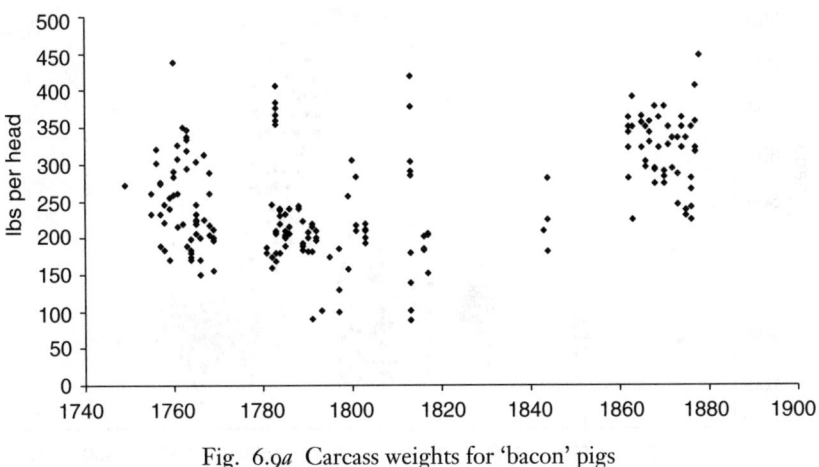

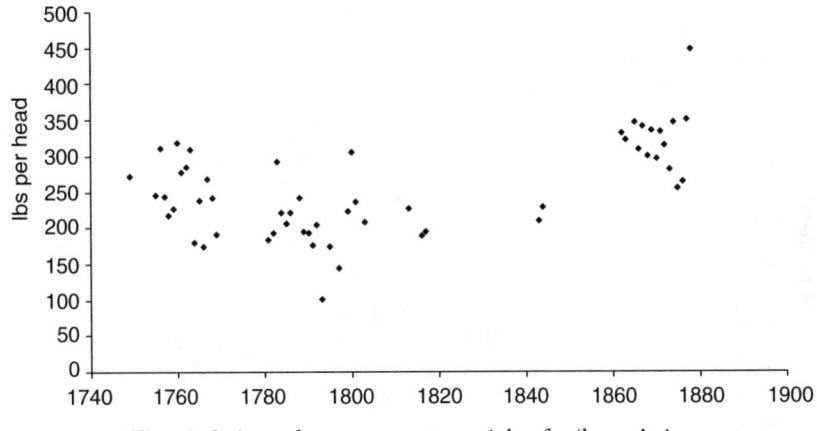

Fig. 6.9b Annual average carcass weights for 'bacon' pigs

There are 787 'hogs' in the sample, and Figs 6.10a and 6.10b show the trend in their carcass weights over time. For the 1740s they arise mainly from a farm in Buckinghamshire; from farms in Worcestershire and Shropshire around the 1760s; Oxfordshire and Devon in the 1780s and 1790s; and mainly from a second Buckinghamshire farm from 1823 onwards. In spite of this geographical variation there is a discernible trend, but it is upset by the extraordinary peak in carcass weights in the period 1805–22. This arises mainly, but not exclusively, from the animals sold off a farm in Swallowfield in Berkshire. Despite this anomaly, the lowest of the weights in that peak are on a clear upward trend from the 1730s to the

Fig. 6.10a Carcass weights for hogs

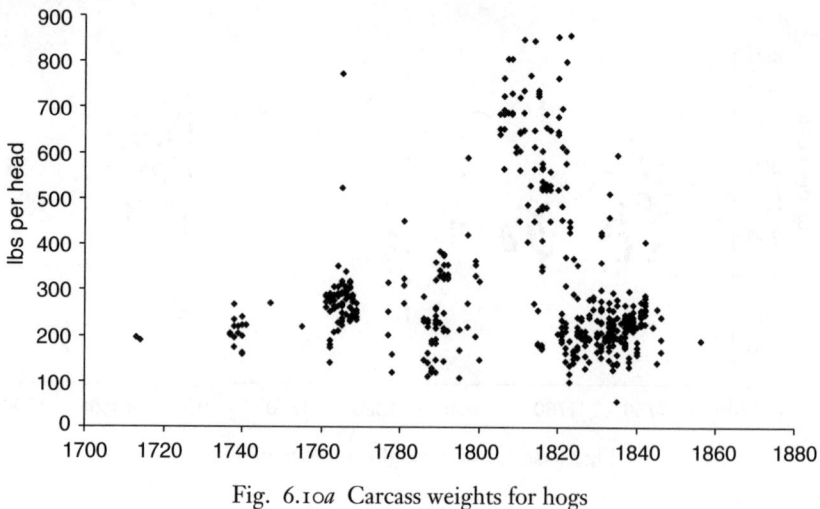

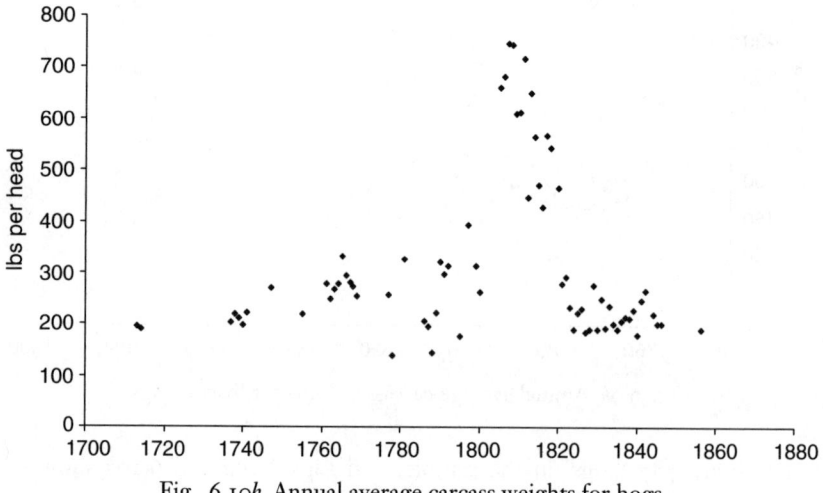

Fig. 6.10b Annual average carcass weights for hogs

1810s. If we discount the Berkshire sample we are left with what appears to be a distorted convex-shaped trend of a steady rise in carcass weights from c.1740 to c.1820 and a dramatic fall thereafter. Conversely, when we isolate the Berkshire farm, internally there was a clear downward trend that began with the peak year in 1807 and a fairly constant fall thereafter.

There are 489 carcasses from sales of pigs classified as porkers. Although there is again a geographical concentration of these sales from particular

Fig. 6.11a Carcass weights for porkers

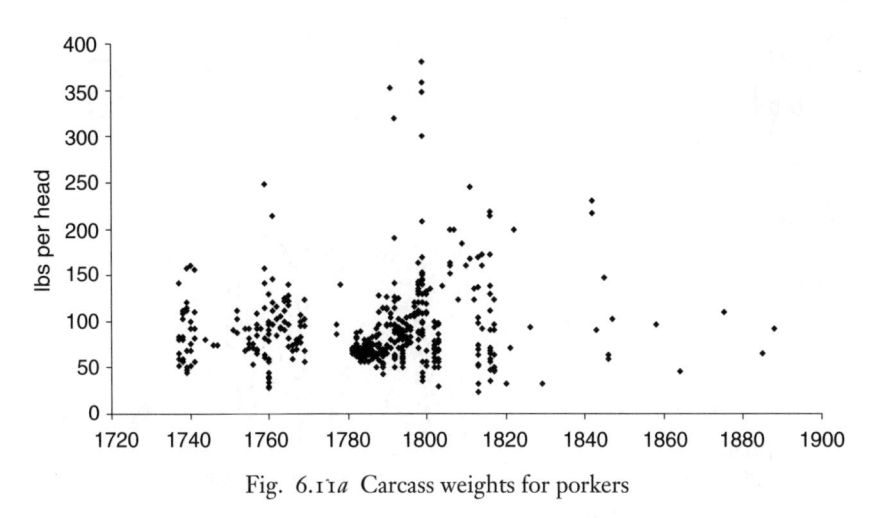

Fig. 6.11b Annual average carcass weights for porkers

farms in particular counties, this concentration is less pronounced than for other animals. For the period 1737–41 the sales come from Buckinghamshire; from 1744–61 from Shropshire; and from 1761–69 a mixture of Shropshire and Worcestershire farms. From 1781–1803 the sales come from Buckinghamshire, Worcestershire, and Devon, from 1806–20 from Buckinghamshire and Berkshire, and from 1826 they come from several counties but only for spot years. Taken together these sales form the trend depicted in Figs. 6.11a and 6.11b. It suggests a marked peak in the size of animals during the French wars. This is illusory and formed entirely from sales from a farm in Swallowfield in Berkshire, the same farm that distorted

Fig. 6.12*a* Carcass weights for 'pigs'

Fig. 6.12*b* Annual average carcass weights for 'pigs'

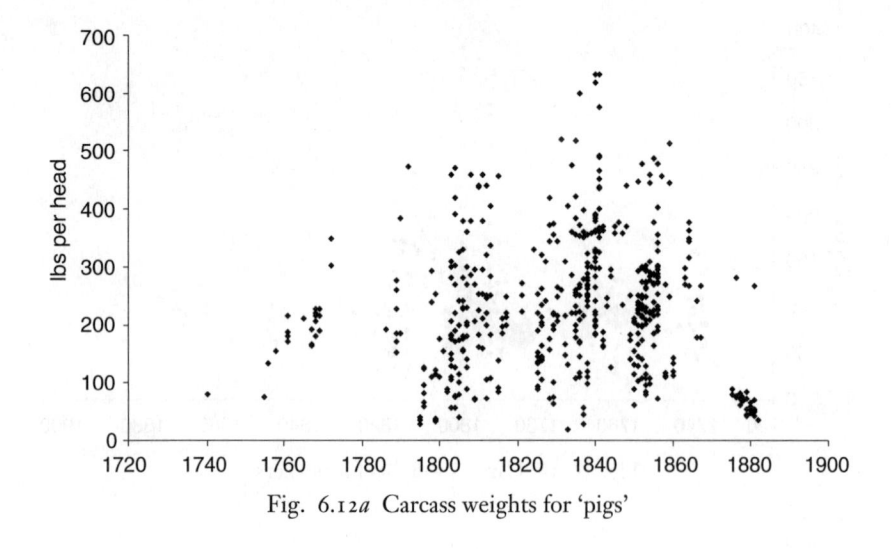

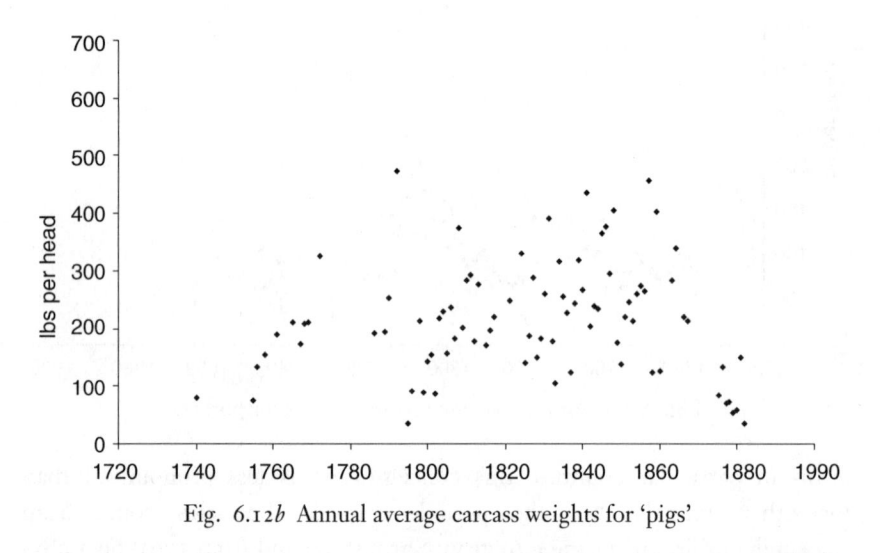

the trend for hogs. If we eliminate this farm and thereby reduce the number of pigs in the sample, the trend from *c.*1740 to *c.*1820 is, if anything, downwards. The pigs presented for sale, on average, were smaller. The Swallowfield pigs were very large animals. When the trends from the five main counties in the sample—Berkshire, Buckinghamshire, Devon, Shropshire, and Worcestershire—are isolated separately, they each show the same pattern. The size of animals became smaller over time, with the single exception of the porkers from a farm in Shropshire which became slightly larger.

Finally we note the trend in carcass weights for 'pigs'. This trend is shown in Figs 6.12a and 6.12b. It arises from the sale of 672 animals. The range of carcass weights at slaughter was large, nevertheless a trend of sorts is apparent. There was an increase in 'pig' carcass weights from about 150 pounds in 1760 to about 300 pounds in c.1840, and then a decline in weight thereafter to just 50 pounds in about 1880. However, that decline is entirely determined by the very much smaller than average size of 'pigs' at Audley End in Essex. Otherwise the trend arises from the slaughter of animals from a very wide range of counties, Buckinghamshire, Devon, Essex, Gloucestershire, Lancashire, Oxfordshire, Somerset, and Warwickshire particularly, and Berkshire, Cambridgeshire, Cheshire, Cumberland, Dorset, Durham, Kent, Lincolnshire, Northumberland, Nottinghamshire, Somerset, Suffolk, West Sussex, Westmorland, Wiltshire, Worcestershire, and East Yorkshire more marginally. If we ignore the Essex pigs at the end of the period, and we recognize the wide range of weights throughout the period, then we do see a rising trend in pig carcass sizes c.1750–c.1860.

Conclusion

The complexities of age, end use, condition, and breed, to mention only four of the possible variables we have encountered in studying the pastoral economy, have made our task difficult. We have brought together a substantial number of observations from farm records, and compiled a database which is in every respect larger than has been achieved in the past. We might still be open to the accusation that even several thousand sheep or cattle is hardly a major database, but from our perspective the slaughter of one animal from a flock or herd (assuming an average-sized animal) was sufficient to add a vital observation to the database. We cannot offer, as we might like to do, data from several thousand sheep flocks or cattle herds, but we are able to present several hundred observations from several dozen farms located in several counties for most of 200 years.[52]

---

[52] When we demonstrated our preliminary findings at the 1998 conference of the British Agricultural History Society, Professor Gordon Mingay suggested that in terms of the number of animals a database of several thousand sheep hardly amounted to the equivalent of one or two flocks at

We settled on animal weights as the proxy for productivity change, despite the many concerns that there must inevitably be about measuring carcass weights, let alone interpreting them. We attempted as far as possible to separate the animals into those with approximately the same ages, and those that were used for approximately the same purposes. So lambs were distinguished from other sheep, calves from cows and heifers, and so on. Thus at slaughter the animals were separated by age and purpose. As far as possible therefore, the lambs (or bacon pigs or others) in the early eighteenth century were more or less equivalent to the age of lambs in the late nineteenth century. The end use destination for the animals was an important element in whether large or not so large animals were reared. Over time, sheep were increasingly raised for their meat, and less so for their manure and wool. This is reflected in our findings. Ideally we would want to know the ages of the generic 'sheep' and thereby say something not only about the size of their stock, but also their flow through the market system. On its own the long-term increase in carcass weights is not of necessity a measure of productivity because quality was as important as size. The trend was towards increasing the edible meat on an animal, and the ratio of carcass to live weight, hence the importance of the work of breeders such as Robert Bakewell who sought to rear animals that gave more meat, smaller bones, and less non-edible parts.

Carcass weights remain the only obvious way of assessing change across time. The size of ewes increased modestly over the period 1720–1900, although because these particular animals may not have come to market at specific ages—spent ewes might be of various ages—the dispersion of their weights about the annual mean was quite wide. The average size of wethers was more or less flat over the period, but the generic 'sheep' increased in size, though modestly. The most noticeable productivity advance, using size as our proxy, was in the carcass weights of lambs. They almost certainly came to slaughter at relatively narrowly defined ages and they increased markedly in size from the 1810s to the 1890s. Similarly there was an increase in the carcass weight of calves, traceable in this case from about 1720 to 1830, with a dip in slaughter weights from then until the 1890s. There was also a long-run increase in the weights of cows and heifers and of adult 'cattle' from the second quarter of the eighteenth century to about

---

any one time from a county like Cambridgeshire. While we appreciated Professor Mingay's viewpoint, it is our belief that it was hardly possible to build a history of the productivity of sheep production based on a single sheep flock, from a single farm, in a single county, in a single year.

1860. In the case of the latter they increased in size from about 500 pounds per head in *c*.1730 to 800 pounds per head in the 1850s. It is less easy to identify trends in the long-run weight of pigs, we think because they came to market at less definable ages and because on one or two farms there were outstandingly large animals. They certainly did not get smaller over time.

The average sizes of most descriptions of animal increased on an incremental basis rather than through any dramatic breakthroughs in breeding. The major problem in taking the analysis further involves the distinction between stocks and flows: it is to do with the number of, as distinct from size of, animals that reached the market. In theory we can take average carcass weights and proceed to an estimate of meat production for the national farm, but we have been unable to establish either the size of the national herd or the flow of animals heading for market at any particular time. If we accept that the number of animals brought to market increased, and that the average size of carcasses grew over time, output must have risen significantly, if only to prevent shortages which would have led to price rises putting meat and dairy products out of the reach of a growing and mainly urban population.

CHAPTER 7

Farm Production and the Agricultural Revolution

The search for an agricultural revolution has preoccupied historians of rural England since the great crisis of the late nineteenth-century depression. Initially it was a matter of description. Population, and particularly urban population, had risen sharply from the later eighteenth century in conjunction with an industrial revolution. Together they had transformed the English economy. The role of the farming community in this process was nothing more nor less than to feed many more mouths, so that the import requirement did not rise to a level which would have crowded out industrial investment. The mechanics of this 'agricultural revolution' were discussed in terms of structures (enclosure, farm sizes, and so forth) and innovations (turnip husbandry, clover, and artificial grasses). Collectively these changes were believed to have raised output, but in turn they relied on progress elsewhere in the economy (notably canals and subsequently railways) to ensure an adequate distribution network. For much of the twentieth century, and for most historians, this explanation was perfectly acceptable, and in any case there seemed to be no obvious alternative given the absence of statistics prior to 1866.

Once questions were raised about the timing and nature of the industrial revolution it was perhaps inevitable that the course of agricultural change would also be scrutinized. In the 1960s the salient features of the agricultural revolution came to be deemed matters ripe for debate. For some historians agriculture had, like industry, made significant progress by the mid-eighteenth century, although few were prepared to go as far as Eric Kerridge and describe the resultant revolution as more or less over by about 1750. Nor is this surprising. However much may have been achieved by the mid-eighteenth century no one could deny that population rose from 8.5 million in c.1770 to 21 million in 1851 without notable recourse to imports. This remarkable record on the part of the English farmer needed to be acknowledged, and in the 1980s quantitative historians sought to do so by finding new ways of measuring agricultural change. Contemporary estimates such as those of Gregory King were manipulated to serve as proxies for the nation's agricultural health, and new methods were developed of extracting data from probate inventories to assess the state of farming c.1700. But the results were still inconclusive. The best 'measures' seemed to indicate strong growth in agricultural output to about 1770, then some falling away before renewed growth after about 1800. When we spliced together all the evidence assembled since the 1960s our conclusion (Table 1.1) was that measured simply in terms of real grain output the agricultural revolution took place in the period 1750–1850. Yet, like everyone else since the 1960s, we recognized that the absence of reliable figures for output was a major stumbling block. Much depended on cutting and recutting a rather meagre database of contemporary material in different ways. Meanwhile no one had systematically trawled the records of working farmers.

For all the discussions of agriculture over the past four decades, surprisingly little attention has been paid to the farmer, and hardly any to the records he generated. Of course there are problems with the records. Except on estate farms they were generally kept only for interest, since there was no pressing need in the form of government census enumeration or the income tax inspector. The quantity and quality inevitably varied a great deal, and with no obvious end use a great many contemporary records must have been destroyed over time. Yet the fact that farmers were encouraged to keep records means that a database has survived, albeit a database which is inconsistent and irregular, and which consists of records kept for all sorts of reasons, not all of which are obvious to us today. We were able to identify over 900 such records, admittedly of differing quality, and of course much of the material was not systematic in the manner of a set of estate rentals. Cost accounting is a relatively modern concept, and farmers were more likely to record the state of the weather than the depreciated value of their stock. What is impressive about the records is not their consistency, but the wealth of data they contain when brought together. The archives of a single farm may prove disappointing taken in isolation—although we found several sets of quite excellent accounts—but aggregated with those of many other enterprises they turn out to offer a remarkable picture of English farming in the eighteenth and nineteenth centuries. In truth it is a magnificent collection which has been virtually neglected in the numerous efforts made to understand the agricultural history of the two centuries or so prior to 1914.

The value of farm records is nowhere clearer than in the light they shed on the practice of farming, and the output of farms. The farmer's output was his annual grain harvest, and his animals (with their related products), and these can be measured in yields and weights. In the process of investigating farm production it is conclusive when that production is complete—the crop is harvested and weighed, the animals are sent to market or disposed of in some other way. We have, in short, the end result, the output, and a line can be drawn under another season. Farm records have proved to be an excellent source from which to establish the position of this line. Less satisfactory has been the measure of productivity. Much of the rhetoric in the analysis and identification of the English agricultural revolution has concerned measures of productivity, and we would have liked to make a more positive analysis of the relationship between inputs and outputs. However, the farm records simply do not provide all the factors of production required to enable us properly to conduct a reliable productivity exercise. For example, farm records do not always record output recycled for internal farm use such as farmyard manure and fodder crops. In the absence of such detail we have not been able to offer an empirical dimension to the otherwise cleverly crafted inferences employed by economic and agricultural historians in relation to productivity.

This has not prevented us from locating what we believe to be key findings about farm production in eighteenth- and nineteenth-century English agriculture, particularly in relation to crop yields. Wheat has long been regarded as the yardstick against which to measure the agricultural economy. Existing estimates suggest that from about 16 bushels an acre in *c*.1680, output rose to about 21 bushels an acre in the 1780s, but that there was little or no rise over the period 1800–50. These findings help to explain why historians have been keen to identify a pre-1770 agricultural revolution, but our new series suggests a rather different picture. We have identified two uneven plateaux of yields from the 1720s to the 1810s and from the 1850s to the 1900s. The first was at or about 20 bushels per acre, and suggests that yields increased by rather less than 10 per cent in the eighteenth century with some falling off in the last two or three decades. In output terms it does not point in the direction of an agricultural revolution prior to 1770, although it does show the slowing down of growth post-1770 identified by Crafts,

Jackson, and others. The second plateau in the later nineteenth century was at or about 25–30 bushels per acre. These trends were particularly noticeable when yields on good-quality soils were isolated from the rest (Fig. 4.2), in which instance there was a decline in yields down to the 1790s followed by a rise to the 1840s. The break between the plateaux occurred in the period c.1820–1840s, when yields rose significantly. If we take the view that the agricultural revolution cannot satisfactorily be divorced from the great rise in population, on the basis of wheat output figures we would place it firmly in the first half of the nineteenth century.

To a significant degree the trend of barley yields mirrors the trend in wheat yields, although there was much more annual and decennial variation. The plateau effect is clear in Table 5.1, although annual variations render the two plateaux less flat than for wheat. The rise in yields to c.1840 is also apparent. The trend in the yield of oats is less clear cut. There was a strong rise in yields from the 1720s to the 1770s, a reversal similar to but weaker than for wheat and barley in the last decades of the eighteenth century, and then relatively strong growth down to the 1840s, followed by a plateau (much as we have identified for the other grains). Our data for other crops is not as full, but Figs 5.3 and 5.4 showed that for both grains and pulses there was a marked upward shift in output in the second quarter of the nineteenth century. This surely is compelling evidence of a real, and sustained, break with the past.

If so, how was it achieved? The difficulty we have faced is that however confident we may feel in our findings regarding grain output, relating these specifically to changes in practice is more or less out of the question. As we showed in Chapter 3, farm records enable us to piece together something of the way in which farmers worked, but we cannot make a straightforward link between the purchase of, for example, a new plough and an increase in output. Yet there are indicators which help to explain shifts in practice which must surely have had an impact on output. These include the spread of the use of clover, which was in use on about 85 per cent of farms in the first half of the nineteenth century (Fig. 3.2); the uptake of turnips and, from the early nineteenth century, other roots (Fig. 3.3); the increasing variety of fertilizers and manures, perhaps particularly in the use of bone manure (Fig. 3.6); an increase in the use of nitrogen-fixing crops (Fig. 3.7); and, towards the middle of the nineteenth century, the introduction of artificials, although much of the rise in output had occurred before these entered widespread use. Also significant was an increase in the efficiency of seeding rates. The long-run seeding rate declined, marginally for wheat, quite dramatically for barley and oats. The yield rate—the output of grain per unit input of seed—increased. Not only does this point to an increase in efficiency through a decrease in the (partial) capital to output ratio, it must also have had an effect on the quantity of grain available to the market. The saving on grain with the decline in seeding rates, along with a long-run rise in yields, points to an increase in the grain available to the market. In addition, greater efficiency allowed farmers more flexibility in mixing and matching their land uses. Finally, we cannot measure with any confidence the impact of structural changes such as enclosure, drainage, and farm sizes, but unless we argue that they had a negative impact—which is surely unlikely—they must have added something to the increasing efficiency of arable agriculture in the period, particularly from about 1790. In the case of enclosure we might infer that the enormous expenditure that took place before 1815 was rewarded in the next thirty years through the increase in output. The precise combination of these changes which brought the rise in output which we have identified is almost impossible to identify. We would suggest that possibly the most important shift was the increased use of a variety of crops in the rotations, especially those which were nitrogen fixing.

Agricultural production is an integrative process. In modern times there might be arable farmers, livestock farmers, milk producers, and so on. In the past this was not the case. Farming, of necessity, was a mix of crop and animal production. Arable crops needed fertilizing, and farmyard manure was a time-tested preferred source of supply. So farmers had to keep livestock. In the warmer months the animals could graze on the meadows and in the pasture. For those times of the climatic year when grass did not grow farmers cultivated fodder crops, cut and stored hay, and retained straw from their grain. In these ways they could overwinter their draught animals, and at least a proportion of their stock. Unfortunately, the livestock side of farming is difficult to analyse. There was something final about the annual grain harvest that was captured in a measurable way, in our case through crop yields. It could also be captured financially by the off-farm sale of the produce in the year after the harvest. The equivalent measure and the equivalent sense of finality regarding the livestock 'harvest' are less easy to capture. We chose to use the trend over time in slaughter rates, and on this basis the most noticeable productivity advance was in the carcass weights of lambs and calves. We also found a long-run increase in the weights of cows and heifers between the 1730s and 1740s, and about 1860. We were not as confident in our findings of the long-run weight of pigs, probably because they reached the market at different ages. Yet even allowing for the problems of measurement, and for the fact that breeding and veterinary science still had far to go, we have no reason to doubt that the output of the livestock sector improved in terms of both numbers and weights. Some of this improvement was a result of better understanding of stocking and pasture management, and the purchase of off-farm feeds (especially by the 1840s).

What do our findings tell us about the timing and nature of the agricultural revolution? We have attempted a long-run measure of change in Table 7.1. It is confined to an analysis of wheat production, partly because barley and oats show a similar pattern, partly because the livestock trends are not sufficiently clear cut, and partly because if we are correct in what we said in Chapter 4 about the centrality and sensitivity of wheat in the nation's diet then it may act as a proxy for the national farm. The table brings together our own findings with data from other sources in an attempt to establish the relationship between output and consumption (or, technically, per capita availability in column 7 since we make no assumptions as to whether or not these figures represent actual consumption). We cannot be certain of all the figures. The wheat acres are only plausible assumptions because there is nothing firmer on which to base conclusions, whereas the population figures benefit both from the nineteenth-century census findings and from the methods of counting for earlier periods developed by Wrigley and Schofield. With these provisos, the table enables us to reinterpret this period in terms of the timing and nature of the agricultural revolution.

In 1750 the population was about 5.8 million, and our per capita availability figure points in the direction of overproduction. The Corn Bounty Act of 1688 introduced subsidies on grain exports in an attempt to keep prices at home above 48s. a quarter and so afford the farming community a measure of protection. This in turn was the result of static and perhaps even falling population since the 1650s. Rents rose marginally in a period of what has often been seen as strong growth, perhaps as farmers sought to counter falling profit margins by pushing up output, but they probably averaged no more than 8s. an acre by 1750. Low grain prices encouraged many farmers on the heavy Midland clays to convert from open-field arable to permanent pasture. Often this was achieved by enclosure.[1] Even so, overproduction

---

[1] TBA 310; M. E. Turner, *English Parliamentary Enclosure: Its Historical Geography and Economic History* (Folkestone: Dawson, 1980), chapter 6; John Broad, 'Alternate Husbandry and Permanent Pasture in the Midlands, 1650–1800', *AgHR* 28 (1980), 77–89; Joan Thirsk, 'Agrarian History, 1540–1950', in Victoria County History, *Leicestershire*, ii (Oxford: Oxford University Press, 1954), 199–264.

predominated, with nearly 400,000 quarters of grain being exported annually during the 1740s.[2] Some parts of the country may have experienced little short of depression conditions in the 1730s and 1740s.[3] Farmers trying to raise output so that they could compensate for falling prices by having more grain to take to market may have innovated, hence the claims of those who see an agricultural revolution in the late seventeenth and early eighteenth centuries. For the most part farmers did not endure the tenurial restrictions of their European counterparts and were free to take whatever steps seemed necessary (within the context of their leases) for survival. Perhaps this is why between the 1690s and 1750s the gross output of cereals is estimated to have grown by 19 per cent while the area of land sown did not increase by much more than 7 per cent.[4] Even so, there is something inherently unsatisfactory about seeing a period of overproduction, relatively static rents, and marginal yield increases in wheat depicted as a period of agricultural revolution.

Are the figures in Table 7.1 plausible? In 1766 Charles Smith drew a distinction between wheat eaters and other grain eaters. Of a population of 6 million or so, Smith suggested that 3.75 million were wheat eaters, a number which 'if any thing' he had 'set rather too low'. These people, representing 62.5 per cent of the population, he judged to consume on average one quarter of grain per annum. In Table 7.1, we estimate that the net output available in 1750 was 34.03 million bushels (or 32.73 using different time periods for estimation). With a population of 5.772 million, this suggests a per capita 'availability' of 5.9 (5.67) bushels per head. If we adopt Smith's assumptions that 62.5 per cent of our population of 5.772 (or 3.61 million) consumed wheat, on our net output figures they each had available between 9 and 9.4 bushels per head. Smith's figures need to be treated with care, but his estimate of consumption (8 bushels) and our estimate of availability (say 9 bushels) are sufficiently close to suggest that we are looking at the right level of magnitude.[5]

Inevitably there is an element of speculation about the figures in the first row of Table 7.1, but we would argue that they provide a reasonable starting

---

[2] D. Ormrod, *English Grain Exports and the Structure of Agrarian Capitalism, 1700–1760* (Hull: Hull University Press, 1985). This was not excessive overproduction. It represented about 2.6 per cent of Charles Smith's estimate of 15.35 million quarters for total grain production in the mid-18th century for England and Wales: quoted in the 'Progress and Present State of Agriculture', *Edinburgh Review*, 62 (Jan. 1836), 321.

[3] J. V. Beckett, 'Regional Variation and the Agricultural Depression, 1730–50', *EcHR* 35 (1982), 35–51.

[4] *AHEW* vi. 444.

[5] Smith's figures taken from P. Deane and W. A. Cole, *British Economic Growth, 1688–1959* (2nd edn. Cambridge: Cambridge University Press, 1969), 63.

point from which we may proceed. From around 1750 a headlong demographic revolution was in train. Population had risen by less than one million over the previous seventy years; now it grew by nearly three million in half a century. The farming community reacted sluggishly. Rents increased only about 10 per cent 1750–90, and yields for all the principal grain crops remained static or even declined. Net output grew by only 18 per cent, and perhaps rather less, and the potential mismatch between supply and demand was met by converting the export surplus into an import deficit.[6] By the 1760s England was a net importer of grain, but farmers still perceived conditions to be adverse, hence the continuing importance of enclosure. From *c*.1760 this was undertaken predominantly by parliamentary Act, and it was still generally associated with the conversion of open-field arable to pasture. The relative sluggishness of agriculture in the second half of the eighteenth century is now well attested.[7]

Then came the 1790s, with its near-lethal combination of rising population, poor harvests, and import restrictions due to war. Fourteen of the twenty-two grain harvests between 1793 and 1814 were deficient in varying degrees. In the desperate conditions of 1795–6 and 1799–1800 the Privy Council instructed the Commissioners of the Victualling Office to experiment with making bread and biscuits out of combinations of 'inferior grains', and also out of potatoes, and even of straw! The Privy Council studied advice on ways of making bread which did not rely entirely on wheat.[8] It solicited reports from across Europe, and especially the Mediterranean, on the purchase of other bread grains.[9] Finally, it made arrangements to purchase rice and maize from North America.[10] For all this effort grain prices spiralled out of control, rents rose rapidly in their wake, sometimes doubling within two decades, and, in a brave attempt to raise the output of grain,

---

[6] Turner, *English Parliamentary Enclosure*, 130.
[7] Mark Overton, *Agricultural Revolution in England: The Transformation of the Agrarian Economy 1500–1850* (Cambridge: Cambridge University Press, 1996), 86–7.
[8] W. M. Stern, 'The Bread Crisis in Britain, 1795–6', *Economica*, 31 (1964), 168–87; W. E. Minchinton, 'Agricultural Returns and the Government during the Napoleonic Wars', *AgHR* 1 (1953), 29–43; M. E. Turner, 'Agricultural Productivity in England in the Eighteenth Century: Evidence from Crop Yields', *EcHR* 35 (1982), 495–6; R. Wells, *Wretched Faces: Famine in Wartime England 1793–1801* (Gloucester: Alan Sutton, 1988), esp. 202–18; PRO, HO/42/52–5; PRO PC Minutes, PC 4/6, 26 Oct. 1795, 2 Nov 1795; PRO PC loose letters including printed recipes for making flour and meal out of potatoes, PC 1/32/A80, Jan. 1796; PRO PC Minutes, PC 4/9, Dec. 1799, fos. 240–4.
[9] PRO PC loose letters, PC 1/31/A73, Dec. 1795, PC 1/32/A80, Jan. 1796; PRO PC Minutes, PC 4/9, Dec. 1799, fos. 209–39.
[10] PRO PC loose letters, PC 1/32/A80, Dec. 1795–Jan. 1796.

Table 7.1 Locating the 'agricultural revolution' in England

(a)

	Wheat acres (million)	Average yield (bushels/acre)	Seeding rate (bushels/acre)	Net yield (bushels/acre)	Net output (million bushels)	Pop. (million)	Per capita availability (bushels/capita)
1750	1.70	22.42 (21.75)	2.4 (2.5)	20.02 (19.25)	34.03 (32.73)	5.772	5.90 (5.67)
1800	2.19	20.98 (18.97)	2.7 (2.5)	18.28 (16.47)	40.03 (36.07)	8.664	4.62 (4.16)
1820	2.55	23.60 (21.17)	2.7 (2.6)	20.90 (18.57)	53.30 (47.35)	11.491	4.64 (4.12)
1850	3.42	27.47 (30.6)	1.6 (1.9)	25.87 (28.7)	88.48 (98.15)	16.736	5.29 (5.86)

(b)

	Index of the value of net output	Index of rent
1750	34.8 (37.1)	66.5
1800	100.0 (100.0)	100.0
1820	122.0 (120.3)	180.5
1850	151.2 (186.1)	175.4

Notes: Subject to rounding errors. Acreage and population approximates to the years specified. Yields and seeding rates refer to the mean decennial yields following the population and acre figures (1750s, 1800s, 1820s, 1850s), and the figures in brackets refer to the yields and seeding rates in the preceding decades (1740s, 1790s, 1810s, 1840s).

Acreages: The wheat acreage for 1850 is based on Caird, and the estimate for 1750 is based on Holderness's assumption that the wheat acreage doubled 1750–1850. The acreage for 1800 is based on the 1801 crop returns in Turner. In the absence of a quotable acreage for 1820, it is assumed that the wheat acreage increased in a linear fashion between c.1800 and c.1850.

Seeding rates: Taken from Table 5.5 above.

Average yields: Taken from the mean yields in Table 4.4 above.

Net yield: Average yield minus seeding rate.

Net output: Net yield multiplied by acreage.

Population: English population figures from Schofield.

Per capita availability: Net output divided by population.

Index of value of net output: Based on net outputs multiplied by annual average wheat prices for 1746–54, 1796–1805, 1816–24, and 1846–54, from Mitchell and Deane.

Index of rent: Based on the annual average rent received per acre for 1746–54, 1796–1805, 1816–24, and 1846–54 from TBA.

Sources: J. Caird, *English Agriculture in 1850–51* (2nd edn. London: Longman, 1852), 522; *AHEW* vi. 129; M. E. Turner, 'Arable in England and Wales: Estimates from the 1801 Crop Returns', *Journal of Historical Geography*, 7 (1981), 294, 296; R. Schofield, 'British Population Change, 1700–1871', in R. Floud and D. McCloskey (eds.), *The Economic History of Britain since 1700*, i: *1700–1860* (2nd edn. Cambridge: Cambridge University Press, 1994), 64; B. R. Mitchell and P. Deane (eds.), *Abstract of British Historical Statistics* (Cambridge: Cambridge University Press, 1962), 487–8; TBA 310–12.

commons, wastes, and other land were turned down to arable production during a further wave of enclosure. The country, so the classical economists thought, stood on the verge of crisis, and this might well be the interpretation we would favour on the basis of the figures in the second row of Table 7.1. Per capita availability had declined by not much short of one-quarter. Living standards more generally may also have declined.

The war years brought dislocation for the farmer, but the combined impact of rising prices through inflation and rising rents was already having an effect on his energies. The response of the rural community to the war had been to increase the wheat acreage, and, among farmers—driven perhaps by rising prices and rising rents—to shift their practices in such a way as to raise yields. In these years enclosure was not designed to change land use as it had been prior to 1793, but instead to intensify arable production. And by moving farming from a communal to an individual basis, and separating the common holdings with hedges and fences, enclosure created the opportunities for individual enterprise. For landlords, enclosure offered the chance to renegotiate lease covenants, which they sometimes took, and to impose rack rents, which they invariably did. The net result of enclosure in terms of output remains contested, but in the longer term it brought improvements in farming practice by making available conditions for adopting innovatory farming techniques, and by altering the arable to pastoral balance, because farmers were no longer limited by the area of commons, meadows, and pasture which was available. It offered farmers freedom of land use to grow a greater variety of crops in order to exploit market opportunities.

All this was happening through the war years, in conjunction with the spread of Norfolk-type rotation systems, improvements in drainage, and other changes we identified in Chapter 3. For the farmer, all the indicators were favourable. Population increased from 7.7 million at the outbreak of war to 10.7 million at its conclusion, an increase of 39 per cent. At the same time imports from the Continent were restricted, leading to rising prices. Grain prices peaked in 1812 at 126s. a quarter. Farmers, in other words, enjoyed immensely favourable conditions, to which they responded so positively that yields were already much higher at the end of the war than at the beginning. Yet the potential for disaster was considerable. Grain prices began to decline even before the end of the war,[11] and no one could predict

---

[11] B. R. Mitchell and P. Deane (eds.), *Abstract of British Historical Statistics* (Cambridge: Cambridge University Press, 1962), 488.

with any certainty whether population would continue to increase, and if so at what rate. For the farming community to operate after the war with higher rents and falling prices could have brought catastrophe, hence the pressure of the landed interest in 1815 to introduce a new Corn Law. In the event, even with this measure of protection, and—as it transpired—a rapidly rising population, conditions remained difficult.

Most of this picture is well known. Growth, it is usually argued, continued after the war, with output reaching new levels, but in the context of falling rents, periods of real distress, and sliding prices which were kept from collapse only by the 1815 Corn Law and its various amendments in the 1820s. Such an explanation has fitted the observable trends, but those trends have been weak precisely where our data are strong, in assessing yields. Our data suggest a real improvement in yields over the first half of the nineteenth century and, of course, this is a further reason why agriculture faced problems in the post-war years. The price of wheat collapsed from 126*s.* a quarter in 1812 to scarcely more than half of this in the post-war period, with a low point of only 39*s.* in 1835.[12] To maintain the level of farm incomes enjoyed during the war landowners and their tenants either had to extend the cultivated acreage, or increase the unit acre yield, or ensure a combination of both. England was cultivated practically to its limits, and so the only way to increase cash crop or wheat production was by changing existing land uses. The popular belief is that the wheat acreage continued to rise in the post-war decades, reaching a maximum extent in the first half of the 1840s.[13] Given the decline in prices after the war there had to be a rise in unit acre yields as well as an extension of the wheat acreage. We have located this rise in yields.

The last time the price of wheat had been as low as in 1835 was in 1776, though more generally the price of wheat in the first half of the 1830s was about the level it had been on the eve of the war. Rents were two or three times higher after the war than before, but they did not move very much, if at all, from 1820 to 1850. Farmers and landlords alike were coming to terms with the significantly changed conditions of these years, which was far from easy. When farmers failed to pay their rents, landlords preferred to give abatements rather than permanent reductions, in the vain hope that an upturn was just around the corner. Farmers were forced to try to pay their

---

[12] Mitchell and Dean, *Abstract*, 488.   [13] *AHEW* vi. 128.

rent, while often leaving other creditors unsatisfied. The result was that they were 'farming under war rents, while ... selling their corn at peace prices'.[14] One after another witness to the three committees of inquiry set up by the government in the 1830s spoke of declining rents and an impoverished agricultural sector.[15] Of course some of this was special pleading of the sort only to be expected before a government committee, and there were reports from some parts of the country which painted a rather different picture. While the Select Committees were regaled with horror stories of collapsing rents, the overall picture for these years is much less gloomy.[16] In 1836 the *Edinburgh Review* carried a review of George Robertson's *Rural Recollections*, and took a similar stance—the depression had been exaggerated, it was localized, and in general agriculture had achieved what it was meant to achieve. It had fed the nation, and in the four years prior to 1836 (incidentally the years when the parliamentary Select Committees sat, inquired, and reported) there had been no resort, or next to no resort, to imports of grain, 'so that it necessarily follows, that all the vast numbers that have been added since 1760 to the population of Great Britain, must be exclusively indebted for their subsistence to the subsequent improvement and extension of agriculture'. At the same time, a modest yet observable increase in carcass weights (such as we have identified through this period) could, in the opinion of the reviewer, have come about only by an increase in the availability of fodder, because 'an improved system of breeding [alone] would [only] have improved the symmetry of the cattle, and increased their aptitude to fatten'. This all came about as a result of the stimulus offered to agriculture by the high prices prevailing during the war, 'its previous progress being not merely maintained, but considerably accelerated'. That progress was attributed particularly to drainage, but also to better crop rotations and, most important of all, the general use of bone manure. We may not agree with all the details of this diagnosis, but we would accept the reviewer's conclusion that the complaints to the Agricultural Committee of 1833 'were certainly very much exaggerated'.[17]

Of course the situation was more complicated than the emotional appeals to government committees were ever likely to suggest, or the responses in

---

[14] P. Roe, 'Norfolk Agriculture, 1815–1915' (University of East Anglia, M.Phil. thesis, 1976), 21.
[15] The relevant evidence is cited in TBA 242–3.   [16] TBA 244–5.
[17] 'Progress and Present State of Agriculture', 319–45, esp. 321, 329, 333, 335, 337–8; George Robertson, *Rural Recollections; or, The Progress of Improvement in Agriculture and Rural Affairs* (Irvine, 1829).

reviews were ever likely to dispel. Without much doubt rents were too high at the end of the war, but because landlords were reluctant to reduce them on a permanent basis they did not reach a realistic economic level as fast as farmers would have liked. Farmers were partially protected from the full force of the market by the Corn Laws, but their real salvation lay in demographic change. Population rose by nearly 3 million 1800–20, and by a further 5 million 1820–50. Urban growth peaked in the 1820s with major English towns growing in a single decade by 40 per cent or more.[18] Increasing demand helped to take up much of the slack brought about by rising yields and output, without which there would surely have been real disaster in the countryside in these decades. This explains the figures in row 4 of Table 7.1. Farmers were meeting the requirements of an expanding population through a process of production which was well in advance of the yields achieved before the war. This was nothing short of an agricultural revolution in terms of yields achieved and population fed.

Precisely what was achieved became clear only with the repeal in 1846 of the Corn Laws. Between 1800 and 1850 population grew by 93 per cent, but output by 121 per cent and perhaps more. Although wheat output kept pace with or even exceeded the growth of population during the period, there was no resurgence of wheat exports. Instead there was an increase in domestic wheat grain bread consumption, closely correlated with urbanization, and a decrease in the use of all other forms of bread grains. Therefore, there was certainly no crisis in the agricultural sector by the 1840s. On the eve of Corn Law repeal wheat imports represented no more than 7 or 8 per cent of total English and Welsh consumption. It was only after repeal that this situation changed. Wheat imports rose to one-fifth on repeal and to one-quarter in the 1850s to reach 40 per cent of the total in the 1860s.[19] But by now the agricultural sector had adjusted to the post-war problems, and was enjoying the benefits not only of increased yields, but of ever growing demand as a

---

[18] M. E. Rose, 'Social Change and the Industrial Revolution', in R. Floud and D. N. McCloskey (eds.), *The Economic History of Britain since 1700, i: 1700–1860* (1st edn. Cambridge: Cambridge University Press, 1981), 257.

[19] Collins points out that in 1800 about 66% of the population of England and Wales consumed wheaten grain, but by 1850 this proportion had risen to 88%, and of course the population had more or less doubled. E. J. T. Collins, 'Dietary Change and Cereal Consumption in Britain in the Nineteenth Century', *AgHR* 23 (1975), 97–115, esp. the table on 114. See also C. Petersen, *Bread and the British Economy c.1770–1870* (Aldershot: Scholar Press, 1995), 205–6 for the equivalent proportions for Britain. For details on the trade in grain see S. Fairlie, 'The Corn Laws and British Wheat Production 1829–76', *EcHR* 22 (1969), 88–116; Mitchell and Deane, *Abstract*, 98–9, 488–9; D. C. Moore, 'The Corn Laws and High Farming', *EcHR* 18 (1965), 545–7.

result of population growth. Once it was obvious that the Corn Laws were not holding together the fabric of agricultural society, farmers and landlords were again able to adjust the terms of agreement between them. In the uncertain conditions since 1815 landlords had been cautious about raising rents. Caird certainly recognized under-renting, and possibly the relatively low level of arrears in mid-century is a further reflection of the way the tenurial terms of trade in farming moved in favour of the tenant.[20] It was not to last, since after the mid-century point landlords were again able to retain part of the profits of farming through rents which increased by *c.*30 per cent 1850–80.

Over the whole period 1750–1850 population grew by about 190 per cent, and net output by 160 per cent (or more, depending on how we cut the figures). Agricultural commentators both at the time and subsequently accept that yields rose.[21] There has been less certainty about timing, but Figs. 4.1, 5.1, and 5.2 point clearly in the direction of yields beginning to rise significantly from around the turn of the century, hence the figures on row 3 of Table 7.1. Had this not occurred the industrial revolution might well have been crowded out by the need to fund grain imports to feed the growing cities.[22] As it was, rising output kept food prices at a level where population growth could be sustained, and the whole economy could move forward.[23] Relatively low grain prices may even have allowed a rise in living standards. The third quarter of the nineteenth century was a golden age for farmers, but it was not all plain sailing. F. M. L. Thompson argued the case for what he called a second agricultural revolution in the period post-1815. His revolution was based on an assessment of external inputs. Our findings suggest that this phase of change has to be treated circumspectly because the new inputs did not necessarily have a cumulative effect in proportion to their usage. As a result, yields reached a plateau in the second half of the century. Had they continued rising the agricultural sector might have been in better shape to respond positively to the cold winds of change which overtook it in the 1880s, particularly the influx of cheap grain from North America.

---

[20] J. Caird, *English Agriculture in 1850–51* (2nd edn. London: Longman, 1852; repr. London: Frank Cass, 1968), 477; TBA 182.

[21] Caird, *English Agriculture*, 474–5; Turner, 'Agricultural Productivity in England', 501, 504; *AHEW* vi. 140.

[22] J. V. Beckett and M. E. Turner, 'Taxation and Economic Growth in Eighteenth-Century England', *EcHR* 43 (1990), 377–403.

[23] Data collected by Dr Susan Fairlie, and published in *AHEW* vi. 1067 suggests that the growth patterns of our putative wheat production estimates agree in large measure with the trend of grain production she reported, particularly for the years after 1820.

As it was, the unprotected grain farmer was caught cold. Rents collapsed, and growth fell back.[24]

In a number of significant ways the real golden age was not 1850–73 but from 1810 or so to the late 1870s. Farmers raised yields while sheltering behind the Corn Laws until the 1840s, and being even more effectively protected against potential disaster by rapidly rising population. Thus in Lincolnshire the Corn Laws provided the confidence to invest in steam pumps during the 1830s and 1840s in order to drain the land for grain production. As a result, the Spalding region soon became a net exporter of wheat, most of it leaving the area by rail.[25] Not surprisingly, having enjoyed the benefits of protective legislation, local farmers were in the forefront of opposition to Corn Law repeal.[26]

Only from about 1850 or thereabouts did the landlords join in this golden age. Once it was clear that contrary to expectations the farmers had not been ruined by repeal, landlords set about extracting what they now recognized as surplus income, hence the 30 per cent increase in rents 1850–80.

The increases in yield were also accompanied by significant changes in other areas of the agricultural economy, not least in labour productivity. All agricultural historians accept that labour productivity grew rapidly, perhaps doubling over the period 1700–1850 with the majority of the improvement after 1800.[27] The number of males aged 20 or over employed in agriculture increased from about 910,000 in 1811 to 1,010,000 in 1851. As a proportion of the male population aged 20–69 this represented a fall from 37.8 to 23.8 per cent.[28] Given the rise in output we have already identified, this must have been a reflection of changes in productivity. In 1760 each agricultural worker's output was capable of feeding 2.7 non-agricultural workers, while over the period 1700–1850 the agricultural labour force probably increased by between 50 and 75 per cent, but labour productivity rose very considerably

---

[24] TBA 149–50, 250–2; M. E. Turner, 'Output and Prices in UK Agriculture, 1867–1914, and the Great Agricultural Depression Reconsidered', *AgHR* 40 (1992), 38–51; F. M. L. Thompson, 'An Anatomy of English Agriculture, 1870–1914', in B. A. Holderness and M. E. Turner (eds.), *Land, Labour and Agriculture, 1700–1920* (London: Hambledon, 1991), 211–40.

[25] T. W. Beastall, *The Agricultural Revolution in Lincolnshire* (Lincoln: History of Lincolnshire Committee, 1978), 70–1.

[26] R. J. Olney, *Rural Society and County Government in Nineteenth-Century Lincolnshire* (Lincoln: History of Lincolnshire Committee, 1979), 153.

[27] G. Clark, 'Labour Productivity in English Agriculture, 1300–1860', in Campbell and Overton (1991), 211–35; E. A. Wrigley, 'Energy Availability and Agricultural Productivity', ibid. 323–39.

[28] E. A. Wrigley, 'Men on the Land and Men in the Countryside: Employment in Agriculture in Early Nineteenth Century England', in L. Bonfield, R. M. Smith, and K. Wrightson (eds.), *The World We Have Gained* (Oxford: Oxford University Press, 1986), 332–5.

given the volume of output by the mid-nineteenth century. As to how it increased, there is less agreement. One obvious explanation would be that technology was widely adopted. Although we found some supporting evidence in the farm records of technology, measuring the pace of change was all but impossible, and most commentators accept that significant productivity improvements as a result of technology date only from the late nineteenth century. Suggestions have been made that English agricultural labourers worked harder than their continental neighbours;[29] that they worked on larger and more efficient farms—although by the late nineteenth century it was widely held that peasant agriculture was more efficient; and that they enjoyed the benefits of a substantially greater input of animal power in proportion to human labour by comparison with France.[30]

Assessing labour productivity from farm records proved almost impossible. The various records which record the employment of farm labour do not usually have anything to say about the work of the farmer and his family, and these were important on the input side of labour productivity. Consequently we have worked in a different way, and Table 7.2 offers evidence on labour productivity derived on the basis of our yield figures. The figures suggest that relatively little increase in productivity occurred while agriculture was making few strides in the second half of the eighteenth century, but much more impressive growth took place between 1800 and 1850. This improvement occurred as the land under cultivation, and particularly the wheat acreage, continued to expand down to the 1840s. Our figures show that the increase in output did not simply keep pace with the rise in acreage, it leapt ahead. This is clear from a comparison of the figures in Table 7.1 for the growth in acreage, the growth in yields, and the economizing in seeding. We give an increase in acreage 1800–50 of 56 per cent, and a net output rise of 120 per cent.[31]

If yields and output were rising, and labour was becoming more productive, what was the role in this of capital? Enclosure and drainage were perhaps the most expensive investments of the period. The greatest expenditure

[29] G. Clark, 'Productivity Growth without Technical Change in European Agriculture before 1850', *JEH* 47 (1987), 419–32.
[30] R. C. Allen, 'The Growth of Labour Productivity in Early Modern English Agriculture', *Explorations in Economic History*, 25 (1988), 117–46; Wrigley, 'Energy Availability'.
[31] Commentators on labour productivity all point roughly in the same direction as our findings, with different orders of magnitude depending on which way they cut the figures. Wrigley put the majority of the increase after 1800, as does Overton, *Agricultural Revolution*, 81–2. Clark, 'Labour Productivity', proposes more modest figures for the 19th century and instead emphasizes the achievements in labour productivity before *c.*1770.

Table 7.2 Labour productivity

	Overton/Wrigley indices[a]		Clark index[b]	New index[c]
c.1750	100	100	100	100 (100)
c.1800	112	127	96	101 (94)
c.1820			98 (104)	123 (114)
c.1830		147		
c.1850	156	154	110	169 (195)

[a] The Overton/Wrigley indices are here rebased on 1750. The first is an index derived from output estimates using the 'population method', and the second is derived from output estimates using the 'volume' method, to both of which are applied Wrigley's rural agricultural population figures to derive productivity estimates. The methodology and the terms are fully explained in Overton.
[b] Clark's original index was related to a scale of grain yields. He also linked those yield figures to specific dates. Since our own yield estimates in Chapter 4 do not entirely agree with Clark for the same periods, we have taken our yields and matched them up with his equivalent labour productivity index. Thus in c.1750 our yield is 22 bushels per acre, at which level Clark's productivity index is 100. In c.1800 our yields are 21 or 19 (depending whether we base them on the decade before or after this date), and at an average yield of 20 Clark's labour productivity index is 96. In c.1820 our yields are 23.6 or 21 (depending on the decade chosen). Clark's pro rata index for 21 bushels would be 98, and his index for 24 bushels is 104. In c.1850 our yields are 27 and 30.6, and Clark's nearest index is 110 representing 28 bushels.
[c] The new labour productivity index is based on gross wheat production which we are taking as a crude proxy for the national farm. It has been derived by applying Wrigley's rural agricultural population estimates as reported in Overton to our own wheat production estimates reported in Table 7.1 (but adjusted to gross outputs). We quote two indices for the same reasons given in Table 7.1. Unfortunately, Wrigley only quotes a population estimate for 1831, and this is the one we have adopted. There is a sense in which there should be a lower population figure for 1820 because the rural agricultural population was almost certainly still rising in the period. The effect of adopting a lower figure is to increase the output per head, and therefore increase the index number. Supposing that instead of 3.38 million people in 1821 in rural agricultural pursuits (which is Wrigley's figure for 1831) there were only 3.25 million. That would change the output figures accordingly and the corresponding productivity index for c.1820 would become 128 (119). In other words, the argument still favours an emphasis on labour productivity after 1820.

Sources Table 7.1 and associated text; M. Overton, *Agricultural Revolution in England: The Transformation of the Agrarian Economy 1500–1850* (Cambridge: Cambridge University Press, 1996), 74–7, 80–2, 86.

for open-field enclosure was 1790–1819, and for the reclamation and enclosure of waste 1790–1809 (the latter at an altogether much higher rate of expenditure than open-field land).[32] Underdrainage, as we saw in Chapter 3, was important long before the advent of pipes in the 1840s. Estimates suggest

---

[32] B. A. Holderness, 'Agriculture, 1770–1860', in C. H. Feinstein and S. Pollard (eds.), *Studies in Capital Formation in the United Kingdom, 1750–1920* (Oxford: Oxford University Press, 1988), 19, 22.

that 35,000 acres were underdrained in the 1770s, 45,000 acres by the 1800s, 70,000 acres by the 1820s, and 120,000 acres in the 1840s.[33] Farm sizes probably increased from the mid-eighteenth to the mid-nineteenth centuries, although Caird noted that they varied according to regional farming and soil characteristics.[34] In any case, heavy investment in farm improvements was a feature of the depression in the 1880s and 1890s rather than earlier periods.[35] The steady elimination of customary tenures and other common rights in the course of the nineteenth century must also have had a significant effect on the efficiency of farming.[36]

Capital investment also implied mechanization. We showed in Chapter 3 that farm records document the spread of implements and machines, but not in a meaningful statistical way. We know that the seed drill was in use before 1800, but harvesting technology was introduced only after the turn of the century. In Oxfordshire and the Welsh borders there is evidence of a take-up in the use of winnowing machines by about 1820, as well as the introduction of turnip cutters and chaff and threshing machines. By 1850 in these areas all three processes in the agricultural cycle—sowing, harvesting, and processing—had enjoyed limited mechanization, although as is well known from the Swing disturbances the introduction of threshing machines was greeted with considerable hostility in many areas of the country.[37] Most agricultural historians have concluded rather hesitantly that abundance of labour slowed down the introduction of machinery,[38] hence Mingay's comment that 'to some extent, at least, the productive achievements of [1750–1850] were secured at the expense of the hardship and deprivation of the more than 900,000 workers who laboured on the farms of England and Wales for meagre rewards'.[39] Farmers who tried to introduce machinery met resistance, either openly as with the Swing riots, or more covertly through

---

[33] Holderness, 'Agriculture', 25; A. D. M. Phillips, *The Underdraining of Farmland in England during the Nineteenth Century* (Cambridge: Cambridge University Press, 1989).

[34] Caird, *English Agriculture*, 481.   [35] *AHEW* vii. 739.

[36] This is the implication of the tenurial adjustments on the Oxbridge colleges' estates during the 19th century, M. E. Turner, 'Corporate Strategy or Individual Priority? Land Management, Income and Tenure on Oxbridge Agricultural Land in the Mid-Nineteenth Century', *Business History*, 42 (2000), 1–26.

[37] J. R. Walton, 'Mechanisation in Agriculture: A Study of the Adoption Process', in H. S. A. Fox and R. A. Butlin (eds.), *Change in the Countryside: Essays on Rural England 1500–1900* (London: Institute of British Geographers, Special Publication no. 10, 1979), 23–42.

[38] Overton, *Agricultural Revolution*, 121–2 writes that 'little mechanisation of farming took place before the mid-nineteenth century, but before then there can be little doubt that small incremental improvements were made to basic farm implements over the centuries'.

[39] *AHEW* vi. 961.

the drip-drip of machine breaking, and the result was that as late as 1870 much the greater part of the British corn harvest was still cut by hand tools, even though other farm operations such as threshing and livestock feed preparation, which mostly occupied the off-peak periods of the farming year, were already highly mechanized.[40]

Consequently we have the odd situation where machinery was least used during some of the most concentrated parts of the farming cycle, largely because of a labour surplus in a sector of the economy which was experiencing considerable productivity savings.

Finally, external changes encouraged the farmer, particularly shifts in marketing and transport structures arising from the turnpiking of roads, the improvement of rivers, the digging of canals, and, from the 1830s, the laying of railways. All of these changes were a feature of the period 1750–1850, and each in its own way aided the process of moving food supplies cheaply. In this way the ability of the farmer to raise output was encouraged by the capacity of the transport network to move his output cheaply and efficiently into the growing urban markets.[41] Of course not all improvements were beneficial to the farmer. The flourishing livestock trade from Scotland and Wales was interrupted by restrictions on wayside pasturage and the cost of turnpike tolls. On the other hand the specialization of land use arising from improved transport systems and wider markets almost certainly played a part in raising productivity. Using the best grass land for stock rearing and milking and the best arable soils for wheat and barley production clearly made sense, and declining urban mortality rates after about 1850 point to an increased efficiency of food supply as one of several material factors.

Collectively these changes point us clearly in the direction of a long period of prosperity for the farmer. Of course it did not always look that way, hence the agonized debates about the Corn Laws, and any claims regarding prosperity looked hollow in the dark days of the 1880s and 1890s. Once the cold winds of unfettered competition brought a reversal of fortunes there was no incentive for farmers to press ahead with experiments designed to improve

---

[40] E. J. T. Collins, 'Harvest Technology and Labour Supply in Britain, 1790–1870', *EcHR* 22 (1969), 455; id. 'The Rationality of Surplus Agricultural Labour: Mechanisation in English Agriculture in the Nineteenth Century', *AgHR* 35 (1987), 36–46.

[41] W. Thwaites, 'The Corn Market and Economic Change: Oxfordshire in the Eighteenth Century', *Midland History*, 16 (1991), 103–25; id., 'Dearth and the Marketing of Agricultural Produce: Oxfordshire c.1750–1800', *AgHR* 33 (1985), 119–31; R. Scola, *Feeding the Victorian City: The Food Supply of Manchester, 1770–1870* (Manchester: Manchester University Press, 1992).

yields. They looked instead to more profitable alternatives (as they had done in the eighteenth century), as rising disposable urban income put a premium on the products of market gardens and town dairies.[42] The farmer who really was in trouble was in the Lincolnshire fens, East Anglia, and similar regions, where he had no easy means of changing track, unless he could find his way into the specialist malting barley market or, in the Lincolnshire fens, the bulb trade.[43]

We now have a plausible answer to that most elusive of questions, when was the agricultural revolution? The original interpretation was of a revolution running alongside its industrial counterpart. Once the timing of the industrial revolution no longer seemed so secure, questions were inevitably asked about the agricultural revolution. One explanation put forward was in terms of a long period of change prior to the eighteenth century, echoing the concept of far-distant causes of the industrial revolution. A second explanation favoured more than one revolution, with a second phase beginning sometime after the Napoleonic wars. Yet, as historians of the industrial revolution have discovered, trying to phase it according to time periods runs foul of the essential seamlessness of history, which simply cannot be fitted into neat compartments. And, just as industrial historians lost faith in the 'great men' approach to innovation, with its emphasis on major players such as James Watt, Richard Arkwright, and Isambard Kingdom Brunel, so agricultural historians recognized that Jethro Tull, Coke of Norfolk, Lord 'Turnip' Townshend, and Arthur Young were not alone as the progenitors of a farming revolution. Breaking down the stereotypes has helped our understanding of the course of change. Once we move away from explanations of the industrial and agricultural revolutions in terms of steam engines or seed drills we must surely conclude that the key moment for identifying the agricultural revolution is when it met the challenge of the headlong growth of population. The critical test is whether it fed that population. It did, and the rise in output relative to mouths, and therefore the location of the agricultural revolution, is firmly in the period from about 1800 to 1850.

---

[42] Joan Thirsk, *Alternative Agriculture* (Oxford: Oxford University Press, 1997).
[43] E. Eagle, 'Some Light on the Beginnings of the Lincolnshire Bulb Industry', *Lincolnshire Historian*, 6 (1950), 220–9.

Appendix 1

Farm Records, 1700–1914

Location	Date	Reference
Bedfordshire		
Billington	1772	Bedfordshire and Luton Archives and Record Service, BO/1334
Eversholt	1802–18	RUL, BED P245/1–2
Sharnbrook	1805–6	RUL, BED 5/2/1
Stopsley	1816–37	RUL, BED 3/1/1–2
Berkshire		
Ardington, Lockinge	1696–1717	RUL, BER43/2/1–2
Bradfield	1778–1817	Berkshire RO, D/E Sv (M) E7/1–5; E8; E10/1–3
Englefield	1762	Berkshire RO, D/E By E70
Hampstead Norreys	1764–93	Berkshire RO, D/Ex 62/2
Radley	1783–1866	RUL, BER 13/3/1–6; 13/5/-; 13/6/1–4; 13/8/1–5
Steventon	1778–1848	RUL, BER 16/1/1–2; 16/2–1
Sulham	1800–31	Berkshire RO, D/E Wi E16–18
Swallowfield	1790–1830	RUL, BER P322/1–3
Buckinghamshire		
Chicheley	1737–86	Buckinghamshire RO, D/C 2/56; 4/30
Little Shardeloes	1802–14	Buckinghamshire RO, D/DR/2/118
Medmenham	1823–36	Buckinghamshire RO, D85/13/1 and 3

APPENDIX I

Location	Date	Reference
Buckinghamshire		
Quainton	1794–1806	RUL, BUC 6/1/1
Quainton	1739–1826	RUL, BUC P321/2–4
Stokenchurch	1849–66	RUL, BUC 7/1/1
Stokenchurch	1875–84	Oxfordshire AO, Fa xxi/16
Cambridgeshire		
Barton	1827–61	Cambridgeshire RO, Cambridge, R65/15/1–2
Chippenham	1793–1811	Cambridgeshire RO, Cambridge, R55/7/8/6, 15, 24–8
Comberton	1805–9	Cambridgeshire RO, Cambridge, R53/16/21
Coton	1830–55	Cambridgeshire RO, Cambridge, R58/9/4/2(a, b, c), 3(a, b), 4, 7
Histon	1864–98	Cambridgeshire RO, Cambridge, 300A/1–36
Histon	1817–35	Cambridgeshire RO, Cambridge, R62/36
Landbeach	1855–6	Cambridgeshire RO, Cambridge, R51/1/17/57
Trumpington	1856–62	Cambridgeshire RO, Cambridge, R58/9/12/1
Cheshire		
Bramhall	1782–1815	RUL, CHE 1/1/1
High Legh	1784–90	Manchester Local Studies Unit Archives, City Library, Misc. 254
Styal	1809–11	Manchester Local Studies Unit Archives, City Library, C5/6/7
Wybunbury	1823–54	RUL, CHE 2/1
Cornwall		
Fentrigan (in Warbstow)	1871–93	RUL, COR 8/1/1
Kilkhampton	1885–91	RUL, COR 6/1/2
Polperro	1845–1903	RUL, COR 2/1/1 and 3

Location	Date	Reference
Cumberland		
Gilsland	1800–20	Cumbria RO, Carlisle, D/CL/P8/17, 58
Holm Cultram	1826–7	Cumbria RO, Carlisle, DX/74/5
Newton Arlosh	1858–66	Cumbria RO, Carlisle, D/Ph
Derbyshire		
Hathersage	1836–74	Sheffield AO, MD 5857/1–5
Thurvaston	1889–1913	RUL, DER 3/2/1
Wormhill, Norton	1772–5	Sheffield AO, OD1518
Devoï		
Braunton	1809–14	Devon RO, 1240 M E1
Buckland Monochorum	1788–1813	Devon RO, 346M E2–23; E53–5
Castle Hill and Filleigh	1752–1809	Devon RO, 1262M/E1/27
Chulmleigh	1798–1821	Devon RO, 1262/M/E3/7
Clayhidon	1824–50	Devon RO, 1061A ZB3
Milton Abbot	1755–88	Devon RO, 2168M E1, E2, E4
Nymet Rowland	1823–32	Devon RO, 1107M E1
Sandford	1820–50	Devon RO, 1283M E1
South Milton	1751–9	RUL, DEV 3/8/1
Tavistock	1740–53	Devon RO, W1258M /LP1/1
Upottery	1764–78	Devon RO, 152 M/box 75 E1
Willand	1764–1866	Devon RO, 181M E1–3, E8, E11, E18
Dorset		
Bloxworth	1781–3	RUL, DOR 8/1/1
Charlton Marshall	1784–90	Dorset RO D 1/5339
Maiden Newton	1744–6	Dorset RO, D 475 E9
Puddletown	1787–1805	Northumberland RO, ZCU 12–27
Shapwick	1749–78	Dorset RO, D 755/1
Stoke Abbot	1841–74	Dorset RO, D SSA E1

Location	Date	Reference
Dorset		
Tarrant Monkton	1858–95	RUL, DOR 5/1/3-4
Winterborne St Martins, Charminster	1791–1818	Dorset RO, D ASH B E20
Wool	1758	Dorset RO, D 406/1
Durham		
Brancepeth	1861–78	Durham RO, D/Br E225
Castle Eden	1890–1903	Durham RO, D/CE/28
Ryton	1844–87	Durham RO, NCB I/ SC/92/6 and 366
Silksworth	1828–33	Durham RO, D/Ph 61/1
Westgate in Weardale	1913–14	RUL, DUR 1/1/1-2
Essex		
Audley End	1822	Essex RO, D/D By A302
Bocking	1794–1888	Essex RO, D/D Ta A15-16, A33
Bocking	1812–44	RUL, ESS 11/1
Bulphan	1740–70	RUL, ESS P 243/2 and 3
Castle Hedingham	1787–1808	Essex RO, D/D Mh E61-4
Elmdon	1907–14	Essex RO, D/DU 508/11-17
Fingringhoe, Peldon, West Mersea	1807–46	Essex RO, D/DU 251/2-7, 93-4
Great Henny	1838–82	Essex RO, D/DU 441/52-75
Halstead	1805–36	Essex RO, D/De A3
Harlow	1807–22	Essex RO, D/DU/676/1
High Easter	1898–1906	Essex RO, D/DU 246/2
High Easter	1827–35	Essex RO, D/DU 392/45-7
Layer Marney	1836–56	RUL, ESS P300
Little Dunmow	1817–45	Essex RO, D/D Zu 214
Loughton	1815–23	Essex RO, D/D Ht A1/1
Rochford	1847–76	RUL, ESS 18/3/1
Rochford	1862–1914	RUL, ESS P264/1
Upminster	1797–9	Essex RO, D/D Jn E5

Location	Date	Reference
Essex		
Upminster	1794–9	RUL, ESS P 243/4
Wickham Bishops, Hatfield Peverel	1819–62	Rothamsted Institute of Arable Crop Research, B/9/3
Wickham St Pauls	1800–25	RUL, ESS P250/2 and 3
Witham	1845–54	Essex RO, D/D Bs A3–6
Gloucestershire		
Arlingham	1851	Gloucestershire RO, D4124/3
Blockley	1835–72	Worcestershire RO, b 705:66 BA4221/42, 52(i–ii)
Longhope	1878–90	Gloucestershire RO, D4380/1/1–5
Lower Swell, Maugersbury	1822–9	RUL, GLO 1/2/1–9
Oddington	1779–1853	RUL, GLO 3/2/1
Snowshill	1854–94	Gloucestershire RO, D2267/E23
Stoke Orchard	1751–78	Gloucestershire RO, D2375/E12
Stonehouse	1711–20	Gloucestershire RO, P316 IN 3/1
Westington	1792–1807	Gloucestershire RO, D5042/A3
Hampshire and Isle of Wight		
Arreton	1730–83	Isle of Wight RO, BRS/J/1–2
Binstead	1820–84	Hampshire RO, 4M51/288–94
Candover	1755	Hampshire RO, 28M82/F1
Godshill, Arreton	1824–67	Isle of Wight RO, BRS/B1–44
Hackwood, Wallop	1860–3	Hampshire RO, 11M49/69/1
Kings Somborne, Compton, Horsebridge	1711–1806	Hampshire RO, 2M37/148–51, 338–42
Kings Somborne, Compton, Horsebridge	1711–1830	Hampshire RO, 35M63/1–43, 69–92
Lockerley	1882–1907	Hampshire RO, 13M74/E1
Meonstoke, Hambledon	1864	Hampshire RO, 53M63/43
Micheldever, East Stratton	1736–60	Hampshire RO, 149M89/R4/6063 and 6040

APPENDIX 1

Location	Date	Reference
Hampshire and Isle of Wight		
Nether Wallop, Pittleworth, Kings, Somborne	1815–67	RUL, HAN 11/1/1 and 11/2/1
North Stoneham	1701	Hampshire RO, 102M71/T133
North Waltham	1786–1823	RUL, HAN P285/1–2
Pibworth, Stockbridge, Bullington, Easton	1872–95	Hampshire RO, 60M77/F1–2
Pittleworth	1850	Hampshire RO, 18M54 coffer 6 box H
Pittleworth, Broughton, Wallop	1825–67	Hampshire RO, 96M88/1–2
South Warnborough	1888	Hampshire RO, 3M48/XIV 308
South Wonston	1855–65	RHC, D 93/10
Whitchurch	1894–1907	Hampshire RO, 83M76/PZ15
Whitchurch	1845–7	RUL, HAN 9/1/1
Yaverland, Great Briddlesford, Godshill	1811–69	Isle of Wight RO, WHT 1–6
Herefordshire		
Moccas	1786–1813	Herefordshire RO, J56/III/116–17
Hertfordshire		
Hatfield	1760–1	RUL, HERT 1/1/1
Hitchin	1898–1913	RUL, HERT 4/1/1
Hitchin, Wellbury	1720–42	Hertfordshire RO, D/ER E94; E101
Huntingdonshire		
Somersham	1771–98	Norfolk RO, MC 64/19
Kent		
Ash	1812–57	RUL, KEN 4/3/1–4 and 9; 4/5/1–2; 4/6/1; 4/7/2
Burmarsh	1696–1775	RUL, KEN 19/1/1

Location	Date	Reference
Kent		
East Sutton	1722–48	Centre for Kentish Studies, U120 A17–19
Milstead	1722–86	Centre for Kentish Studies, U593 A3, 7, and 10
Rodmersham	1829–44	Rothamsted Institute of Arable Crop Research, C/38/1
Tudeley	1744–58	RUL, KEN 13/1/1
Lancashire		
Atherton	1800–29	Lancashire RO, DD Li Box 92
Bashall (formerly Yorkshire, West Riding)	1807–34	Lancashire RO, DDX 8/95
Bury	1805–7	Manchester Local Studies Unit Archives, City Library, CL, M39/13/17–20
Didsbury	1791–1838	Manchester Local Studies Unit Archives, City Library, M62 1/1–3; 1/2/1–2
Dumplington	1708–74	Manchester Local Studies Unit Archives, City Library, MS 942.72 H 43
Little Crosby	1748–70	Lancashire RO, DD Bl 54/11
nr. Liverpool	1811–13	RUL, LAN 1/1
Rufford	1821–6	Lancashire RO, DD He 62/25–8
Scarisbrick	1788	Lancashire RO, DD SC 78/3
Leicestershire		
Beaumont Leys	1804–30	RUL, LEI 4/1/1–2
Breedon-on-the-Hill	1759–1820	Leicestershire, Leicester and Rutland RO, DE 1563/135, 152
Breedon-on-the-Hill, Tonge, Wilson	1787–1810	Leicestershire, Leicester and Rutland RO, DE 41/1/197/4, 6
Burton on the Wold, Cotes, Prestwold, Hoton, Loughborough	1774	Leicestershire, Leicester and Rutland RO, DE 258/D/3

Location	Date	Reference
Leicestershire		
Chadwell	1755–90	Leicestershire, Leicester and Rutland RO, 11 D51
Owston	1786	Leicestershire, Leicester, and Rutland RO, DG 7/1/80
Lincolnshire		
Burwell, Sudbrooke	1835–1914	Lincolnshire AO, Scorer Farm 1/1–47; 2/1–10
Dunholme, Washingborough	1860–92	Lincolnshire AO, WEST 1–30
Greetham	1832–61	Sheffield AO, EM 823–25
Holton le Moor, Thornton, Nettleton, Riby, Searby	1755–1864	Lincolnshire AO, Dixon 4/1–3; 5/2/15; 5/5/1; 5/6/2
Knaith	1810	Lincolnshire AO, 2 T d'E/D/6
Langton	1690–1727	Lincolnshire AO, MASS 28/1–3
Leverton	1826	Lincolnshire AO, PSJ 8/D/4
Nocton Rise	1826–45	RUL, LIN P323
Redbourne	1854–86	Lincolnshire AO, RED 3/1/4/6/1
Norfolk		
Clenchwarton	1810–11	Norfolk RO, MS 576
Clenchwarton	1816–17	Norfolk RO, NRS 21432 39C
Costessey	1817–27	Norfolk RO, BR 126/1
Felbrigg	1802–9	Norfolk RO, WKC 5/233
Garboldisham	1792–1824	Norfolk RO, BR 149/1
Gaywood	1825–30	Norfolk RO, BL XII k/10–12
Geldeston	1810–79	Suffolk RO, Lowestoft, AR 270
Great Witchingham	1825–82	Norfolk RO, MC 561/47–51
Kettlestone	1795–1804	Norfolk RO, PD 610/22
Kirstead cum Langhale	1742–8	Norfolk RO, PD 300/14
Langham	1768–1855	Norfolk RO, MC 120/1–82

Location	Date	Reference
Norfolk		
Ludham	1806–18	RUL, NORF 19/1
Morningthorpe	1798–1827	Norfolk RO, MC 150/51–2
Moulton	1823–34	RUL, NORF 23/1/1
Thurgarton	1818–34	Norfolk RO, MC 259/69/72
Wacton	1795–1800	Norfolk RO, BR 187/1
Wymondham	1794–9	Norfolk RO, MC 216
Northamptonshire		
Ashby St Ledgers	1747–77	Northamptonshire RO, ASL 1227
Grafton	1726	Northamptonshire RO, G3883
Little Houghton	1853–96	Northamptonshire RO, ML 1226–9
Towcester, Abthorpe	1864–84	RUL, OXF 2/2/4
Wickham, Deanshanger	1843–50	Northamptonshire RO, XYZ 1336
Wittering	1836–69	RUL, NORTHAM 1/1/1
Northumberland		
Bamburgh, Middleton	1676–1857	Berwick-upon-Tweed RO, ZSI/1–4, 47, 75–7, 81–2
Castle Heaton	1904–14	Northumberland RO, 302/21
Castle Heaton	1906–12	Northumberland RO, 4539/1
Coldstream, Ford	1760–4	Northumberland RO, 2/DE/19/3–5, 17
East and West Brunton	1823–7	Berwick-upon-Tweed RO, 1978 3/1
Ford	1784–1867	Berwick-upon-Tweed RO, ZBE 4
Ford	1760–4	Northumberland RO, ZDE 19/4/3, 17
Hexham	1826–56	Northumberland RO, ZHE 34
Hexham	1822–30	Northumberland RO, ZMD 81
Newbiggin	1861–8	Northumberland RO, 3480/1
Park End	1859–65	Northumberland RO, 414/7–14
Thornbrough	1853–1903	RUL, NORTHUM 2/3/1–2

Location	Date	Reference
Nottinghamshire		
Bingham	1892–1901	Nottinghamshire AO, DD 1405/2–3
Chilwell	1827–33	Nottinghamshire AO, DD 1089/5
Collingham	1748–62	RUL, NOT 4/1/1–5
Cropwell Butler	1846–7	Nottinghamshire AO, DD 872/16
Nottingham	1830	Nottinghamshire AO, DD TB 4/1/34/1
Shelford	1789–1849	Nottinghamshire AO, DD 69/1–2
Strelley	1795–1804	Nottinghamshire AO, DDE 1/12; 3/24
West Bridgford	1847	Nottinghamshire AO, DD T 4/1/33/1
Oxfordshire		
Adderbury	1868–1914	RUL, OXF 19/1/19; 19/2/1–3
Adderbury, Deddington	1852–67	RUL, OXF 16/1
Bloxham	1827–45	Warwickshire County RO, CR 1635/125, 126, 128, 135
Ditchley	1781–6	Oxfordshire AO, DIL/C/22a
Kingham	1851–6	Oxfordshire AO, Chap XIV/i/2a
Shilton	1896–1907	RUL, OXF 17/2/1
Sibford Ferris	1774–1893	RUL, OXF 14/2/1–10; 14/3/1–5
Sibford Ferris	1807–37	RUL, OXF 20/1/1
Standlake	1785–94	Oxfordshire AO, P4/2/F3/1
Rutland		
Preston	1812–1901	Leicestershire, Leicester, and Rutland RO, DE3858/1–6
Shropshire		
Bridgnorth	1744–69	RUL, SAL 5/1
Ellesmere, Houlston (sometimes Houghton or Haughton)	1849–50	Shropshire Records and Research Centre, 611, bundle 337
Moreton Say	1747–64	Shropshire Records and Research Centre, 2125/1–2

Location	Date	Reference
Shropshire		
West Drayton	1824–33	Shropshire Records and Research Centre, 2600/1–3
Whitchurch	1829–37	Shropshire Records and Research Centre, 212, bundle 361
Wrockwardine	1873–93	RUL, SAL P430/4
Somerset		
Bishop's Lydeard, Uphill	1793–1802	Somerset AO, DD/FS 5/1–5, 7–8; 6/1–2; 7/1–3, 5
Crowcombe	1740	Somerset AO, SS/TB 14/5
Dulverton	1826–48	Somerset AO, DD/X/PPP
Dunster Castle	1718–1803	Somerset AO, DD/L 1/5/16–17
Middlezoy	1812–40	Devon RO, 880M E4
Nettlecombe, Bishop's Lydeard	1806–15	Somerset AO, DD/WO 51/9, 11–12, 19
South Petherton	1783–1809	Somerset AO, DD/MR 110
Taunton	1828–86	Somerset AO, T/PH/BGE/1–3
Trull	1770–1861	Somerset AO, DD/CT 34–8
Staffordshire		
Abbots Bromley	1775–1851	RUL, STA P262/1–2
Litchfield	1730–1	Staffordshire RO, D661/8/2/3
Milwich	1831	Staffordshire RO, D864/1/2/1
Suffolk		
Badmondisfield	1787–1843	Suffolk RO, Bury St Edmunds, T924/1341/518
Bulcamp, Darsham, Henham	1793–1868	Suffolk RO, Ipswich, HA11 C6/28, C24/4/1 and C36/2/1
Buxhall	1862–70	Suffolk RO, Bury St Edmunds, 1198
Coddenham	1798–1800	Suffolk RO, Ipswich, HA24/50/19/3.26
Flempton	1829–35	Suffolk RO, Bury St Edmunds, FB 76 C1/9

APPENDIX I

Location	Date	Reference
Suffolk		
Kessingland	1839–53	Suffolk RO, Ipswich, JA1/59
Mettingham (Bungay)	1837–48	Suffolk RO, Lowestoft, 1057/2/6 and 7
Playford	1814–33	Suffolk RO, Ipswich, HA2 A3/1 and 5
Playford	1806–25	RUL, SUF 5/5/2
Redgrave	1796–1817	Suffolk RO, Ipswich, HD79/AD4/2/1 and HD79/AD4/3/1
Rickinghall, Swilland, Wortham	1860–1914	Suffolk RO, Ipswich, HC 423 B1/1
Stansfield	1766–80	Suffolk RO, Bury St Edmunds, FL 627/3/18, 21
Stanton All Saints	1713–21	Suffolk RO, Bury St Edmunds, E1/11
Sudbourne	1851–6	Suffolk RO, Ipswich, HA28/50/23/4.4 (2)
Surrey		
Pyrford	1814–18	RUL, SUR 2/1/1–2
Shirley	1852–3	RUL, SUR 3/1/2
Sussex		
Chailey	1736–49	East Sussex RO, HOOK 16/2–3
Goring	1864–95	West Sussex RO, Add. MSS 22777–9
Hurstpierpoint	1727–50	East Sussex RO, DAN 2199 and 2201
Sompting	1840–94	West Sussex RO, Add. MSS 22768–73; 22776
West Dean	1753–1831	West Sussex RO, MP 1477
Warwickshire		
Castle Bromwich	1778–98	Warwickshire RO, 2374/1
Eathorpe	1824–55	RUL, WAR 2/2

Location	Date	Reference
Warwickshire		
Weethley	1803–52	Warwickshire County RO, CR 394/37–8
Wootton	1806–7	RUL, WAR 1/1/1
Westmorland		
Kirkby Thore	1806–60	RUL, WES P242/1–2
Wiltshire		
Aldbourne, Baydon	1727–1914	RUL, WIL11/2/2–13, 16, 18–26; 11/4/1–8
Bishops Cannings	1825–8	Wiltshire and Swindon RO, 2574/4
Bratton	1824–76	Wiltshire and Swindon RO, 929/22–24, 31–2
Crudwell	1827–44	Wiltshire and Swindon RO, 392/17/6
Enford	1729–37	Wiltshire and Swindon RO, 415/89
Kingston Deverill	1867–9	RUL, WIL 6/2/2; 6/4/1–3
Wingfield	1871–1914	RUL, WIL 9/1/1
Wishford	1803–20	Wiltshire and Swindon RO, 1785/1
Worcestershire		
Castlemorton	1824–89	Worcestershire RO, 705:351/(iv); BA 2044
Church Lench	1845–73	Worcestershire RO, 705:379; BA 3565
Church Lench	1866–73	Worcestershire RO, 705:47/3 BA 1551
Clifton	1755–70	Northamptonshire RO, T (KEL)123–4, 126
Ombersley	1761–1801	Buckinghamshire RO, D/DA/170
Wolverley	1847	Worcestershire RO, b705:198; BA 1290
Yorkshire		
Bradfield	1803–8	Sheffield AO, MD 3518
Bramley	1780–9	Sheffield AO, MD 6913

Location	Date	Reference
Yorkshire		
Burton Constable	1806	East Riding of Yorkshire AO, DD CC (2) 19/6
Campsell	1767–8	Sheffield AO, BFM 1284
Cleveland	1799–1810	RUL, YOR 13/1/1
North Frodingham	1831–2	East Riding of Yorkshire AO, DD SA 1056
Heslington	1858–88	Rothamsted Institute of Arable Crop Research, D/90/6
High Hoyland	1794–1853	RUL, YOR P338/1
Hook	1752	East Riding of Yorkshire AO, DD SE (2) 30/9
Houndhill (in Darfield)	1802–17	Sheffield AO, EM 345
Loftus	1890	RUL, YOR 4/1/1; 3/1
Londesborough	1894–1914	RUL, YOR 17/1/1
North Holme	1819–51	RUL, YOR P389/1
Rudby	1821–4	RUL, YOR 10/1/1
Saltmarshe	1801–46	East Riding of Yorkshire AO, DD SA 1203/1–6
Shelley	1901–7	RUL, YOR 8/1/1–3; 8/2/1
Wentworth, Badsworth, Hoyland, Tankersley, Wath upon Dearne	1767–77	Sheffield AO, WWM A/1380, A/1493
Worsbrough Bridge	1803–20	Sheffield AO, EM 685

Appendix 2

Measurement and Weighting Problems in Farm Records

One of the most complex problems we faced in using farm records was inconsistency of unit, whether in terms of area or measure.

The size of the acre and the volume measure used varied over time and space. Although the acre was officially standardized by Edward I at 40 roods by 4 roods, or 4,840 square yards, there were many customary variations. In Lancashire in the eighteenth century, a customary acre of 7,866 sq. yards was used.[1] The same sized acre, and known as the Cheshire acre, was used in Cheshire at the beginning of the nineteenth century at the time when the 1801 crop returns were collected.[2] At West Dean in Sussex an acre of 107 rods, or two-thirds of a statute acre, was used.[3] In Dorset, the pre-enclosure customary acre in use at Charlton Marshall was *c*.63 per cent of a statute acre.[4] At other locations in Dorset and Wiltshire, the customary acre was 3,630 sq. yards or 75 per cent of an Imperial acre.[5] According to Primrose McConnell's *Agricultural Note-Book*, customary acres in England ranged from just under 48 per cent of an Imperial acre in Leicestershire to nearly 212 per cent in Cheshire and Staffordshire.[6]

Crop volumetric measures also varied over time and location, and they also varied from crop to crop. Wheat might be measured in bushels, but the size of the bushel varied. In 1824, the Act for Ascertaining and Establishing Uniformity of Weights and Measurements replaced the 7.76-Imperial

---

[1] Lancashire RO, DD Bl 54/11.
[2] See M. E. Turner (ed.), *Home Office Acreage Returns: (HO67) List and Analysis Part 1 Bedfordshire–Isle of Wight* (London: List and Index Society, 189, 1982), 71–3.
[3] West Sussex RO, MP1477.
[4] Dorset RO, D1/5339.
[5] P. McConnell, *Note-Book of Agricultural Facts and Figures for Farmers and Farm Students* (7th edn. London, 1904), 32.
[6] Ibid. 32.

APPENDIX 2

Table A2.1 Variations in measurements

Region	Unit	Equivalent
Worcestershire, Oxfordshire	bag	3 bushels
Cambridgeshire, Suffolk	coomb	4 bushels
Hampshire, Wiltshire, Essex	sack	4 bushels
Kent	seam	5½ bushels
Kent	tovet	½ bushel
Ford, Northumberland	boll	2 bushels
Middleton, Northumberland	boll	6 bushels
Northumberland	old boll	8 bushels
Sheffield area	load	3 bushels
Hampshire, Isle of Wight, Sussex	load	5 quarters
Nottinghamshire	strike	1 bushel

Sources Worcestershire RO, BA 1551 705:41/3; Warwickshire RO, CR 1635/128; Cambridgeshire RO, R55/7/8/15; Suffolk RO, Ipswich, HC 423/B1/1; Wiltshire RO, 929; Essex RO, D/DU 251; Centre for Kentish Studies, U593/A3; Northumberland RO, 414/14; Berwick-upon-Tweed RO, ZSI/82; Northumberland RO, ZDE 19/4/17; Sheffield AO, MD3518 and WWM A 1380; Hampshire RO, 2M37/342; Isle of Wight RO, BRS/J/1; East Sussex RO, Hook; Nottinghamshire AO, DD 872/16.

gallon, Winchester bushel, with the 8-gallon Imperial bushel. From this time the Imperial bushel became the only legal bushel, though other-sized bushels had been commonly used.[7] At West Dean in Sussex, for example, the bushel measure for wheat changed from 9 to 8 gallons in 1789.[8] In some locations, the bushel was given a weight equivalent. For example, in Newton, Cumberland, the bushel weight for wheat was 14 stone.[9] Many other measures were used beside the bushel. These were often local and not standardized around the country. Table A2.1 lists some of the units found in the archives compared with the Imperial equivalent.[10] The variations can be

---

[7] C. M. Watson, *British Weights and Measures as Described in the Laws of England from Anglo-Saxon Times* (London: John Murray, 1910), 85, 97.

[8] West Sussex RO, MP1477.

[9] Cumbria RO, Carlisle, DX/408/1.

[10] See also an indication of the variations in these volumetric measures at the time of the 1801 crop returns in M. E. Turner, J. V. Beckett, and B. Afton, 'Taking Stock: Farmers, Farm Records, and Agricultural Output in England, 1700–1850', *AgHR* 44 (1996), 30–1.

confusing. For example, in one document from Ford in Northumberland, both the old boll and the new boll were used.[11]

The measurement of livestock weights was also complex, as we show in Chapter 6.

There are several ways to resolve the problems caused by the different units of measure. Often individual documents provided clues. For livestock weights, the unit could often be found by comparing the price per pound with the total price, and hence deriving the total weight of the animal. With corn, the unknown unit could sometimes be calculated in much the same way. At North Waltham, Hampshire, for example, grain was measured in loads, quarters, and sacks. The maximum number of quarters used before a load was registered was four.[12] This suggests that there were five quarters in a load. The resulting calculation can be checked using contemporary printed sources. Appendix A to the *Second Report of the Parliamentary Commissioners on Weights and Measures* (1820) provides an index to the various units in use in the late eighteenth century and early nineteenth. These were largely extracted from the *General Views* of the various counties published by the Board of Agriculture.[13]

Many of the nineteenth-century cyclopedias and notebooks have similar, though generally less complete lists. One of the best can be found in John Morton's *Cyclopedia of Agriculture*.[14] As useful as these lists and indexes are, it would have been a mistake to rely solely on them since the most important evidence was found within individual documents. A single example can demonstrate this. In Dorset, according to the Index from the Commissioners on Weights and Measures, a customary acre was generally 134 perches or 83.75 per cent of a statute acre. According to Primrose McConnell, the acre in Wiltshire and Dorset was 75 per cent of a statute acre. However, to use either of these rates to convert the area found in a document from Charlton Marshall in Dorset would result in a considerable error in yield. On several pages of the document the farmer shows the statute equivalent relative to the customary acre used on the farm. The amount used varied slightly but his measure of the customary acre was approximately 63 per cent of a statute acre.[15]

---

[11] Northumberland RO, ZDE 19/4/17.   [12] RUL, HAN P 285.
[13] *Second Report of the Commissioners on Weights and Measures*, BPP, VII (1820), appendix A. Reproduced in *AHEW* vi. 1117–55.
[14] J. C. Morton (ed.), *Cyclopedia of Agriculture* (London, 1855), 417–18, 720–7, 1123–7.
[15] *AHEW* vi. 1117; McConnell, *Note-Book*, 32; Dorset RO, D1/5339.

Bibliography

I. Archive sources not otherwise listed in Appendix 1

Public Record Office, Kew

Home Office Papers: Domestic Correspondence, George III, Letters and Papers:
 HO/42/52; October 1800.
 HO/42/53–4; November 1800.
 HO/42/55; December 1800.
 HO/67; Parish Acreage Returns, 1801.
Privy Council and Privy Council Office: Minutes and Associated Papers;
 PC 4/6; August–November 1795.
 PC 4/9; 1799.
Privy Council and Privy Council Office: Miscellaneous Unbound Papers:
 PC 1/31/73–80; includes papers on scarcity of corn, December 1795–January 1796.

Rural History Centre, Reading University

The Working Papers of J. R. Bellerby—Agricultural Economist:
 Bellerby MSS, D84/8/1–24, esp. 17.

II. Parliamentary Papers (BPP) (in date order)

Second Report of the Commissioners on Weights and Measures. VII (1820), appendix A. Reproduced in *AHEW* vi. 1117–55.
Census of Ireland for the Year 1841. XXIV (1843).
Census of Great Britain, 1851: Population Tables. LXXXIII (1852–3).
Report from the Select Committee Appointed to Inquire into the Present State of Agriculture. [612], V (1833).
First Report from the Select Committee on the State of Agriculture: with Minutes of Evidence and Appendix. [79], VIII, part 1 (1836).

Second Report from the Select Committee on the State of Agriculture. [189], VIII, part 1 (1836).
Third Report from the Select Committee on the State of Agriculture: with Minutes of Evidence, Appendix and Index. [465], VIII, part 2 (1836).
Report from the Select Committee of the House of Lords on the State of Agriculture in England and Wales. [464], V (1837).
Reports by Poor Law Inspectors on Agricultural Statistics (England), 1854. LIII (1854–5). Reproduced in *AHEW* vi. 1042–4; *AHEW* vii. 1768–9.
Accounts and Papers, *Annual Agricultural Statistics*. From 1884 for crop yields.
Royal Commission on Agriculture, *Garstang and Glendale: Report by Assistant Commissioner Mr Wilson Fox*. C. 7334, XVI, part 1 (1894).
——*Andover and Maidstone: Report by Assistant Commissioner Dr W. Fream*. C. 7365, XVI, part 1 (1894).
——*Frome and Stratford-on-Avon: Report by Assistant Commissioner Mr Jabez Turner*. C. 7372, XVI, part 1 (1894).
——*Isle of Axholme and Essex: Report by Assistant Commissioner Mr R. Hunter Pringle*. C. 7374, XVI, part 1 (1894).
——*Suffolk: Report by Assistant Commissioner Mr A. Wilson Fox*. C. 7755, XVI (1895).
——*Lincolnshire: Report by Assistant Commissioner Mr A. Wilson Fox*. C. 7671, XVI (1895).
Royal Commission on Agriculture, *Final Report of Her Majesty's Commissioners Appointed to Inquire into the Subject of Agricultural Depression*. C. 8540, XV (1897).
The Agricultural Output of Great Britain, 1908. Cd. 6277, X (1912–13).
Second Report by Mr A. Wilson Fox on the Wages Earnings and Conditions of Employment of the Agricultural Labourers in the United Kingdom. Cd. 2376, XCVII (1913).

III. Contemporary printed sources, pre-1914

ACLAND, T. D., 'On the Farming of Somersetshire', *JRASE* 1st ser. 11 (1850), 666–764.
BEAR, W. E., 'The Future of Agricultural Competition', *JRASE* 3rd ser. 2 (1891), 742–71.
—— 'The Survival in Farming', *JRASE* 3rd ser. 2 (1891), 257–75.
—— 'Agricultural Competition', *JRASE* 3rd ser. 70 (1909), 151–63.
BEDFORD, Duke of, *A Great Agricultural Estate* (London: John Murray, 1897).
BELL, T. G., 'A Report upon the Agriculture of the County of Durham', *JRASE* 1st ser. 17 (1856), 86–123.
BLITH, W., *The English Improver Improved or the Survey of Husbandry* (London: J. Wright, 1652).

BROWN, R., *The Compleat Farmer; or, The Whole Art of Husbandry* (London: J. Coote, 1759).

BROWN, T., *General View of the Agriculture of the County of Derbyshire* (London, 1794).

BURKE, J. F., *British Husbandry: Exhibiting the Farming Practices in Various Parts of the United Kingdom*, 3 vols. (Library of Useful Knowledge, Farmer's Series: London: Baldwin & Craddock, 1834-40).

CAIRD, J., *English Agriculture in 1850-51* (2nd edn. London: Longman, 1852; repr. London: Frank Cass, 1968).

—— 'On the Agricultural Statistics of the United Kingdom', *JRSS* 31 (1868), 127-45.

CHRISTISON, J., *New Tables for Computing the Weight of Cattle by Measurement* (Edinburgh; Stirling Kennedy & Co., 1839).

CLARKE, J. A., 'Practical Agriculture', *JRASE* 2nd ser. 14 (1878), 445-642.

COBBETT, W., *Rural Rides*, i (Everyman edn. London: J. M. Dent & Sons, 1912).

COLEMAN, J., 'Farm Accounts', *JRASE* 1st ser. 19 (1858), 122-43.

—— 'The Breeding and Feeding of Sheep', *JRASE* 1st ser. 24 (1863), 623-38.

COMBER, W. T., *An Inquiry into the State of National Subsistence, as Connected with the Progress of Wealth and Population* (London: T. Cadell & W. Davies, 1808).

The Complete Farmer; or, General Dictionary of Agriculture and Husbandry, 2 vols. (5th edn. London: R. Baldwin, 1807).

CRAIGIE, P. G., 'Taxation as Affecting the Agricultural Interest', *JRASE* 2nd ser. 14 (1878), 385-424.

—— 'Statistics of Agricultural Production', *JRSS* 46 (1883), 1-58.

—— 'On the Production and Consumption of Meat in the United Kingdom', *Report of the British Association for the Advancement of Science* (London, 1884), Section F, 841-7.

CRAWFORD, R. F., 'An Inquiry into Wheat Prices and Wheat Supply', *JRSS* 58 (1895), 75-111.

CURWEN, J. C., *Hints on Agricultural Subjects, and on the Best Means of Improving the Condition of the Labouring Classes* (2nd edn. London: J. Johnson, B. Crosby & Co., 1809).

DAVIS, T., Sr., *General View of the Agriculture of the County of Wiltshire* (London, 1794).

DAVIS, T., Jr., *General View of the Agriculture of the County of Wiltshire* (London, 1811).

DICKSON, R. W., *Practical Agriculture; or, A Complete System of Modern Husbandry with the Best Methods of Planting and the Improved Management of Livestock*, 2 vols. (2nd edn. London: Richard Phillips, 1814).

DONALDSON, J., *General View of the Agriculture of the County of Northampton* (London, 1794).

ELLIS, W., *The Modern Husbandman; or, The Practice of Farming*, 4 vols. (London: T. Osborne, M. Cooper, 1742-4).

EVERSHED, H., 'The Agriculture of Staffordshire', *JRASE* 2nd ser. 5 (1869), 263-317.

EVERSHED, S., 'On the Improved Methods of Cultivating and Cropping Light Land', *JRASE* 1st ser. 14 (1853), 79-96.

Farmers Magazine, 2 (1801).

FREAM, W., *The Complete Grazier* (14th edn. London: Crosby, Lockwood & Son, 1900).

FRERE, P. H., 'On the Feeding of Stock', *JRASE* 1st ser. 21 (1860), 218–58.

GARNIER, R. M., 'The Introduction of Forage Crops into Great Britain', *JRASE* 3rd ser. 7 (1896), 77–97.

GILBERT, J. H., *Results of Experiments at Rothamsted, on the Growth of Root Crops for Many Years in Succession on the Same Land* (London, 1887).

—— and LAWES, J. B., *The Rothamsted Experiments* (Edinburgh, 1895).

GUILHAUD DE LAVERGNE, L. G. L., *The Rural Economy of England, Scotland and Ireland* (Edinburgh, 1855).

HAGGARD, H. RIDER, *Rural England*, 2 vols. (2nd edn. London: Longmans, 1906).

HALE, T., *A Compleat Body of Husbandry* (London: T. Osborne & J. Shipton, 1756).

HALL, A. D., *The Book of the Rothamsted Experiments* (London: John Murray, 1905).

—— *Fertilizers and Manures* (London: John Murray, 1912).

—— *A Pilgrimage of British Farming, 1910–1912* (London: John Murray, 1913).

Hansard's Parliamentary Debates, various.

HAXTON, J., 'Essay on the Cultivation of Oats', *JRASE* 1st ser. 12 (1851), 105–33.

—— 'On Light-Land Farming', *JRASE* 1st ser. 15 (1854), 88–124.

HERBERT, R., 'Statistics of Live Stock and Dead Meat for Consumption in the Metropolis', *JRASE* 1st ser. 20 (1859), 473–81.

HILLYARD, C., *Practical Farming and Grazing, with Observations on the Breeding and Feeding of Sheep and Cattle* (4th edn. London, 1844).

HORNE, T. H., *The Complete Grazier; or, Farmer and Cattle-Dealer's Assistant . . . By a Lincolnshire Grazier* (3rd edn. London: B. Crosby & Co., 1808).

—— *The Complete Grazier and Farmer's and Cattle-Breeder's Assistant: A Compendium of Husbandry* (10th edn. rev. W. Youatt, London; Simpkin, Marshall & Co., 1851).

—— *The Complete Grazier and Farmer's and Cattle-Breeder's Assistant: A Compendium of Husbandry* (12th edn., enlarged by R. S. Burn, London, 1877).

JACOB, G., *The Complete Court-Keeper or Land Steward's Assistant* (London, 1713).

JAMES, W., and MALCOLM, J., *General View of the Agriculture of the County of Surrey* (London, 1794).

JONAS, S., 'On the Farming of Cambridgeshire', *JRASE* 1st ser. 7 (1846), 35–72.

KENT, N., *Hints to Gentlemen of Landed Property* (London, 1775).

LAURENCE, E., *The Duty of a Steward to his Lord* (London: John Shuckburgh, 1727).

LAWES, J. B., 'Report of Experiments on the Comparative Fattening Qualities of Different Breeds of Sheep', *JRASE* 1st ser. 12 (1851), 414–45.

—— 'Experiments on the Comparative Fattening Qualities of Different Breeds of Sheep', *JRASE* 1st ser. 16 (1855), 45–87.

—— and GILBERT, J. H., 'On the Composition of Oxen, Sheep, and Pigs, and of their Increase whilst Fattening', *JRASE* 1st ser. 21 (1860), 433–88.

LAWES, J. B., and GILBERT, J. H., 'Home Produce, Imports, Consumption, and Price of Wheat, over Forty Harvest-Years, 1852–53 to 1891–92', *JRASE* 3rd ser. 4 (1893), 77–133.

—— —— 'The Feeding of Animals for the Production of Meat, Milk and Manure, and for the Exercise of Force', *JRASE* 3rd ser. 6 (1895), 47–146.

The Library of Agricultural and Horticultural Knowledge, with a Memoir by Mr Ellman (3rd edn. Lewes: J. Baxter, 1834).

LISLE, E., *Observations in Husbandry*, 2 vols. (2nd edn. London, 1757).

LOUDON, J. C., *An Encyclopaedia of Agriculture* (2nd edn. London, 1831).

Low, D., *Elements of Practical Agriculture* (Edinburgh, 1834).

LUCCOCK, J., *The Nature and Properties of Wool* (Leeds, 1805).

—— *An Essay on Wool* (London, 1809).

MCCONNELL, P., *Note-Book of Agricultural Facts and Figures for Farmers and Farm Students* (7th edn. London: Crosby, Lockwood & Son, 1904; 10th edn. 1922).

MCCULLOCH, J. R., *A Statistical Account of the British Empire*, 2 vols. (London: The Society for the Diffusion of Useful Knowledge, 1837).

MALDEN, W. J., *Sheep Raising and Shepherding* (London: Gill, 1899).

MARSHALL, W., *The Rural Economy of Norfolk*, 2 vols. (London, 1781, 1787).

—— *The Rural Economy of Yorkshire*, 2 vols. (London, 1788).

—— *The Rural Economy of Gloucestershire*, 2 vols. (Gloucester, 1789).

—— *The Rural Economy of the Midland Counties*, i (London: G. Nicol, 1790).

—— *The Rural Economy of the West of England* (London: G. Nicol, 1796).

—— *The Rural Economy of the Southern Counties*, 2 vols. (London: G. Nicol, 1798).

—— *Minutes, Experiments, Observations, and General Remarks, on Agriculture, in the Southern Counties, a New Edition. To which is Prefixed a Sketch of the Vale of London, and an Outline of its Rural Economy*, i (London: G. Nicol, 1799).

MEATS, T. W., *Agricultural Accounts* (4, The Accountants Library London: Gee, 1901).

MECHI, J. J., *How to Farm Profitably* (4th edn. London, 1864).

MONK, J., *General View of the Agriculture of the County of Leicester* (London, 1794).

MORDANT, J., *The Complete Steward; or, The Duty of a Steward to his Lord*, 2 vols. (London, 1761).

MORTON, J. C., 'On Increasing our Supplies of Animal Food', *JRASE* 1st ser. 10 (1849), 341–79.

—— (ed.), *Cyclopedia of Agriculture* (London, 1855).

—— *Handbook of Farm Labour* (2nd edn. London, 1868).

—— *Labour on the Farm* (London, 1887).

PELL, A., 'On Weighing Live-Stock', *JRASE* 2nd ser. 25 (1889), 447–72.

'Progress and Present State of Agriculture', *Edinburgh Review*, 62 (Jan. 1836), 319–45.

PUSEY, P., 'On the Progress of Agricultural Knowledge during the Last Four Years', *JRASE* 1st ser. 3 (1842), 169–217.

PUSEY, P., 'Evidence on the Antiquity, Cheapness and Efficacy of Thorough-Draining, or Land-Ditching as Practised Throughout the Counties of Suffolk, Hertford, Essex, and Norfolk, with some Notice of Improved Machines for Tile-Making', *JRASE* 1st ser. 4 (1843), 23–49.

RANDELL, C., FRANKISH, W. M., and WARREN, R. A., 'Report of the Judges of Book-Keeping', *JRASE* 2nd. ser. 19 (1883), 693–702.

READ, C. S., 'On the Farming of Oxfordshire', *JRASE* 1st ser. 15 (1854), 189–276.

—— 'Recent Improvements in Norfolk Farming', *JRASE* 1st ser. 19 (1858), 265–311.

REW, R. H., 'An Inquiry into the Statistics of the Production and Consumption of Milk and Milk Products in Great Britain', *JRSS* 55 (1892), 244–86.

—— 'Production and Consumption of Meat and Milk: Second Report', *JRSS* 67 (1904), 368–84.

—— 'Production and Consumption of Meat and Milk: Third Report', *JRSS* 67 (1904), 385–412.

—— 'Observations on the Production and Consumption of Meat and Dairy Products', *JRSS* 67 (1904), 413–27.

—— 'The Nation's Food Supply', *JRSS* 76 (1912), 98–105.

ROBERTSON, G., *Rural Recollections; or, The Progress of Improvement in Agriculture and Rural Affairs* (Irvine, 1829).

Royal Society, 'Georgicall Account of Devonshire and Cornewall', *Philosophical Transactions of the Royal Society*, 10 (3)/12 (1667).

SMITH, C., *Three Tracts on the Corn-Trade and Corn-Laws* (2nd edn. London, 1766).

SOMERVILLE, Lord, 'The Farmer's, Grazier's and Butcher's Ready-Reckoner; a Short Table, by which the Weight of Stock, According to the Different Usages in England, can be Ascertained; and the Value of Stock of Any Size, with the Difference, at once Discovered', Letters and Papers on Agriculture, *Journal of the Bath and West of England Society*, 9 (1799), 289–91.

SPOONER, W. C., *The History, Structure, Economy, and Diseases of the Sheep* (5th edn. London, 1888).

STEPHENS, H., *The Book of the Farm* (2nd edn. Edinburgh, 1851).

STONE, T., *General View of the Agriculture of the County of Lincoln* (London, 1794).

THOMPSON, H. S., 'Agricultural Progress and the Royal Agricultural Society', *JRASE* 1st ser. 25 (1864), 1–52.

TULL, J., *The Horse-Hoeing Husbandry, to which is Prefixed, an Introduction by William Cobbett* (London: William Cobbett, 1829).

TURNBULL, R. C., 'The Household Food Supply of the United Kingdom', *Transactions of the Highland and Agricultural Society*, 15 (1903), 197–211.

TURNOR, C., *Land Problems and National Welfare* (London: The Bodley Head, 1911).

—— *The Land and its Problems* (London: Methuen, 1921).

VANCOUVER, C., *General View of the Agriculture of the County of Cambridgeshire* (London, 1794).

VOELCKER, J. A., 'On the Commercial Value of Artificial Manures', *JRASE* 1st ser. 23 (1862), 277–86.

WATSON, C. M., *British Weights and Measures as Described in the Laws of England from Anglo-Saxon Times* (London: John Murray, 1910).

WEBB, S. H., *The Practical Farmers' Yearly Account Book* (various edns. London and Norwich, 1845 et seq.).

WEBSTER, C., 'On the Farming of Westmorland', *JRASE* 2nd ser. 4 (1868), 1–37.

WEDGE, T., *General View of the Agriculture of the County Palatine of Cheshire* (London, 1794).

WILKINSON, J., 'The Farming of Hampshire', *JRASE* 1st ser. 22 (1861), 239–346.

WILSON, J., 'On the Various Breeds of Sheep in Great Britain, Especially with Reference to the Character and Value of their Wool', *JRASE* 1st ser. 16 (1855), 222–49.

—— *Our Farm Crops*, 2 vols. (London, 1860).

—— *British Farming: A Description of the Mixed Husbandry of Great Britain* (Edinburgh, 1862).

WILSON, J. M. (ed.), *The Rural Cyclopedia; or, A General Dictionary of Agriculture*, 4 vols. (Edinburgh, 1847–9).

WILSON FOX, A., 'Agricultural Wages in England and Wales during the Last Half Century', *JRSS* 66 (1903), 273–348, repr. in W. E. Minchinton, (ed.), *Essays in Agrarian History*, ii (Newton Abbott: British Agricultural History Society, 1968), 121–98.

WORLIDGE, J. (alias J. W. Gent), *Systema Agriculturae* (4th edn. London, 1687).

WREN HOSKYNS, C., 'On Agricultural Statistics', *JRASE* 1st ser. 16 (1855), 554–606.

WRIGHTSON, J., *Fallow and Fodder Crops* (London: Chapman Hall, 1889).

—— *The Principles of Agricultural Practices as an Instructional Subject* (3rd edn. London: Chapman & Hall, 1893).

—— *Sheep: Breeds and Management* (2nd edn. 1895).

YOUATT, W., *Sheep: Their Breeds, Management, and Disease* (London: Library of Useful Knowledge, Farmers' Series, 1837).

YOUNG, A., *A Six Weeks Tour through the Southern Counties* (London: W. Strahan, 1769).

—— *A Six Months Tour through the North of England*, 4 vols. (London: W. Strahan, 1770).

—— *The Farmer's Tour through the East of England*, i (London: W. Strahan, 1771).

—— 'Experiments in Weighing Fatting Cattle Alive', *Annals of Agriculture*, 14 (1790), 140–63.

IV. Books and articles, post-1914

AFTON, B., 'The Great Agricultural Depression on the English Chalklands: The Hampshire Experience', *AgHR* 44 (1996), 191–205.

AFTON, B., 'Mixed Farming on the Hampshire Downs, 1837–1914' (University of Reading, Ph.D. thesis, 1993).
—— 'Investigating Agricultural Production and Land Productivity: Opportunities and Methodologies Using English Farm Records', paper for the Twelfth International Economic History Congress Session C54 'Production et productivité agricoles dans le monde occidental (XIIe–XXe siècles)' (Madrid, 24 to 28 Aug. 1998).
—— 'Land Productivity in a Lightland Agricultural System: The Hampshire Downs, 1835–1914', in B. J. P. van Bavel and E. Thoen (eds.), *Land Productivity and Agrosystems in the North Sea Area, Middle Ages–Twentieth Century: Elements for Comparison* (Turnhout: Brepols, 1999), 325–36.
—— and TURNER, M. E., 'The Statistical Base of Agricultural Performance in England and Wales, 1850–1914', *AHEW* vii (2000), 1755–2140.
ALLEN, R. C., 'The Efficiency and Distributional Consequences of Eighteenth-Century Enclosures', *Economic Journal*, 92 (1982), 937–53.
—— 'The Growth of Labour Productivity in Early Modern English Agriculture', *Explorations in Economic History*, 25 (1988), 117–46.
—— 'Inferring Yields from Probate Inventories', *JEH* 48 (1988), 117–25.
—— 'The Two English Agricultural Revolutions, 1450–1850', in Campbell and Overton (1991), 236–54.
—— *Enclosure and the Yeoman: The Agricultural Development of the South Midlands 1450–1850* (Oxford: Oxford University Press, 1992).
—— 'Agriculture during the Industrial Revolution', in R. Floud and D. McCloskey (eds.), *The Economic History of Britain since 1700*, i: *1700–1860* (2nd edn. Cambridge: Cambridge University Press, 1994), 96–122.
—— 'Tracking the Agricultural Revolution in England', *EcHR* 52 (1999), 209–35.
—— and Ó GRÁDA, C., 'On the Road again with Arthur Young: English, Irish and French Agriculture during the Industrial Revolution', *JEH* 48 (1988), 91–116.
AMERY, G. D., 'The Writings of Arthur Young', *JRASE* 3rd ser. 85 (1924), 175–205.
ARMSTRONG, A., *Farmworkers in England and Wales: A Social and Economic History, 1770–1980* (London: Batsford, 1988).
ASHCROFT, L. (ed.), *Vital Statistics* (Kendal: Cumberland and Westmorland Antiquarian and Archaeological Society, 1992).
ASHLEY, W., *The Bread of our Forefathers* (Oxford: Oxford University Press, 1928).
ATKINS, P. J., 'The Retail Milk Trade in London, c.1790–1914', *EcHR* 33 (1980), 522–37.
—— 'The Growth of London's Railway Milk Trade, c.1845–1914', *Journal of Transport History*, 4 (1978), 208–26.
BEASTALL, T. W., *The Agricultural Revolution in Lincolnshire* (Lincoln: History of Lincolnshire Committee, 1978).
BECKETT, J. V., 'The Peasant in England: A Case of Terminological Confusion?', *AgHR* 32 (1981), 113–23.

BECKETT, J. V., 'Regional Variation and the Agricultural Depression, 1730-50', *EcHR* 35 (1982), 35-51.
—— *The Agricultural Revolution* (Oxford: Blackwell, 1990).
—— 'Estate Management in Eighteenth-Century England: The Lowther–Spedding Relationship in Cumberland', in J. A. Chartres and D. Hey (eds.), *English Rural Society, 1500–1800* (Cambridge: Cambridge University Press, 1990), 55-72.
—— 'Agricultural Landownership and Estate Management', *AHEW* vii (2000), 693-758.
—— and HEATH, J. E., *Derbyshire Tithe Files 1836-50* (Chesterfield: Derbyshire Record Society 22, 1995).
—— and TURNER, M. E., 'Taxation and Economic Growth in Eighteenth-Century England', *EcHR* 43 (1990), 377-403.
—— 'Freehold from Copyhold and Leasehold: Tenurial Transition in England between the Sixteenth and Nineteenth Centuries' (forthcoming, Leiden, 2001).
BENNETT, M. K., 'British Wheat Yield Per Acre for Seven Centuries', *Economic History*, 3 (1937), 12-29.
BIDDICK, K., 'Agrarian Productivity on the Estates of the Bishopric of Winchester in the Early Thirteenth Century', in Campbell and Overton (1991), 95-123.
BONFIELD, L., SMITH, R. M., and WRIGHTSON, K. (eds.), *The World We Have Gained* (Oxford: Oxford University Press, 1986).
BOWDEN, P. J., 'Agricultural Prices, Wages, Farm Profits, and Rents', *AHEW* v (2) (1985), 1-118.
—— 'Statistics', *AHEW* v (2), (1985), 827-902.
BOWIE, G. G. S., 'New Sheep for Old: Changes in Sheep Farming in Hampshire, 1792-1879', *AgHR* 35 (1987), 15-24.
—— 'Watermeadows in Wessex: A Re-evaluation of the Period 1640-1750', *AgHR* 35 (1987), 151-8.
—— 'Northern Wolds and Wessex Downlands: Contrast in Sheep Husbandry and Farming Practice, 1770-1850', *AgHR* 38 (1990), 117-26.
BRASSLEY, P., 'Farming Techniques', *AHEW* vii (2000), 495-593.
—— 'Agricultural Science and Education', *AHEW* vii (2000), 594-649.
BROAD, J., 'Alternate Husbandry and Permanent Pasture in the Midlands, 1650-1800', *AgHR* 28 (1980), 77-89.
BROWN, J., *Agriculture in England: A Survey of Farming 1870-1947* (Manchester: Manchester University Press, 1987).
—— 'The Malting Industry', *AHEW* vi (1989), 501-19.
—— and Beecham, H. A., 'Arable Farming: Farming Practices', *AHEW* vi (1989), 275-95.
BRUNT, L., 'Nature or Nurture? Explaining English Wheat Yields in the Agricultural Revolution', *University of Oxford Discussion Paper in Economic and Social History*, 19 (1997).

CAMPBELL, B. M. S., 'Land, Labour, Livestock, and Productivity Trends in English Seignorial Agriculture, 1208–1450', in Campbell and Overton (1991), 144–82.

——and Overton, M. (eds.), *Land, Labour and Livestock: Historical Studies in European Agricultural Productivity* (Manchester: Manchester University Press, 1991).

————'Norfolk Livestock Farming, 1250–1740', *Journal of Historical Geography*, 18 (1992), 377–96.

————'A New Perspective on Medieval and Early Modern Agriculture: Six Centuries of Norfolk Farming, c.1250–c.1850', *Past and Present*, 141 (1993), 38–105.

CHALKLIN, C. W., and HAVINDEN, M. A. (eds.), *Rural Change and Urban Growth*, 1500–1800 (London: Longman, 1974).

CHAMBERS, J. D., and MINGAY, G. E., *The Agricultural Revolution, 1750–1880* (London: Batsford, 1966).

CHARTRES, J. A. 'The Marketing of Agricultural Produce', *AHEW* v (2) (1985), 406–502.

——and HEY, D. (eds.), *English Rural Society, 1500–1800* (Cambridge: Cambridge University Press, 1990).

CHORLEY, G. P. H., 'The Agricultural Revolution in Northern Europe, 1750–1880: Nitrogen, Legumes and Crop Production', *EcHR* 34 (1981), 71–93.

CLARK, C., *The British Malting Industry since 1830* (London: Hambledon Press, 1998).

CLARK, G., 'Productivity Growth without Technical Change in European Agriculture before 1850', *JEH* 47 (1987), 419–32.

——'Labour Productivity in English Agriculture, *1300–1860*', in Campbell and Overton (1991), 211–35.

——'Yields Per Acre in English Agriculture 1250–1860: Evidence from Labour Inputs', *EcHR* 44 (1991), 445–60.

——'Agriculture and the Industrial Revolution, 1700–1850', in J. Mokyr (ed.), *The British Industrial Revolution: An Economic Perspective* (Oxford: Westview Press, 1993), 227–66.

——and VAN DER WERF, Y., 'Work in Progress? The Industrious Revolution', *JEH* 58 (1998), 830–43.

COLEMAN, D. C., *The Economy of England 1450–1750* (Oxford: Oxford University Press, 1977).

COLLINS, E. J. T., 'Historical Farm Records', *Archives*, 7 (1966), 143–9.

——'Harvest Technology and Labour Supply in Britain, 1790–1870', *EcHR* 22 (1969), 453–73.

——'The Diffusion of the Threshing Machine in Britain, 1790–1880', *Tools and Tillage*, 2 (1972), 16–33.

——'Dietary Change and Cereal Consumption in Britain in the Nineteenth Century', *AgHR* 23 (1975), 97–115.

——'The Rationality of Surplus Agricultural Labour: Mechanisation in English Agriculture in the Nineteenth Century', *AgHR* 35 (1987), 36–46.

COLLINS, E. J. T., (ed.), *The Agrarian History of England and Wales*, vii: *1850–1914* (Cambridge: Cambridge University Press, 2000).

COOKE, G. W., 'Advice on Using Fertilisers, 1861–1967', *JRASE* 128 (1967), 107–24.

COPPOCK, J. T., 'The Statistical Assessment of British Agriculture', *AgHR* 4 (1956), 4–21.

——and BEST, R. H., *The Changing Use of Land in Britain* (London: Faber & Faber, 1962).

COPUS, A. K., 'Changing Markets and the Development of Sheep Breeds in Southern England, 1750–1900', *AgHR* 37 (1989), 36–51.

CRAFTS, N. F. R., *British Economic Growth during the Industrial Revolution* (Oxford: Oxford University Press, 1985).

CULLEN, L. M., and SMOUT, T. C. (eds.), *Comparative Aspects of Scottish and Irish Economic and Social History 1600–1900* (Edinburgh: University of Edinburgh Press, 1977).

DAUNTON, M. J., *Progress and Poverty: An Economic and Social History of Britain 1700–1850* (Oxford: Oxford University Press, 1995).

DAVIS, S. J. M., and BECKETT, J. V., 'Animal Husbandry and Agricultural Improvement: The Archaeological Evidence from Animal Bones and Teeth', *Rural History*, 10 (1999), 1–17.

DEANE, P., and COLE, W. A., *British Economic Growth 1688–1959* (2nd edn. Cambridge: Cambridge University Press, 1969).

DODD, J. P., 'The Agricultural Statistics for 1854: An Assessment of their Value', *AgHR* 35 (1987), 159–70.

EAGLE, E., 'Some Light on the Beginnings of the Lincolnshire Bulb Industry', *Lincolnshire Historian*, 6 (1950), 220–9.

ENGLISH, B., 'On The Eve of the Great Depression: The Economy of the Sledmere Estate 1869–1878', *Business History*, 24 (1982), 23–47.

FAIRLIE, S., 'The Corn Laws and British Wheat Production 1829–76', *EcHR* 22 (1969), 88–116.

FEINSTEIN, C. H., 'Pessimism Perpetuated: Real Wages and the Standard of Living in Britain during the Industrial Revolution', *JEH* 58 (1998), 625–58.

——and Pollard, S. (eds.), *Studies in Capital Formation in the United Kingdom, 1750–1920* (Oxford: Oxford University Press, 1988).

FLOUD, R., and MCCLOSKEY, D. (eds.), *The Economic History of Britain since 1700*, i: *1700–1860* (1st edn. Cambridge: Cambridge University Press, 1981).

————(eds.), *The Economic History of Britain since 1700*, i: *1700–1860* (2nd edn. Cambridge: Cambridge University Press, 1994).

Fox, H. S. A., 'Devon and Cornwall', *AHEW* iii (1991), 303–23.

——and BUTLIN, R. A. (eds.), *Change in the Countryside: Essays on Rural England 1500–1900* (London: Institute of British Geographers, Special Publication no. 10, 1979).

Fox, N. E., 'The Spread of the Threshing Machine in Central Southern England', *AgHR* 26 (1978), 26-8.
Fussell, G. E., 'The Size of English Cattle in the Eighteenth Century', *Agricultural History*, 3 (1929), 160-81.
—— 'Population and Wheat Production in the Eighteenth Century', *History Teachers' Miscellany*, 7 (1929), serialized, 65-8, 84-8, 108-11, 120-7.
—— (ed.), *Robert Loder's Farm Accounts, 1610-1620* (London: Royal Historical Society, Camden Society 3rd ser. 53, 1936).
—— 'The Collection of Agricultural Statistics in Great Britain: Its Origin and Evolution', *Agricultural History*, 18 (1944), 161-7.
—— *Crop Nutrition: Science and Practice before Liebig* (Lawrence, Ks.: Colorado Press, 1971).
—— *The Farmer's Tools: The History of British Farm Implements, Tools and Machinery, AD 1500-1900* (London: Bloomsbury, 1985).
—— and Goodman, C., 'Eighteenth-Century Estimates of British Sheep and Wool Production', *Agricultural History*, 4 (1930), 131-51.
Gardner, H. W., and Gardner, H. V., *The Use of Lime in British Agriculture* (London: Farmer and Stock-Breeder Publications, 1953).
Gerrard, F., *Meat Technology* (3rd edn. London: Leonard Hill, 1964).
Glennie, P., 'Measuring Crop Yields in Early Modern England', in Campbell and Overton (1991), 255-83.
Goddard, N., 'Information and Innovation in Early-Victorian Farming Systems', in B. A. Holderness and M. Turner (eds.), *Land, Labour and Agriculture, 1700-1920* (London: Hambledon, 1991), 165-90.
—— 'The Development and Influence of Agricultural Periodicals and Newspapers, 1780-1880', *AgHR* 31 (1983), 116-31.
—— *Harvests of Change: The Royal Agricultural Society of England 1838-1988* (London: Quiller Press, 1988).
Gourvish, T. R., and Wilson, R. G., *The British Brewing Industry, 1830-1980* (Cambridge: Cambridge University Press, 1994).
Grantham, G., and Leonard, C. S. (eds.), *Agrarian Organization in the Century of Industrialization* (*Research in Economic History*, 5, 1989).
Griffiths, T., Hunt, P. A., and O'Brien, P. K., 'Inventive Activity in the British Textile Industry, 1700-1800', *JEH* 52 (1992), 881-906.
Havinden, M. A., 'Agricultural Progress in Open-Field Oxfordshire', *AgHR* 9 (1961), 73-83.
—— 'Lime as a Means of Agricultural Improvement: The Devon Example', in C. W. Chalklin and M. A. Havinden (eds.), *Rural Change and Urban Growth, 1500-1800* (London: Longman, 1974), 104-34.
Healy, M. J. R., and Jones, E. L., 'Wheat Yields in England, 1815-59', *JRSS* 125 (1962), 574-9.

HOLDERNESS, B. A., 'Productivity Trends in English Agriculture, 1600–1850: Observations and Preliminary Results', unpublished paper (Edinburgh: International Economic History Conference, 1978).

—— 'Agriculture, 1770–1860', in C. H. Feinstein and S. Pollard (eds.), *Studies in Capital Formation in the United Kingdom, 1750–1920* (Oxford: Oxford University Press, 1988), 9–34.

—— 'Prices, Productivity, and Output', *AHEW* vi (1989), 84–189.

—— and Turner, M. E. (eds.), *Land, Labour and Agriculture, 1700–1920* (London: Hambledon Press, 1991).

HOLMES, G. S., 'Gregory King and the Social Structure of Pre-industrial England', *Transactions of the Royal Historical Society*, 27 (1977), 41–68.

HUGGINS, M. J., 'Thoroughbred Breeding in the North and East Ridings of Yorkshire in the Nineteenth Century', *AgHR* 42 (1994), 115–25.

JACKSON, R. V., 'Growth and Deceleration in English Agriculture, 1660–1790', *EcHR* 38 (1985), 333–51.

JOHN, A. H., 'The Course of Agricultural Change, 1660–1760', in L. S. Pressnell (ed.), *Studies in the Industrial Revolution* (London: Athlone Press, 1960), 125–55.

—— 'Aspects of Economic Growth in the First Half of the Eighteenth Century', *Economica*, 28 (1961), 176–90.

—— 'Agricultural Productivity and Economic Growth in England, 1700–1760', *JEH* 25 (1965), 19–34.

—— 'Statistical Appendix', *AHEW* vi (1989), 972–1155.

JONES, E., *Accountancy and the British Economy 1840–1980* (London: Batsford, 1981).

JONES, E. L., 'Agriculture and Economic Growth in England 1660–1750: Agricultural Change', *JEH* 25 (1965), 1–18.

—— 'Agriculture, 1700–80', in R. Floud and D. McCloskey (eds.), *The Economic History of Britain since 1700*, i: *1700–1860* (1st edn. Cambridge: Cambridge University Press, 1981), 66–86.

—— and COLLINS, E. J. T., 'The Collection and Analysis of Farm Record Books', *Journal of the Society of Archivists*, 3 (1965), 86–9.

JONES, M. J., 'The Accounting System of Magdalen College, Oxford, in 1812', *Accounting, Business and Financial History*, 2 (1991), 141–61.

KAIN, R. J. P., *An Atlas and Index of the Tithe Files of Mid-Nineteenth-Century England and Wales* (Cambridge: Cambridge University Press, 1986).

—— and PRINCE, H. C., *The Tithe Surveys of England and Wales* (Cambridge: Cambridge University Press, 1984).

KERRIDGE, E., *The Agricultural Revolution* (London: Allen & Unwin, 1967).

KUSSMAUL, A., *Servants in Husbandry in Early Modern England* (Cambridge: Cambridge University Press, 1981).

LEE, G. A., and OSBORNE, R. H., 'The Account Book of a Derbyshire Farm of the Eighteenth Century', *Accounting, Business and Financial History*, 4 (1994), 147–61.

LINDERT, P. E., and WILLIAMSON, J. G., 'Revising Britain's Social Tables, 1688–1913', *Explorations in Economic History*, 19 (1982), 385–408.

MCCLOSKEY, D. N., 'The Economics of Enclosure: A Market Analysis', in W. N. Parker and E. L. Jones (eds.), *European Peasants and their Markets* (Princeton: Princeton University Press, 1975), 73–119.

MACDONALD, S., 'The Diffusion of Knowledge among Northumberland Farmers, 1780–1815', *AgHR* 27 (1979), 30–9.

—— 'The Role of George Culley of Fenton in the Development of Northumberland Agriculture', *Archaeologia Aeliana*, 3 (1974), 131–41.

—— 'The Progress of the Early Threshing Machine', *AgHR* 23 (1975), 63–77.

—— 'Further Progress with the Early Threshing Machine: A Rejoinder', *AgHR* 26 (1978), 29–32.

MACLEOD, C., 'Strategies for Innovation: The Diffusion of New Technology in Nineteenth-Century British Industry', *EcHR* 45 (1992), 285–307.

MAFF, *A Century of Agricultural Statistics: Great Britain 1866–1966* (London: HMSO, 1968).

MALCOLMSON, R., and MASTORIS, S., *The English Pig: A History* (London: Hambledon Press, 1998).

MATHEW, W. M., 'Peru and the British Guano Market, 1840–70', *EcHR* 23 (1970), 112–28.

—— 'Marling in British Agriculture: A Case of Partial Identity', *AgHR* 41 (1993), 97–110.

MATHIAS, P., 'The Social Structure in the Eighteenth Century: A Calculation by Joseph Massie', *EcHR* 10 (1957), 30–45.

—— *The First Industrial Nation: An Economic History of Britain 1700–1914* (London: Methuen, 1969).

MILLER, E. (ed.), *The Agrarian History of England and Wales*, iii: *1348–1500* (Cambridge: Cambridge University Press, 1991).

MINCHINTON, W. E., 'Agricultural Returns and the Government during the Napoleonic Wars', *AgHR* 1 (1953), 29–43.

—— (ed.), *Essays in Agrarian History*, ii (Newton Abbott: British Agricultural History Society, 1968).

—— (ed.), *Agricultural Improvement: Medieval and Modern* (Exeter: University of Exeter Press, 1981).

MINGAY, G. E. (ed.), *Arthur Young and his Times* (London: Macmillan, 1975).

—— *Rural Life in Victorian England* (London: Heinemann, 1976).

—— (ed.), *The Victorian Countryside*, 2 vols. (London: Routledge & Kegan Paul, 1981).

—— 'Agricultural Productivity and Agricultural Society in Eighteenth-Century England', in G. Grantham and C. S. Leonard (eds.), *Agrarian Organization in the Century of Industrialization* (*Research in Economic History*, 5, 1989), 31–48.

MINGAY, G. E. (ed.), *The Agrarian History of England and Wales*, vi: *1750–1850* (Cambridge: Cambridge University Press, 1989).
—— 'The Progress of Agriculture, 1750–1850', *AHEW* vi (1989), 938–71.
—— 'The Farmer', *AHEW* vii (2000), 759–809.
MITCHELL, B. R., AND DEANE, P. (eds.), *Abstract of British Historical Statistics* (Cambridge: Cambridge University Press, 1962).
MOKYR, J. (eds.), *The Economics of the Industrial Revolution* (London: Allen & Unwin, 1985).
—— (ed.), *The British Industrial Revolution: An Economic Perspective* (Oxford: Westview Press, 1993).
MOORE, D. C., 'The Corn Laws and High Farming', *EcHR* 18 (1965), 544–61.
MOORE-COLYER, R. J., *The Welsh Cattle Drovers: Agriculture and the Welsh Cattle Trade before and during the Nineteenth Century* (Cardiff: University of Wales, 1976).
—— 'Livestock', *AHEW* vi (1989), 313–60.
MORGAN, R., *Dissertations on British Agrarian History* (Reading: University of Reading, 1981).
—— 'Supplements to the Bibliography of Theses on British Agrarian History', *AgHR* 30 (1982), 150–5; 37 (1989), 89–97; 42 (1994), 168–85.
Museum of English Rural Life, *G. E. Fussell: A Bibliography of his Writings on Agricultural History* (Reading: University of Reading, 1967).
NAPIER, C. J., 'Aristocratic Accounting: The Bute Estate in Glamorgan, 1814–1880', *Accounting and Business Research*, 21 (1991), 163–74.
NAPOLITAN, L., 'The Centenary of the Agricultural Census', *JRASE* 127 (1966), 81–96.
NEESON, J. M., *Commoners: Common Right, Enclosure and Social Change in England, 1700–1820* (Cambridge: Cambridge University Press, 1993).
O'BRIEN, P. K., 'Agriculture and the Industrial Revolution', *EcHR* 30 (1977), 166–81.
—— 'Agriculture and the Home Market for English Industry 1660–1820', *English Historical Review*, 100 (1985), 773–800.
—— 'British Incomes and Property in the Early Nineteenth Century', *EcHR* 12 (1959), 255–67.
OLNEY, R. J., *Rural Society and County Government in Nineteenth-Century Lincolnshire* (Lincoln: History of Lincolnshire Committee, 1979).
OLSEN, M., *The Economics of the Wartime Shortage* (Durham, NC: University of North Carolina Press, 1963).
ONIONS, C. T. (ed.), *The Oxford Dictionary of English Etymology* (Oxford: Oxford University Press, 1966).
ORMROD, D., *English Grain Exports and the Structure of Agrarian Capitalism, 1700–1760* (Hull: Hull University Press, 1985).
OVERTON, M., 'Estimating Yields from Probate Inventories: An Example from East Anglia, 1585–1735', *JEH* 39 (1979), 363–78.

OVERTON, M., 'Agricultural Productivity in Eighteenth-Century England: Some Further Speculations', *EcHR* 37 *(1984)*, 244–51.

—— 'The Determinants of Crop Yields in Early Modern England', in Campbell and Overton *(1991)*, 284–322.

—— 'Re-estimating Crop Yields from Probate Inventories', *JEH* 50 *(1990)*, 931–5.

—— *Agricultural Revolution in England: The Transformation of the Agrarian Economy 1500–1850* (Cambridge: Cambridge University Press, 1996*)*.

—— 'Re-establishing the English Agricultural Revolution', *AgHR* 44 *(1996)*, 1–20.

—— and CAMPBELL, B. M. S., 'Statistics of Production and Productivity in English Agriculture, 1086–1871', in B. J. P. van Bavel and E. Thoen (eds.), *Land Productivity and Agro-systems in the North Sea Area, Middle Ages–Twentieth Century: Elements for Comparison* (Turnhout: Brepols, 1999*)*, 189–208.

OWEN, C. C., *'The Greatest Brewery in the World': A History of Bass, Ratcliff & Gretton* (Chesterfield: Derbyshire Record Society 19, 1992).

PARKER, R. A. C., *Coke of Norfolk: A Financial and Agricultural Study, 1707–1842* (Oxford: Oxford University Press, 1975).

PARKER, W. N., and JONES, E. L. (eds.), *European Peasants and their Markets* (Princeton: Princeton University Press, 1975).

PERKINS, J. A., 'Tenure, Tenant Right, and Agricultural Progress in Lindsey, 1780–1850', *AgHR* 23 *(1975)*, 1–22.

PERREN, R., *The Meat Trade in Britain 1840–1914* (London: Routledge & Kegan Paul, 1978*)*.

PERRY, P. J., *British Farming in the Great Depression, 1870–1914* (Newton Abbot: David & Charles, 1974).

PETERSEN, C., *Bread and the British Economy c.1770–1870* (Aldershot: Scolar Press, 1995).

PHILLIPS, A. D. M., 'Agricultural Land Use and Cropping in Cheshire around 1840: Some Evidence from Cropping Books', *Transactions of the Lancashire and Cheshire Antiquarian Society*, 84 *(1987)*, 46–63.

—— *The Underdraining of Farmland in England during the Nineteenth Century* (Cambridge: Cambridge University Press, 1989).

—— 'Agriculture and Land Use, Soils and the Nottinghamshire Tithe Surveys, circa 1840', *East Midland Geographer,* 6 *(1976)*, 284–301.

POLLARD, S., *The Genesis of Modern Management* (London: Edward Arnold, 1965).

PRESSNELL, L. S. (ed.), *Studies in the Industrial Revolution* (London: Athlone Press, 1960).

PROTHERO, R. E. (LORD ERNLE), *English Farming Past and Present* (6th edn. London: Heinemann, 1961*)*.

Reading University Library, *Historical Farm Records: A Summary Guide to Manuscripts and Other Material in the University Library and Collected by the Institute of Agricultural History and the Museum of English Rural Life* (Reading: University of Reading Library, 1973).

RICH, E. E., and WILSON, C. H. (ed.), *The Cambridge Economic History of Europe*, v: *The Economic Organization of Early Modern Europe* (Cambridge: Cambridge University Press, 1977).

ROBINSON, D. H. (ed.), *Fream's Elements of Agriculture* (14th edn. London: John Murray, 1962).

ROBINSON, G. W., 'The Use of Lime', *JRASE* 104 (1943), 284–301.

ROE, P., 'Norfolk Agriculture, 1815–1915' (University of East Anglia, M.Phil. thesis, 1976).

ROSE, M. E., 'Social Change and the Industrial Revolution', in R. Floud and D. N. McCloskey (eds.), *The Economic History of Britain since 1700*, i: *1700–1860* (1st edn. Cambridge: Cambridge University Press, 1981), 253–75.

ROWE, D. J., 'The Culleys, Northumberland Farmers 1767–1813', *AgHR* 19 (1971), 156–74.

Rural History Centre, 'Survey of Farm Records in Other Repositories and in Private Hands', covering Bedfordshire to Gloucestershire (Reading: Rural History Centre, 1972).

RUSSELL, E. J., *Soil Conditions and Plant Growth* (11th edn. ed. A. Wild, Harlow: Longman, 1988).

RUSSELL, N., *Like Engend'ring Like: Heredity and Animal Breeding in Early Modern England* (Cambridge: Cambridge University Press, 1986).

Saorstát Éireann; *Agricultural Statistics 1847–1926: Reports and Tables* (Dublin: Department of Industry and Commerce, 1930).

SCHOFIELD, R., 'British Population Change, 1700–1871', in R. Floud and D. McCloskey (eds.), *The Economic History of Britain since 1700*, i: *1700–1860* (2nd edn. Cambridge: Cambridge University Press, 1994), 60–95.

SCOLA, R., *Feeding the Victorian City: The Food Supply of Manchester, 1770–1870* (Manchester: Manchester University Press, 1992).

SHEAIL, J., 'Elements of Sustainable Agriculture: The UK Experience', *AgHR* 43 (1995), 178–92.

—— 'Town Wastes, Agricultural Sustainability and Victorian Sewage', *Urban History*, 23 (1996), 189–210.

SHIEL, R. S., 'Improving Soil Productivity in the Pre-fertiliser Era', in Campbell and Overton (1991), 51–77.

SILL, M., 'Using the Tithe Files: A County Durham Study', *Local Historian*, 17 (1986), 205–11.

SIRAUT, M., 'A Somerset Farming Account Book', *Somerset Archaeology and Natural History*, 129 (1984–5), 161–70.

SLICHER VAN BATH, B. H., *The Agrarian History of Western Europe A.D. 500–1850* (London: Edward Arnold, 1963).

—— 'Yield Ratios, 810–1820', *AAG Bijdragen*, 10 (1963).

—— 'Agriculture in the Vital Revolution', in E. E. Rich and C. H. Wilson (eds.), *The Cambridge Economic History of Europe*, v: *The Economic Organization of Early Modern Europe* (Cambridge: Cambridge University Press, 1977), 42–122.

SPEDDING, C. R. W. (ed.), *Fream's Agriculture* (16th edn. London: Murray, 1983).

STAMP, Sir JOSIAH, *British Incomes and Property* (London: P. S. King & Sons, 1927).

STAMP, L. D., 'Fertility, Productivity, and Classification of Land in Britain', *Geographical Journal*, 96 (1940), 389–412.

—— *The Land of Britain: Its Use and Misuse* (London: Longman, 1948).

STEPHEN, L., and LEE, S. (eds.), *Dictionary of National Biography*, vi (Oxford: Oxford University Press, 1921–2).

STERN, W. M., 'The Bread Crisis in Britain, 1795–6', *Economica*, 31 (1964), 168–87.

STONE, L. (ed.), *Social Change and Revolution in England, 1540–1640* (London: Longman, 1966).

STOVIN, J. (ed.), *Journals of a Methodist Farmer, 1871–1875* (London: Croom Helm, 1982).

SULLIVAN, R. J., 'The Revolution of Ideas: Widespread Patenting and Invention during the English Industrial Revolution', *JEH* 50 (1990), 349–62.

TAYLOR, D., 'The English Dairy Industry, 1860–1930', *EcHR* 29 (1976), 585–601.

THIRSK, J., 'Agrarian History, 1540–1950', in Victoria County History, *Leicestershire*, ii (Oxford: Oxford University Press, 1954), 199–264.

—— (ed.), *The Agrarian History of England and Wales*, iv: *1500–1640* (Cambridge: Cambridge University Press, 1967).

—— (ed.), *The Agrarian History of England and Wales*, v: *1640–1750*, 2 vols. (Cambridge: Cambridge University Press, 1984/5).

—— 'Agricultural Policy: Public Debate and Legislation', *AHEW* v (2) (1985), 298–388.

—— *England's Agricultural Regions and Agrarian History, 1500–1750* (Basingstoke: Macmillan, 1987).z

—— *Alternative Agriculture* (Oxford: Oxford University Press, 1997).

THOMAS, E., 'The June Returns One Hundred Years Old', *Agriculture*, 73 (1966), 245–9.

THOMPSON, F. M. L., 'The Social Distribution of Landed Property in England since the Sixteenth Century', *EcHR* 19 (1966), 505–17.

—— 'The Second Agricultural Revolution, 1815–80', *EcHR* 21 (1968), 62–77.

—— 'An Anatomy of English Agriculture, 1870–1914', in B. A. Holderness and M. E. Turner (eds.), *Land, Labour and Agriculture, 1700–1920* (London: Hambledon, 1991), 211–40.

THORNTON, C., 'The Determinants of Land Productivity on the Bishop of Winchester's Demesne of Rimpton, 1208 to 1403', in Campbell and Overton (1991), 183–210.

THWAITES, W., 'Dearth and the Marketing of Agricultural Produce: Oxfordshire c.1750–1800', *AgHR* 33 (1985), 119–31.
—— 'The Corn Market and Economic Change: Oxfordshire in the Eighteenth Century', *Midland History*, 16 (1991), 103–25.
TITOW, J. Z., *Winchester Wheat Yields: A Study in Medieval Agricultural Productivity* (Cambridge: Cambridge University Press, 1972).
TOLLEY, B. H., 'M. J. R. Dunstan and the First Department of Agriculture at University College, Nottingham, 1890–1900', *Transactions of the Thoroton Society*, 87 (1983), 71–9.
TROW-SMITH, R., *A History of British Livestock Husbandry to 1700* (London: Routledge & Kegan Paul, 1957).
—— *The History of British Livestock Husbandry, 1700–1900* (London: Routledge & Kegan Paul, 1959).
TURNER, M. E., *English Parliamentary Enclosure: Its Historical Geography and Economic History* (Folkestone: Dawson, 1980).
—— 'Arable in England and Wales: Estimates from the 1801 Crop Returns', *Journal of Historical Geography*, 7 (1981), 291–302.
—— (ed.), *Home Office Acreage Returns: (HO67) List and Analysis Part 1 Bedfordshire–Isle of Wight* (London: List and Index Society 189, 1982).
—— 'Agricultural Productivity in England in the Eighteenth Century: Evidence from Crop Yields', *EcHR* 35 (1982), 489–510.
—— 'Agricultural Productivity in Eighteenth-Century England: Further Strains of Speculation', *EcHR* 37 (1984), 252–7.
—— 'English Open Fields and Enclosures: Retardation or Productivity Improvements', *JEH* 46 (1986), 669–92.
—— 'Corn Crises in Britain in the Age of Malthus', in M. E. Turner (ed.), *Malthus and his Time* (London: Macmillan, 1986), 112–28.
—— 'Output and Prices in UK Agriculture, 1867–1914, and the Great Agricultural Depression Reconsidered', *AgHR* 40 (1992), 38–51.
—— 'Weighing the Fat Pig: Agricultural History and the National Income Accounts', Inaugural Lecture, University of Hull (25 Jan. 1993).
—— *After the Famine* (Cambridge: Cambridge University Press, 1996).
—— 'Counting Sheep: Waking up to New Estimates of Livestock Numbers in England, c.1800', *AgHR* 46 (1998), 142–61.
—— 'Corporate Strategy or Individual Priority? Land Management, Income and Tenure on Oxbridge Agricultural Land in the Mid-Nineteenth Century', *Business History*, 42 (2000), 1–26.
—— 'Agricultural Output, Income and Productivity', *AHEW* vii (2000), 224–320.
—— and BECKETT, J. V., 'The Lingering Survival of Ancient Tenures in English Agriculture in the Nineteenth Century', in F. Galassi, K. Kauffman, and J. Liebowitz (eds.), *Land, Labour and Tenure: The Institutional Arrangements of*

Conflict and Cooperation in Comparative Perspective (Madrid: Fundación Fomento de la Historia Económica, 1998), 97–114.

—— and AFTON, B., 'Taking Stock: Farmers, Farm Records, and Agricultural Output in England, 1700–1850', *AgHR* 44 (1996), 21–34.

———— *Agricultural Rent in England, 1690–1914* (Cambridge: Cambridge University Press, 1997).

VAN BAVEL, B. J. P., and THOEN, E. (eds.), *Land Productivity and Agro-systems in the North Sea Area, Middle Ages–Twentieth Century: Elements for Comparison* (Turnhout: Brepols 1999).

VENN, J. A., *The Foundations of Agricultural Economics* (Cambridge: Cambridge University Press, 1933).

WADE MARTINS, S., *A Great Estate at Work: The Holkham Estate and its Inhabitants in the Nineteenth Century* (Cambridge: Cambridge University Press, 1980).

—— and Williamson, T., 'Floated Water-Meadows in Norfolk: A Misplaced Innovation', *AgHR* 42 (1994), 20–37.

—— —— (eds.), *The Farming Journal of Randall Burroughes (1794–1797)* (Norwich: Norfolk Record Society 58, 1995).

—— —— 'Labour and Improvement: Agricultural Change in East Anglia, c.1750–1870', *Labour History Review*, 62 (1997), 275–95.

—— —— *Roots of Change: Farming and the Landscape in East Anglia, c.1700–1870* (Exeter: British Agricultural History Society, 1999).

WALTON, J. R., 'Mechanisation in Agriculture: A Study of the Adoption Process', in H. S. A. Fox and R. A. Butlin (eds.), *Change in the Countryside: Essays on Rural England 1500–1900* (London: Institute of British Geographers, Special Publication no. 10, 1979), 23–42.

—— 'Pedigree and the National Cattle Herd, circa 1750–1950', *AgHR* 34 (1986), 149–70.

—— 'Varietal Innovation and the Competitiveness of the British Cereals Sector, 1760–1930', *AgHR* 47 (1999), 29–57.

WELLS, R., *Wretched Faces: Famine in Wartime England 1793–1801* (Gloucester: Alan Sutton, 1988).

WHETHAM, E. H. (ed.), *The Agrarian History of England and Wales*, viii: *1914–39* (Cambridge: Cambridge University Press, 1978).

WILKES, A. R., 'Adjustments in Arable Farming after the Napoleonic Wars', *AgHR* 28 (1980), 90–103.

WILKES, R., 'The Diffusion of Drill Husbandry, 1731–1850', in W. Minchinton (ed.), *Agricultural Improvement: Medieval and Modern* (Exeter: University of Exeter Press, 1981), 65–94.

WOOD, T. B., *Animal Nutrition* (London: W. B. Clive, 1924).

WOODWARD, D., 'A Comparative Study of the Irish and Scottish Livestock Trade in the Seventeenth Century', in L. M. Cullen and T. C. Smout (eds.), *Comparative*

Aspects of Scottish and Irish Economic and Social History 1600–1900 (Edinburgh: University of Edinburgh Press, 1977), 147–64.

WORDIE, J. R., 'The Chronology of English Enclosure, 1500–1914', *EcHR* 36 (1983). 504–5.

WRIGLEY, E. A., 'Men on the Land and Men in the Countryside: Employment in Agriculture in Early Nineteenth Century England', in L. Bonfield, R. M. Smith, and K. Wrightson (eds.), *The World We Have Gained* (Oxford: Oxford University Press, 1986), 295–336.

—— 'Energy Availability and Agricultural Productivity', in Campbell and Overton (1991), 323–39.

—— and SCHOFIELD, R. S., *The Population History of England, 1541–1871* (London: Edward Arnold, 1981).

Index

Abbots Bromley, Staffs. 241
Abthorpe, Northants. 239
Accountant's Library 40–1
accounting systems 44–51
accounts, *see* farm accounts
acidity, *see* soil acidity
acres, *see* customary measures; statute acres
Act for Ascertaining and Establishing Uniformity of Weights and Measurements 245–6
Adderbury, Oxon. 98, 240
ages of animals:
   and carcass weights 174–5
   at slaughter 195
agistment 60, 96, 103
agrarian revolution, *see* agricultural revolution
agricultural accounts 40–1
agricultural change 138–9, 215–30
agricultural depression:
   of late nineteenth century 38, 68, 112–13, 134–5, 141, 151–2
   of post-French wars 222
   of second quarter of eighteenth century 216
agricultural growth 210–11
   growth rates 14–17, 19–20, 21–3
Agricultural Holdings Act, 1875 37–8
agricultural labour force 28, 128, 225, 227
   see also labour
agricultural output 3, 5, 8–26, 223
   proxied by wheat 224–7
agricultural production 9–26, 214–16
   see also farm production
agricultural productivity 9–26
   see also labour
agricultural protection, *see* Corn Bounty; Corn Laws
agricultural returns, *see* agricultural statistics

agricultural revolution 1, 8–9, 12, 23, 180, 223, 230
   definitions of 10–14
   and farm production 210–30
   measurement of 10, 21–2
   timing of 14–25
   and wheat yields 122, 137–41, 223
agricultural statistics:
   1790s inquiries 3, 22
   1801 crop returns 11, 14, 22
   1830s 222
   1854 returns 4
   in Ireland 3
   June Returns 4, 25, 37
   in the livestock sector 173–4
   official crop yields 4, 151, 160–2
   re. parliamentary debates 1–5
   in Scotland 4 n. 9
   wheat yields 126, 136–7, 141 n. 38
   see also Overton, M.
AHEW 24, 26
   on agricultural output 11–12
   on agricultural statistics 18
   on farm production 12
   on farmers 29
   on grain yields 120–1, 127, 131, 133, 172
   on seeding rates 168
   on sheep carcass weights 183
Aldbourne, Wilts. 243
   Brown family 36–7, 58–9
   grain yields 42
   use of lime 82
Allen, R.C.:
   on agricultural growth 14–15, 19, 21–3
   on agricultural revolution 15, 138
   on wheat yields 117, 122–4

allowances, *see* perquisites
American wheat 79
Andover, Hants.:
  for livestock supplies 108
  report on in RC 50
Angier family of Coton and Histon, Cambs.,
    see Coton; Histon
animals:
  ages and carcass weights 174–5
  breeding 58, 99
  breeds and carcass weights 7, 174
  carcass weights 58, 174–5
  dung as fertilizer 84–5, 87
  feeds in a Norfolk type rotation 71
  health 106–7
  numbers 4, 18 n. 38, 37
  output 20–3
  stocking rates in Norfolk-type 19
  urine as fertilizer 83–4
  see also breed by name; named animals
Annals of Agriculture:
  and Arthur Young 2
  on sheep carcass weights 181
Anstey Mill, Leics. 48
arable:
  acreage 140
  conversion to pasture 55
  farmers and farming 57
  farming practice 214
  as livestock feed 101–2
  rotations 55, 70
  see also farmers; named crops
archives, *see* farm records
Ardington, Berks. 231
Arlingham, Glos. 235
Armstrong, Alan:
  on the farmer 29
Arreton, Isle of Wight 235
  barley yields 155 n. 13
  wheat yields 127, 131–3
artificial feeds 102–3
artificial fertilizer 82, 86–7
ash 54, 57
  on John West's farm 112
  as manure and fertilizer 83–4, 87
Ash, Kent 37, 236
Ashby St Ledgers, Northants. 239
Ashley, W. 150
Atherton, Lancs. 94, 237

Audley End, Essex 234
  carcass weights 192, 207

bacon pigs:
  carcass weights 186–7, 202–3
Badmondisfield, Suffolk 241
Badsworth, Yorks. 244
bag:
  customary measure 246
bagging hook 92
Bakewell, Robert, animal breeder of
    Dishley, Leics.:
  on carcass weights 195
  on selective breeding 96
ballards 190
Bamburgh, Northumb. 239
bare fallow, *see* fallow
Barker, William of Upton Farm, Sompting,
    Sussex 113
barley 42, 54, 56, 75, 83, 86, 154
  acreages 18, 151
  in agricultural output 20
  in arable rotations 66–8
  as bread grain 150
  as a catch crop 68–9
  as feed 58, 60, 101–2, 150, 155
  for malting 34, 68, 79, 150, 154–6, 230
  in a Norfolk-type rotation 71–2, 74–6, 89,
    111, 156
  as seed 78
  for straw 156
  seed varieties 79–80
  seeding rates 167–72, 214
  on Upton Farm 113
  yields 7, 42, 79, 150–6, 158, 160–5, 213
Barnard, William, of Harlowbury Farm,
    Harlow, Essex 91, 96–7
Barton, Cambs. 232
  seed drills 92
Bashall, Lancs. 237
Bass Breweries 113
Baydon, Wilts. 243
beans 23 n. 51
  in arable rotations 66–9
  fallowing 75
  as feed 60
  as nitrogen fixers 69, 85–6, 108
  in a Norfolk-type rotation 74–6
  as seed 77–8

INDEX

seed varieties 79–80
   yields 162, 164, 165, 166
beasts:
   carcass weights 199–201
Beaumont Leys, Leics. 48, 98, 237
   crop varieties 79
   livestock farming 97
   ploughs 91
   seed drills 92
Bedfordshire 62, 231
   malting barley 155 n. 12
   wheat yields 146
   see also Billington; Eversholt; Leighton
      Buzzard; Sharnbrook; Stopsley
beef cattle:
   carcass weights 186–7, 199–201
Beeke, Henry:
   on agricultural statistics 12 n. 11
Bellerby, J.R.:
   on agricultural output 21–2
   on wheat yields 126
Berkshire 62, 231
   lamb carcass weights 185
   pig carcass weights 207
   porker carcass weights 205–6
   wheat yields 133–5, 146
   see also Ardington; Bradfield;
      Englefield; Finchampstead;
      Hampstead Norreys; Lockinge;
      Radley; Reading; Steventon;
      Sulham; Swallowfield
Berkshire cattle 98
Billington, Beds. 231
Bingham, Notts. 240
   reapers 93
Binstead, Hants. 235
birds:
   as pests 93
Bishops Cannings, Wilts. 243
Bishop's Lydeard, Som. 241
Black cattle 98
Black Siberian oats 79
Blith, Walter, agricultural writer:
   on manure and fertilizer 83
Blockley, Glos. 235
   estate accounts 34
blood:
   on John West's farm 112
   as manure and fertilizer 54, 83

Bloxham, Oxon. 58, 240
   crop varieties 79
Bloxworth, Dorset 233
Board of Agriculture 2
   on cattle carcass weights 195
   on farmers 28
   on weights and measures 247
   on wheat yields 127, 134–5
   on yield rates 170
Board of Trade:
   on agricultural statistics 2
Bocking, Essex 234
   double-entry accounting 50
   Tabor family 56
boll:
   customary measure 246
bones 57, 83–4, 193, 213, 222
bone dust:
   on John West's farm 112
bookkeeping 44–51
Boswell, George, Dorset farmer:
   on seeds 78, 80
   on water-borne diseases 107
   on water meadows 104
Bowden, P.:
   on cattle carcass weights 195
   on seeding rates 168–9
boxplots:
   of barley yields 152
   of oats yields 157
   of wheat yields 130–1
Bradfield, Berks. 231
Bradfield, Yorks. 98, 243
   underdraining 90
Bradford, Yorks.:
   wheat yields 126
Bramhall, Ches. 232
   land management 106
Bramley, Yorks. 243
Brancepeth, Dur. 234
   bullock carcass weights 200–1
   cow and heifer carcass weights 197, 199
Brandwood, Shropshire:
   Price's Farm 106
Brassley, P.:
   on fertilizers 88
   on pest control 95 n. 72
Bratton, Wilts. 243
Braunton, Devon 92, 233

brawning (birth of piglets) 58
bread 116, 223
breeding 222
   breeding books 58
   and carcass weights 174
   of cattle 95–6
   in farming systems 100
Breedon-on-the-Hill, Leics. 237
   land management 105–6
   underdraining 105
Bridgewater Estate, Whitchurch, Shropshire:
   land management 106
   purchased seeds 78
   underdraining 105
Bridgnorth, Shropshire 98, 240
   cattle carcass weights 200–1
   Coton Hall Farm 63–4, 108–11, 177–8
   cow and heifer carcass weights 199
   ewe carcass weights 185
   lamb carcass weights 185
Britannia wheat 80
Broad clover 79
broadcast sowing 91
   seeding rates 171–2
Broughton, Hants. 236
Brown family, Aldbourne, Wilts. 36–7, 58–9
Brown, J.:
   on malting barley 155 n. 12
Brunt, L.:
   on seed drills 172
Brunton, Northumb. 75
Buckinghamshire 62, 231
   bacon pig carcass weights 202
   calf carcass weights 196
   cow and heifer carcass weights 198
   hog carcass weights 203
   lamb carcass weights 185
   pig carcass weights 207
   porker carcass weights 205–6
   sheep carcass weights 192
   threshing machines 93
   wethers carcass weights 189
   see also Chicheley; Little Shardeloes; Medmenham; Quainton; Stokenchurch; Vale of Aylesbury
Buckland Abbey, Devon:
   Drake Estate 35
Buckland Monochorum, Devon 233
buckwheat:
   in a Norfolk-type rotation 74

bulb trade:
   of the Lincolnshire fens 230
Bulcamp, Suffolk 241
Bullington, Hants. 236
bullocks:
   carcass weights 194, 199–201
   feed for 103
bulls:
   carcass weights 199–201
Bulphan, Essex 234
Bungay, Suffolk:
   purchased feeds from 103
   see also Mettingham
Burke, J.F.:
   on carcass weights 179–80
   on steelyards 178–9
Burmarsh, Kent 98, 236
   Forestall Farm 60
   livestock farming 96
Burroughs, Randall, of Wymondham, Norfolk:
   on underdraining 90
Burton Constable, Yorks. 244
Burton on the Wold, Leics. 237
Burwell wheat 79
Burwell, Lincs. 238
Bury St Edmunds, Suffolk:
   for malting barley 154, 156
Bury, Lancs. 237
bushels:
   definitions of 245–6
Buxhall, Suffolk 241

cabbages 75
   as livestock feed 101–2
   special varieties 79–81
Caird, James, MP and agricultural lobbyist:
   on agricultural statistics 3–5, 12
   on farm sizes 228
   on rents 224
   on wheat yields 127
cake:
   as feed 60, 102, 112
cake crusher 92
calves 42, 57
   carcass weights 186–7, 196–7, 214
calving 7, 47, 58
Cambridgeshire 62, 232
   barley yields 155
   customary measures 246

INDEX

pig carcass weights 207
wheat yields 146
see also Barton; Chippenham; Comberton; Coton; Histon; Landbeach; Trumpington
Campbell, B.M.S.:
see Overton, M.
Campsell, Yorks. 244
canals:
and marketing 229
Candover, Hants. 235
capital 7, 36, 226
investment 9, 17–18, 35, 228
capitalist agricultural systems 9–10, 15
carcass weights 7–8, 42, 58, 64, 100–1, 214–15
general principles of 174–80
selective breeding and 97, 99
see also named animals
cash accounts 39
cash crops:
in a Norfolk-type rotation 72
see also named crops
Castle Bromwich, War. 242
Castle Eden, Dur. 234
ewe carcass weights 185
lamb carcass weights 185
sheep carcass weights 192
Castle Heaton, Northumb. 239
Castle Hedingham, Essex 234
see also Castle Park Farm
Castle Hill, Devon 233
land management 105–6
Castle Park Farm, Castle Hedingham, Essex:
land management 106
Castlemorton, Worcs. 243
catch crops 101–2
cattle 20, 37, 48, 57, 95–7, 100, 194
breeding 95–6
breeds of 98
carcass weights 175, 184–7, 194–201
droving 97
fat cattle 113, 176–8
feeding of 54, 111–12
at Langham, Norfolk-type 89
ratio of meat to offal 177
tuberculosis 107
see also named animals
cattle distemper 106
cattle plague (rindepest) 4
cereals production 24

see also grain output; named crops
Chadwell, Leics. 238
chaff (machines) 38, 92, 228
Chailey, Sussex 242
chalk:
as a soil conditioner 82
Chambers, J.D.:
on agricultural revolution 10–11
on food dearth 16
charge and discharge system 44–5
Charlton Marshall, Dorset 61, 233
customary measures 245, 247
Charminster, Dorset 98, 234
crop varieties 79
Chartres, J.A.:
on agricultural output 11–12, 20–1, 23
on barley exports 154 n. 11
on wheat yields 127
Cheshire 62, 232
customary measures 245
lamb carcass weights 185
pig carcass weights 207
water meadows 104
see also Bramhall; Chester; High Legh; Style; Wybunbury
Chester, Ches.:
wheat yields 126
Chevalier barley 79–80
Cheviot sheep 98
ewe carcass weights 185
Chicheley, Bucks. 231
beef cattle carcass weights 200
sheep carcass weights 192
sheep sales 176
Chilgrove Farm, West Dean, Sussex:
grain yields 42
Chilwell, Notts. 240
underdraining 90
Chippenham, Cambs. 98, 232
Chipping Norton, Oxon.:
wheat yields 126
Chulmleigh, Devon 233
land management 106
Church Lench, Worcs. 243
Churchill oats 79
Clare, William of Coton Hall Farm, nr Bridgnorth, Shropshire, *see* Coton Hall Farm
Clark, G.:
on agricultural growth 16

Clark, G. (*cont.*)
  on agricultural revolution 24 n. 53, 138
  on labour productivity 227
classical economists:
  on harvest crisis 220
  see also Malthus; Ricardo; McCulloch
clay:
  as a soil conditioner 82–3
Clayhidon, Devon 233
  ploughs 91
Clenchwarton, Norfolk-type 238
  crop varieties 79
  underdraining 90
Cleveland, Yorks. 244
Clifton, Worcs. 243
clover 55, 70–1, 213
  in arable rotations 67, 70, 75
  in East Anglia 19 n. 42
  at East Sutton, Kent 108
  and ley farming 70, 101
  as livestock feed 101
  as a nitrogen fixer 69–70, 85–6, 108
  in Norfolk-type 19
  in a Norfolk-type rotation 71, 74–6, 89
  with oats 159–60
  purchased seeds 77–8
  red clover 56, 70
  re. soil conditioners 83
  special varieties 79–80
  see also trefoil
Cobham wheat 79
Coddenham, Suffolk 241
Coke of Holkham (Norfolk) 230
Coldstream, Northumb. 239
  Ford Estate 34
Cole, W.A., *see* Deane, P.
Coleman, John, agricultural writer:
  on double-entry 49
  on farm accounts 39–40
coleseed 75
  as livestock feed 101–2
Collingham, Notts. 240
Collins, E.J.T.:
  on farm records 6–7
  on rye bread 68 n. 2
  on wheaten bread 223 n.19
Colquhoun, Patrick, political arithmetist:
  on the farmer 28, 31
Colworth Farm, Sharnbrook, Beds. 54

comb:
  customary measure 246
Comberton, Cambs. 232
Commissioners of the Victualling Office 217
Compton, Hants. 235
condition (of animals):
  re. carcass weights 174–5
conditioners 81–3
  see also named conditioners; named crops
conies 20 n. 46
consumption 14 n. 18, 17–18
Continent, the:
  grain imports from 220
copyhold 30–1
corn 35
  as feed 59–60, 102, 112
  re. fertilizer 87
  in a Norfolk-type rotation 71
  seeds 77–81, 113
Corn Bounty Act (1688) 215
Corn Laws 3, 221, 223–5, 229
corn merchants 126
Cornwall 62, 232
  ewe carcass weights 185
  manure and fertilizer 83
  see also Fentrigan (in Warbstow);
    Kilkhampton; Polperro
Costessey, Norfolk-type 238
Cotes, Leics. 237
Coton, Cambs. 60, 232
  threshing machines 93
  underdraining 90
Coton Hall Farm, Bridgnorth, Shropshire:
  farm accounts 63–4, 108–11
  on ratio of meat to offal 177–8
Cotswold sheep 57, 181–2
cottage pigs 202
Coventry, War.:
  wheat yields 126
cows 47–8, 60, 197–8
  breeding 58
  carcass weights of cows and heifers 186–7, 197–8
  feeding 60, 103
Crafts, N.F.R.:
  on agricultural growth 15, 138 n. 32, 212–13
Craigie, P.G., agricultural writer:
  on grain yields 161–2

on sheep carcass weights 183–4
on wheat yields 127, 135–6
crops:
  diseases 93–4
  dressing of 54, 57
  output 43
  rotations 2, 7, 35, 55, 222
  varieties 7, 79–80
  see also under named crops
crop returns (1801):
  re. customary measures 246 n. 10
  wheat yields 117, 125
crop yields 4, 8, 19, 24, 36, 42–3, 60, 64, 212
  see also under named crops
cropped fallow:
  in a Norfolk-type rotation 71
cropping books 27, 36, 55
Cropwell Butler, Notts. 240
Crowcombe, Som. 241
Crudwell, Wilts. 243
Culley, George, Northumb. livestock farmer:
  on the production of seeds 78
  on water meadows 104
cultivated area 16–17, 24–5, 221
Cumberland 62, 233
  pig carcass weights 207
  see also Gilsland; Holm Cultram; Newton Arlosh
Cumbrian probate inventories 31
customary measures 245–7
customaryhold 30–1
Cyclopedia of Agriculture:
  on weights and measures 247

dairying 39, 113, 230
  at East Sutton, Kent 108
Danny Estate, Sussex 56
Darfield, Yorks., *see* Houndhill
Darsham, Suffolk 241
Daunton, M.:
  on agricultural production 24–5
Davis, Thomas, agricultural writer:
  on enclosure 89
Day, James of Whitstock, Som. farmer 38
daybooks 57
deadstock weights 42
  see also carcass weights
Deane, P. and Cole, W.A.:
  on cattle carcass weights 194

on wheat yields 137–8
Deanshanger, Northants. 239
  Norfolk-type rotation 75–6
Deddington, Oxon. 240
Denbighshire cattle 98
depression, *see* agricultural depression
Derbyshire 62, 233
  carcass weights 195
  see also Hathersage; Norton; Thurvaston; Wormhill
Devon 62, 233
  cattle carcass weights 195–6
  cow and heifer carcass weights 198
  ewe carcass weights 185
  hog carcass weights 203
  lamb carcass weights 185
  manure and fertilizer 83
  pig carcass weights 207
  porker carcass weights 205–6
  sheep carcass weights 192
  wethers carcass weights 189
  wheat yields 145
  see also Braunton; Buckland Monochorum; Castle Hill; Chulmleigh; Clayhidon; Exeter; Filleigh; Lesant; Milton Abbot; Nymet Rowland; Sandford; South Milton; Tavistock; Upottery; Willand
diaries, *see* farm diaries
dibbling 91, 171–2
Didsbury, Lancs. 237
  crop varieties 79
  land management 106
diseases, *see* named diseases
Dishley, Leics.:
  livestock farming 96–7
  see also Robert Bakewell
Ditchley, Oxon. 240
Doggetts Hall Farm, Rochford, Essex:
  fallowing 75
  pipe drainage 91
Dorset 62, 234–5
  customary measures 245, 247
  George Boswell, farmer 78, 80
  Norfolk-type rotations 73
  pig carcass weights 207
  threshing machines 93
  water meadows 104
  wheat yields 147

Dorset (*cont.*)
  see also Bloxworth; Charlton Marshall; Charminster; Frampton; Maiden Newton; Puddletown; Shapwick; Stoke Abbot; Tarrant Monkton; Winterborne St Martins; Wool
Dorset Down sheep 98
  ewe carcass weights 185
double-entry bookkeeping 48–51
drainage 18, 89–91, 105, 214, 220, 222, 226–8
  in Lincolnshire 225
  pipe drains 56
  see also underdrainage
Drake Estate, Buckland Abbey, Devon: estate accounts 35
dressing (of crop), see crops
drills, see seed drills
droving (of cattle) 97
Duke of Bedford 32
Duke of Grafton, see Grafton Estate
Dulverton, Som. 241
Dumplington, Lancs. 237
dunging:
  in a Norfolk-type rotation 89
Dunholme, Lincs. 238
  crop varieties 80
  and John West 111–13
  steam cultivation 93
Dunster Castle, Som. 241
  feeding animals 60
Durham 62, 234
  bacon pig carcass weights 202
  bullock carcass weights 200–1
  cow and heifer carcass weights 199
  pig carcass weights 207
  seed drills 91
  wethers carcass weights 189
  see also Brancepeth; Castle Eden; Ryton; Silksworth; Westgate in Weardale
Dutch clover 80–1

Earl of Leicester, of Holkham, Norfolk-type 35
Earl of Stamford and Warrington, of Breedon, Leics. 106
Early Anglesea oats 80
Early Sugar Loaf cabbage 79
earths:
  as manure and fertilizer 84
East Anglia:
  agricultural depression 230
  barley yields 154–5
  crop rotations 69
  ploughs 91
  underdraining 90
  see also Norfolk; Suffolk
East Brunton, Northumb. 239
East Stratton, Hants. 235
East Sutton Court Farm, East Sutton, nr Maidstone, Kent 107–8, 237
  cow and heifer carcass weights 197
  drainage 89
  livestock breeds 98
  manure 85
  Norfolk-type rotation 73–4
  use of lime 82
Easton, Hants. 236
  steam cultivation 93
Eathorpe, War. 242
  cow and heifer carcass weights 199
  seed drills 92
  sheep sales 184
Edinburgh Review:
  on post-French wars agricultural depression 222
Ellesmere, Shropshire 240
Ellis, John of Beaumont Leys, Leics.:
  on his accounting methods 48–9
  on livestock farming 97
  see also Beaumont Leys
Elmdon, Essex 234
enclosure 2, 7, 9–10, 16–18, 214–15, 217, 220, 226–7
  and crop yields 88–9
  during the French wars 139
Enford, Wilts. 243
Englefield, Berks. 231
English land utilization survey (1930s) 142–5, 148 n. 43
English, B.:
  on accounting systems at Sledmere, Yorkshire 45
environment:
  re. wheat yields 142–9
Ernle, Lord, see Prothero
Essex 61–2, 234
  barley yields 155
  customary measures 246
  lamb carcass weights 185
  pig carcass weights 207

INDEX

report on in RC 50
sheep carcass weights 191-2
threshing machines 93
wheat yields 128-9, 134-5, 146
see also Audley End; Bocking; Bulphan; Castle Hedingham; Doggett's Hall Farm; Elmdon; Fingringhoe; Great Henny; Halstead; Harlow; Hatfield Peverel; High Easter; Layer Marney; Little Dunmow; Loughton; Peldon; Rayleigh, Lord; Rochford; Strutt, Edward; Upminster; West Mersea; Wickham Bishops; Wickham St Pauls; Witham
estate farms 56
  accounting systems on 44-5
  records 34-6, 41-2
Europe:
  animal breeding 100
Eversholt, Beds. 231
ewe carcass weights 184-8
Exeter, Devon:
  clover sales 70
exports:
  of grain 16, 215-16

fallow 2, 55-6, 70 n. 6
  in arable rotations 66, 69-70, 75
  in a Norfolk-type rotation 71-2, 74-6
family consumption:
  of own produce 57-8
family labour, *see* labour
farm accounts 7, 12, 27, 43-53, 63-5
farm diaries 7, 27, 36, 55, 61, 75
farm incomes 221
farm labour, *see* labour
farm output 2-3, 5, 8, 54
farm production 6, 8, 12, 28
  re. the agricultural revolution 210-30
  see also agricultural production
farm records 6-8, 33-65, 231-44
  carcass weights 174-80
  enclosure 88-9
  labour productivity 226
  mechanization 91-2, 228
  weights and measures 245-7
  see also under individual places; named counties
farm size 10, 29, 36-7, 214, 228
farm tenure 216

farmers 9, 27-33, 38
  as businessmen 27, 46-50, 86
  and mechanization 228-9
  and output 7
  their records 7, 33-65, 211-12
  and rents 221-3
farming:
  cycle 66-79
  practice 7, 66-115, 214, 220
  prosperity 229-30
  systems 7, 54, 56
  techniques 7, 66-115
  see also Norfolk-type rotations
farms in hand 35-6
farmyard manure, *see* manure
fat cattle, *see* cattle
fattening 95-6, 100, 108
feed 54, 60, 101-3, 175
  carcass weights 180
  on John West's farm 112-13
feed crops:
  in a Norfolk-type rotation 71-2
Felbrigg, Norfolk-type 238
Fen wheat 79
Fens, the:
  oats yields 158
Fentrigan (in Warbstow), Corn. 232
fertilizer 35, 38, 56-7, 81, 83-8, 128, 213
  crop yields 36
  farming practice 214
  imported artificials 139-41
  on John West's farm 112-13
field books 56
field draining, *see* drainage
Filleigh, Devon 233
Filmer Estate, *see* East Sutton Court Farm
Finchampstead, Berks.:
  for purchase of seeds 77
finger and toe (disease in root crops) 83
Fingringhoe, Essex 234
fleeces:
  re. sheep breeds 181-2
fish:
  as manure and fertilizer 83
Flempton, Suffolk 241
fodder 20, 101, 222
  see also under named crops
folding:
  re. manuring 85

food supply 1, 3–4, 17
    see also imports
foot and mouth disease 107
forage crops:
    as livestock feed 101
    as nitrogen fixers 69 n. 4
Ford Estate, near Coldstream, Northumb.:
    customary measures 246–7
    estate accounts 34
Ford, Northumb. 239
Forestall Farm, Burmarsh, Kent:
    feeding animals 60
four course rotation,
    see Norfolk-type rotation
Frampton, Dorset:
    sheep carcass weights 183
Fream, Dr W., RC assistant commissioner 50
freeholders 30–1
French wars 14, 16, 139
    grain yields 148–9, 153–4, 160
    harvest crises 217, 220
    porker carcass weights 205
Frome, Som.:
    report on in RC 50
fuller's earth 83
Fussell, G.E., agricultural writer:
    on cattle carcass weights 194–5, 201
    on farmers 29
    on sheep carcass weights 181–2
    on wheat yields 137–8

gallon:
    customary measure 245–6
Garboldisham, Norfolk-type 98, 238
    underdraining 90
Garstang, Lancs.:
    report on in RC 50
Gaywood, Norfolk-type 238
Geldeston, Norfolk-type 238
General Views:
    on sheep carcass weights 181
    on slaughter of animals 175
    on weights and measures 247
    on wheat yields 127, 134–5
Georgicall Inquiries:
    on manure and fertilizer 83
Gilbert, J.H., see Lawes, J.B.
Gilsland, Cumb. 233
Glendale, Lancs.:
    report on in RC 50

Glennie, P.:
    on wheat yields 117, 122–4, 132
Globe turnip 79–80
Gloucestershire 62, 235
    calf carcass weights 196
    pig carcass weights 207
    threshing machines 93
    water meadows 104
    wheat yields 145–6
    see also Arlingham; Blockley; Longhope;
        Lower Swell; Maugersbury;
        Oddington; Snowshill; Stoke Orchard;
        Stonehouse; Westington
Godshill, Isle of Wight 235–6
Golden Drop wheat 79–80
Golden Swan oats 79
Goodman, C., see Fussell, G.E.
goose dung 83
Goring, Sussex 242
Goss Hall Farm, Kent:
    accounting method employed 50
government inquiries, see agricultural
        statistics
Grafton Estate, Northants. 239
    purchased seeds 77
    use of hayseed 70
grain:
    exports of 16, 215–16
    imports of 18, 215, 217, 220, 222, 224
    output of 20–3
    see also prices
grain crops:
    acreages 37
    in arable rotations 70
    for beer/malting, see barley
    on mixed farms 69
    in a Norfolk-type rotation 72
grain production:
    according to Charles Smith 216 n. 2
grain yields 160–2, 165, 217, 221, 223
    in a Norfolk-type rotation 72
    see also named crops
grannary books 56–7
grass:
    in arable rotations 67
    as feed 59–60, 101
    in a Norfolk-type rotation 71–2, 75
    seeds 77–8
    and soil conditioners 83
grass mower 93

INDEX

grassland management 103–5
Gravesend, Kent:
   wheat yields 126
grazing 34, 48, 69, 96
Great Briddlesford, Isle of Wight 236
Great Depression (of late nineteenth
   century), *see* agricultural depression
Great Henny, Essex 234
Great Witchingham, Norfolk-type 238
Greene King brewery 154
Greetham, Lincs. 238
Grey Partridge peas 79
grey-faced sheep 98
guano 86, 89
   on John West's farm 112
gypsum:
   as a soil conditioner 82–3

Hackwood, Hants. 235
hair:
   as manure and fertilizer 83
Hall, Daniel, director of Rothamsted 33
   on keeping accounts 46
Hallett's barley 80
Hallett's wheat 56
Halstead, Essex 234
Hambledon, Hants. 235
Hampshire 61–2, 235–6
   customary measures 246
   Lord Bolton's estate 35
   Norfolk-type rotation 73
   water meadows 104
   wheat yields 117, 122–5, 128–9, 134–5, 146
   see also Andover; Binstead; Broughton;
      Bullington; Candover; Compton; East
      Stratton; Easton; Hackwood;
      Hambledon; Horsebridge; Kings
      Somborne; Lockerley; Meonstoke;
      Micheldever; Nether Wallop; North
      Stoneham; North Waltham; Pibworth;
      Pittleworth; South Warnbrough;
      South Wonston; Stockbridge; Wallop;
      Whitchurch
Hampshire Down sheep 57, 98–9
Hampstead Norreys, Berks. 231
   harvests of 1772, 1777 42
hand tools 92
Harding's Farmers Account Book 39
Harlow, Essex 234
   animal health 107

fallowing 75
ploughs 91
Harlowbury Farm, Harlow, Essex:
   selective breeding 96–7
harrows 91
Hartley Short Tap swede 80
harvest crises (1794–1801) 217, 220
   wheat yields 117, 133–4, 138
harvesting 17, 52, 60, 92, 228–9
Hatfield Peveril, Essex 235
Hatfield, Herts. 236
Hathersage, Derby. 233
Haughton, Shropshire, *see* Houlston
Haxton, John, Prize Essayist 159
hay 34, 37–8, 101, 103, 214
   in a Norfolk-type rotation 72
   seed 77–8
   yields 163, 165
haymaking machine 92
Healy, M.R. (and Jones, E.L.):
   on wheat yields 117–19, 126, 128, 136–7
Heath and Reach, Leighton Buzzard, Beds.:
   in a Norfolk-type rotation 71–2
Heath sheep:
   carcass weights 182
heifers 57, 60
   for carcass weights, *see* cows
hen dung:
   as manure and fertilizer 83
Hensham, Suffolk 241
Herbert, Robert, agricultural writer:
   on cattle carcass weights 201
   on lamb carcass weights 188
   on sheep carcass weights 183
Hereford cattle 98
Hereford, Herefordshire:
   wheat yields 126
Herefordshire 62, 236
   barley yields 155
   carcass weights 177, 195
   sheep carcass weights 182
   wheat yields 129, 134–5, 145
   see also Hereford; Moccas
Hertfordshire 236
   seeding rates 168
   wheat yields 117, 123, 125, 128, 132–3
   see also Hatfield; Hitchin; Ware;
      Welbury
Heslington, Yorks. 244
Hexham, Northumb. 239

hides 42, 176–8, 193
High Easter, Essex 234
High Farming 141
  in Lincolnshire 111–13
  wheat yields 134, 139
High Hoyland, Yorks. 244
High Legh, Ches. 232
  crop varieties 79
hiring books 47, 51
Histon, Cambs. 60, 232
  crop varieties 79
  ploughs 91
  underdraining 90
Historical Manuscripts Commission:
  on farm records 6
Hitchin, Herts. 98, 236
  reaper-binders 93
  wheat yields 128, 132
hoggets 190–1
hogs:
  carcass weights 186–7, 203–4
  fattening 60
Holdernes, B.A.:
  on agricultural output 12, 20–3, 138 n. 32
  on barley yields 151
  on cattle carcass weights 194
  on cattle numbers 194
  on farm accounts 46
  on farmers 29
  on grain yields 172
  on oats yields 157
  on pig numbers 202
  on sheep carcass weights 183
  on wheat yields 127, 133
Holkham, Norfolk-type 35
Holm Cultram, Cumbria 233
Holton le Moor, Lincs. 238
home farms 34–5
Hook Estate, Sussex 34
Hook, Yorks. 244
horn:
  as manure and fertilizer 83
horse feed 23 n. 51
horse-hoes 91
Horsebridge, Hants. 235
horses 20 n. 37, 54, 195
  at East Sutton, Kent 108
Hoton, Leics. 237
Houghton, Shropshire, *see* Houlston

Houlston (sometimes Houghton or Haughton), Shropshire 240
  Middle Wood Farm 106
Houndhill (in Darfield), Yorks. 55, 244
Hoyland, Yorks. 244
Hull, Yorks.:
  for supplying manure 87
Hunter Pringle, R., RC assistant commissioner 50
Huntingdonshire 62, 236
  carcass weights 195
  wheat yields 133
  see also Somersham
Hurstpierpoint, Sussex 242
  seeding rates 169–70
  yield rates 169–70

Ickworth, Suffolk:
  sheep carcass weights 183
Ilminster, Som.:
  carcass weights 177
Imperial measures 245–6
implements, *see* named implements
imports:
  of food 14 n. 18
  of grain 215, 217, 220, 222–224
income:
  and agricultural growth 14 n. 18
  redistribution of 1, 15
income tax 3, 5, 7, 37
industrial revolution 1, 9–10
  and grain yields 224
innovation 7, 35, 220
  in the open fields 88–9
inorganic fertilizer 87
inputs 128, 224
  at East Sutton, Kent 108
  see also under named inputs
Ireland:
  famine and pig stocks 201
  grain yields 160–2
  store cattle 98, 194
Isle of Wight:
  customary measures 246
  wheat yields 146
  see also Arreton; Godshill; Great Briddlesford; Yaverland

Jackson, R.V.:
  on agricultural growth 15, 213

John, A.H. 10
  on agricultural statistics 12
  on wheat yields 127
Jones, Edgar:
  on accountancy 47 n. 75
Jones, Eric L. 10
  on agricultural production 16
  on farm records 6
  see also Healy, M.R.
JRASE:
  on animal health 107
  on farmers 28
  on sheep carcass weights 183
  Prize Essays 2
  see also named agricultural writers
Juliana wheat 80
June Returns, *see* agricultural statistics

kale:
  as livestock feed 101
Kent 62, 236–7
  customary measures 246
  cow and heifer carcass weights 198–9
  East Sutton Court Farm 73–4, 85, 89, 107–8
  ewe carcass weights 185
  Goss Hall Farm 50
  lamb carcass weights 185
  in a Norfolk-type rotation 73
  pig carcass weights 207
  sheep carcass weights 192
  steer cattle carcass weights 200
  wethers carcass weights 189
  wheat yields 128, 147
  see also Ash; Burmarsh; East Sutton; Gravesend; Maidstone; Milstead; Rodmersham; Tonbridge; Tudeley
Kent, Nathaniel, land steward:
  on the farmer 28
Kentish wheat 79
Kerridge, E.:
  on the agricultural revolution 11, 138, 210–11
Kessingland, Suffolk 242
Kettleston, Norfolk-type 238
Kilkhampton, Corn. 232
King, Gregory, political arithmetist 13–14, 20, 211
  on bread 50
  on carcass weights 194–5
  on cattle numbers 194
  on farmers 28, 30–1
  on malting barley 154
  on pig numbers 202
  on sheep carcass weights 181
Kingham, Oxon. 240
Kings Somborne, Hants. 235–6
  crop varieties 79
Kingslake, John, Som. farmer 38
Kingston Deverill, Wilts. 59, 243
  sheep breeding 99
Kirkby Thore, Westmorland 47, 243
Kirstead cum Langdale, Norfolk-type 238
Knaith, Lincs. 238
kohlrabi:
  as livestock feed 101

labour 17–18, 28–9, 51–3, 228–9
  costs 51–2, 60, 72, 104
  and crop yields 36
  family labour 52–4
  productivity 18, 24, 64–5, 92, 225–7
labour records 35, 39, 51–4, 64, 92
  and crop yields 43
labourers 7, 9, 28, 32, 51–3, 92
lambing 58
lambing rates 7
lambs 7, 58–9
  carcass weights 185–9, 214
Lancashire 62, 237
  customary measures 245
  ewe carcass weights 185
  lamb carcass weights 185
  in a Norfolk-type rotation 73
  pig carcass weights 207
  wheat yields 133
  see also Atherton; Bashall; Bury; Didsbury; Dumplington; Garstang; Glendale; Little Crosby; Liverpool; Rufford; Scarisbrick; Warrington
land management 103–6
land productivity 24, 116–49
land quality:
  re. wheat yields 148–9
land utilization, *see* English land utilization
Landbeach, Cambs. 232
landlords 9–10, 15, 29, 30, 32
  re. rents 221–3, 225
landowners 1, 28–9, 35

Langham Farm, Langham, Norfolk:
  farm diaries 63
  open fields and enclosures 88–9
Langham, Norfolk-type 88–9, 238
  crop varieties at 79
Langton, Lincs. 238
Lawes, J.B. (and J.H. Gilbert), agricultural writers:
  on carcass weights 175
  on grain yields 161–2
  on wheat yields 4, 119–20, 126–8, 136–7
Layer Marney, Essex 234
leasehold 30–1
leases:
  covenants regarding manure 84, 87
legumes:
  as livestock feed 101–2
  and manure 85
  on mixed farms 69
  as nitrogen fixers 101
  in Norfolk-type 19
  in a Norfolk-type rotation 75–6
  see also named crops
Leicester sheep 98
  carcass weights 182–3
  scab in 107
Leicestershire 62, 92, 237–8
  customary measures 245
  in a Norfolk-type rotation 73
  see also Anstey Mill; Beaumont Leys; Breedon-on-the-Hill; Burton on the Wold; Chadwell; Cotes; Dishley; Hoton; Loughborough; Owston; Prestwold; Riby; Tonge; Wilson
Leighton Buzzard, Beds.:
  Heath and Reach 71–2
Lesant, Devon 57
Leverton, Lincs. 238
ley and ley grasses 55, 71
  in arable rotations 70
  at East Sutton, Kent 108
  in a Norfolk-type rotation 72–4, 111
  re. and oats 160
lime and liming 105–6
  on John West's farm 112
  in a Norfolk-type rotation 89
  as a soil conditioner 81–2
Lincoln sheep 98
Lincolnshire 62, 238
  agricultural depression 230

barley yields 154–5
  carcass weights 182
  Corn Laws 225
  pig carcass weights 207
  report on in RC 50
  Scorer Farm 63
  West Farm 63, 103
  wheat yields 117, 123, 125, 132–5, 146–7
  see also Burwell; Dunholme; Greetham; Holton le Moor; Knaith; Langton; Leverton; Nettleton; Nocton Rise; Redbourne; Riby; Searby; Sleaford; Spalding; Sudbrooke; Thornton; Turnor, Christopher; Washingborough
linseed dressing:
  as manure and fertilizer 87
Litchfield, Staffs. 241
Little Crosby, Lancs. 237
Little Dunmow, Essex 234
Little Fois swede 80
Little Houghton, Northants. 239
  Little Houghton Farm and fallowing 75
Little Shardeloes, Bucks. 231
liverfluke 107
Liverpool, Lancs. 237
  wheat yields 126
livestock:
  re. agricultural practices 95–107
  at East Sutton, Kent 108
  and manure 56
  see also named animals
livestock farming 57, 214–15
livestock feed 59, 87
  in arable rotations 70
  in a Norfolk-type rotation 72
livestock production 58–9, 173–209
livestock records 57
livestock trade:
  from Scotland and Wales 229
livestock weights 42, 247
liveweight:
  ratio of live to slaughter weight 176–80
load:
  customary measure 246
Lockerley, Hants. 235
Lockinge, Berks. 231
Loftus, Yorks. 244
Londesborough, Yorks. 244
London market:

carcass weights 177
  for livestock 96–7
  for malt 155 n. 12
  for seed merchants 108
  Smithfield 194
  re. Upton Farm 113
long wool sheep, *see* fleeces
Longhope, Glos. 235
Longhorn cattle 98
Lord Bolton's Estate, Hants. 35
Lord Northwick of Blockley, Glos.
  estate accounts 34
Lord Rayleigh, estate owner from
    Essex 46
Lotus, *see* trefoil
Loudon, J.C., agricultural writer:
  on barley yields 152–3
  on double-entry 50
  on farm accounts 39–40
Loughborough, Leics. 237
  land management 105
  selective breeding 96
Loughton, Essex 234
Lower Swell, Glos. 235
Luccock, John, wool stapler from Leeds:
  on sheep numbers 174
lucerne 71
  in ley farming 70
  as a nitrogen fixer 85–6, 108
  in a Norfolk-type rotation 73–4
Ludham, Norfolk-type 239

McConnell, Primrose, agricultural writer:
  on customary measures 245, 247
  on weights 165–6
McCulloch, J.R., economist:
  on agricultural statistics 12
  on wheat yields 127
  on yield rates 171
machinery 91–3
Maiden Newton, Dorset 233
Maidstone, Kent 107–8
  report on in RC 50
maize:
  imports of 217
  as purchased feeds 102–3
Malden, W.J.:
  on livestock feed 101
malt dust:
  as manure and fertilizer 83, 87

Malthus, T.R., economist 1, 23
malting barley 34, 68, 79, 150, 154–6, 230
  on John West's farm 112–13
mangle drill 92
mangolds 73, 83
  in arable rotations 67
  as livestock feed 101–2
  in a Norfolk-type rotation 71–2
  seeds 77
  yields 56
manure 35, 56, 81, 84–5, 87, 105–6, 113, 175,
    213–14
  and crop yields 36
  at East Sutton, Kent 108
  on John West's farm 112
  in Norfolk-type 19
  in a Norfolk-type rotation 72
mares:
  breeding of 58
Mark Lane Express:
  on wheat yields 127
market gardens 230
marketing 7, 10, 57, 229
Market Weighton, Yorks. 57–8
Markets and Fairs (Weighing of Cattle) Act,
    1887 178
marl:
  as a soil conditioner 82–3
Marlborough Fair, Wilts. 59
Marlborough Grees peas 79
Marrow Fat peas 79
Marsh sheep 98
Marshall, William, agricultural writer:
  on farmers 28
  on selective breeding 97
  on sheep carcass weights 181
  on straw from oats 159
  on yield rates 171
Masham sheep 98
Massie, Joseph, social commentator:
  on farmers 28, 31
master and steward system 44–5
Maugersbury, Glos. 235
meadows, *see* pasture
meal:
  as purchased feeds 103
mechanization 228–9
  see also capital; farming; innovations; under
    named items
Medicago, *see* trefoil; lucerne

medieval grain yields 116 n. 1, 166, 168
medieval seeding rates 170
Mediterranean:
  imports from 217
Medmenham, Bucks. 231
  Norfolk-type rotation 75
Meikle, Andrew, inventor of a threshing
    machine 93
memoranda books 7, 27, 51, 53, 55–61
Meonstoke, Hants. 38, 235
Mettingham (Bungay), Suffolk 242
mice:
  as pests 93
Micheldever Down, Hants. 235
  land managament 105
Middle Wood Farm, Houlston,
    Shropshire:
  land management 106
Middleton, John, agricultural writer:
  on agricultural statistics 12 n. 11
Middleton, Northumb. 239
  customary measures 246
Middlezoy, Som. 241
Midland Agricultural College 33
milk production 34, 214
  milk yields 7
Milner Gibson, T., MP and free trader:
  on agricultural statistics 3
Milstead, Kent 54, 237
Milton Abbot, Devon 57, 233
Milwich, Staffs. 241
Mingay, G.E.:
  on agricultural revolution 10–11, 13
  on farm accounts 46
  on farm labour 228
  on farmers 29
  on statistical sampling 207–8 n. 52
mixed farming and farming practice 214
Moccas, Herefordshire 54, 236
Mokyr, J.:
  on food consumption 16–17
Moore-Colyer, R.:
  on sheep carcass weights 183
Moreton Say, Shropshire 240
  cattle distemper 106
Morfe sheep:
  carcass weights 183
Morningthorpe, Norfolk-type 239
mortality rates in animals 58
Morton, John:

  on weights and measures 247
Moulton, Norfolk-type 239
mustard:
  as livestock feed 101
mutton, see sheep

Napoleonic Wars, see French wars
National Register of Archives 6
Nether Wallop, Hants. 236
  crop varieties 80
Nettlecombe, Som. 241
Nettleton, Lincs. 238
  crop varieties 79
Newbiggin, Northumb. 239
Newton Arlosh, Cumbria 233
  customary measures 246
night soil:
  on John West's farm 112
nitrate of soda 87, 89
nitrogen 88, 91
  and barley 113, 155–6
  fixing of 85–6, 108, 213
  in a Norfolk-type rotation 71
  and oats 160
  and yields 69
Nocton Rise, Lincs. 238
Norfolk-type 19, 24, 62, 79, 83, 90, 98, 238
  medieval estate accounts 43–4
  medieval grain yields 166, 168
  probate inventories 31
  seed drills 91
  sheep breeding 99
  sowing rates 168
  water meadows 104
  wheat yields 15, 117, 123, 125, 129, 132,
    134–5, 146–7
  see also Clenchwarton; Costessey;
    Felbrigg; Garboldisham; Gaywood;
    Geldeston; Great Witchingham;
    Holkham; Kettleston; Kirstead cum
    Langdale; Langham; Ludham;
    Morningthorpe; Moulton;
    Norwich; Thurgarton; Wacton;
    Wymondham
Norfolk-type rotation 70–3, 75, 111, 156,
    220
  at Langham, Norfolk-type 89
  and sheep carcass weights 182
  see also named crops

INDEX

Norman, William, labourer on Colworth Farm 54
North America:
  grain imports 217, 224
North Frodingham, Yorks. 244
North Holme, Yorks. 244
North Stoneham, Hants. 236
North Waltham, Hants. 236
  customary measures 247
Northamptonshire 62, 239
  sheep breeding 99
  yield rates 169
  see also Abthorpe; Ashby St Ledgers; Deanshanger; Grafton; Little Houghton; Sibbertoft; Towcester; Wickham; Wittering
Northumberland 34, 62, 239
  calf carcass weights 196
  cow and heifer carcass weights 198–9
  ewe carcass weights 185
  lamb carcass weights 185, 188
  pig carcass weights 207
  seed drills 91
  wethers carcass weights 190
  see also Bamburgh; Brunton; Castle Heaton; Coldstream; Culley, George; East Brunton; Ford; Hexham; Middleton; Newbiggin; Park End; Thornbrough; West Brunton
Norton, Derby. 51–2, 233
Norwich, Norfolk-type:
  wheat yields 126
Nottingham, Notts. 240
Nottinghamshire 62
  cow and heifer carcass weights 199
  customary measures 246
  ewe carcass weights 185
  Norfolk-type rotation 73
  pig carcass weights 207
  technical training in agriculture 33
  see also Bingham; Chilwell; Collingham; Cropwell Butler; Shelford; Strelley; West Bridgford
nutrients 69–70, 83–5, 88
Nymet Rowland, Devon 233
  sheep sales 176

Oakes Farm, Norton, Derby. 51–2
oats 34, 54–6, 86, 150–1, 158, 160
  in agricultural output 20, 23 n
  in arable rotations 66–8
  as a catch crop 68–9
  as feed 68, 101–2, 150, 159, 180
  and nitrogen 160
  in a Norfolk-type rotation 71, 74, 89
  seeding rates 166–72, 214
  seeds 78
  special varieties 79–80
  straw 81, 158–60
  at Upton Farm 113
  yields 7, 42, 54, 81, 150–1, 156–63, 165, 213
O'Brien, P.K.:
  on agricultural prices 22
  on capital formation in agriculture 25–6
Oddington, Glos. 235
offal 176–7
oilcake 38, 86–7, 89, 103, 112
old boll:
  customary measure 246
Ombersley, Worcs. 243
  bullock carcass weights 200
  cow and heifer carcass weights 199
  oxen carcass weights 200
Onobychis, *see* sainfoin
open fields 10, 15, 69, 88, 215
output 6, 35, 57 n. 99, 217
  see also agricultural output
Overton, M.:
  on agricultural growth 15, 20
  on agricultural output 23–4
  on agricultural revolution 24
  on agricultural statistics 26
  on labour productivity 227
  on mechanization 228 n. 38
  on Norfolk-type agriculture 19
  on wheat yields 15, 117, 122–4
owner-occupier 30–1, 36–7
Owston, Leics. 238
  underdraining 105
Ox Moble potatoes 79
oxen:
  carcass weights 176, 195, 199–200
  cost of feeding 60
  on the ratio of meat to offal 177
Oxford 39
Oxford Down sheep 57, 99
Oxfordshire 62, 240
  customary measures 246

Oxfordshire (*cont.*)
  hog carcass weights 203
  mechanization 228
  pig carcass weights 207
  wheat yields 117, 123, 125, 132
  see also Adderbury; Bloxham; Chipping Norton; Deddington; Ditchley; Kingham; Shilton; Sibford Ferris; Standlake
oyster shells:
  as manure and fertilizer 83
paring and burning:
  as manure and fertilizer 84
Park End, Northumb. 239
  cow and heifer carcass weights 199
  crop records 43
pasture 37, 39, 55, 160, 180
peas 23 n. 51, 55, 69, 102
  in arable rotations 66–9
  at East Sutton, Kent 108
  as feed 60
  as nitrogen fixers 69, 85–6, 108
  in a Norfolk-type rotation 74–6, 89
  seeds 77–8
  special varieties 79–80
  yields 164–6
peat:
  as manure and fertilizer 84
Peldon, Essex 234
perquisites 52–3
Perry, P. J.:
  on farmers as businessmen 50
Peru:
  guano imports 86
pests 83, 93, 95 n. 72
  see also named pests
Pibworth, Hants. 236
pigeon dung 56, 83–4, 89
pigs 20, 37, 47, 100, 202
  breeding 58, 201–2
  breeds 98
  carcass weights 184–7, 200–7, 214–15
  at East Sutton, Kent 108
  at Langham, Norfolk-type 89
  ratio of meat to offal 177
  see also named animals
pipe drainage 90–1
Pipe Rolls 43
Pittleworth, Hants. 236
plant varieties 35, 79–80

  see also named crops
Playford, Suffolk 242
ploughing 54, 69
ploughs 2, 91
Poland oats 79–80
Pollard, S.:
  on accounting systems 44
Polperro, Corn. 232
population 3, 211, 215–17
  and agricultural growth 14 n. 18, 218–21, 223–4
  agricultural population 225
  and agricultural revolution 213, 230
  feeding of 1–2, 10, 13–14, 16–17, 25
porkers:
  carcass weights 186–7, 204–7
  fattening 60
Portland sheep 98
potato blight 94
Potato oats 79–80
potatoes:
  in a Norfolk-type rotation 72
  special varieties 79
  and wire worm 94
poultry 20
  and wire worm 94
Pratt, Thomas of Ansty Mill, Leics., client of John Ellis 48
Preston, Rutland 98, 240
  crop varieties 80
  machinery 93
Prestwold, Leics. 237
  land management 105
prices 1, 5, 7, 12, 14, 17, 25
  in French wars 55, 82, 222
  of grain 20–1, 215–17, 220–1
  of livestock 97
  O'Brien's agricultural price index 22
  Rousseaux's agricultural price index 22
Price's Farm, Brandwood, Shropshire:
  land management 106
Privy Council 217
Prize Essays 2
  see also under named agricultural writers
probate inventories 14, 28, 31, 211
  derivation of crop yields from 5, 122–8, 132
productivity 5, 11–12, 24
  in grain crops 166–71
  see also labour productivity

protection, *see* Corn Bounty; Corn Laws
Prothero, R.E. (Lord Ernle):
  on agricultural revolution 9–10
Puddletown, Dorset 233
pulse crops 37
  as fodder 162
  as nitrogen fixers 69 n
  yields 166
purchased feeds 102–3
purchased seeds 77–80, 113
Purple Top swede 80
Pyrford, Surrey 242

Quainton, Bucks. 98, 232
Quantock, James of South Petherton, Som.:
  on carcass weights 42, 177

rabbits 20 n. 46
rack rent, *see* rent and rentals
Radley, Berks. 231
rags:
  as manure and fertilizer 83
railways:
  and marketing 113, 229
rams 58
Ransome's plough 91
rape:
  as livestock feed 101–2
rape dust:
  as manure and fertilizer 87
ratio of live to slaughter weight 176–80
ratio of meat to offal 176–80, 193
rats:
  as pests 93
Read, C.S., MP and agricultural writer 5
  on barley yields 156
Reading, Berks.:
  for the purchase of seeds 77
reap hook 92
reaper-binder 92–3
Red Champion potatoes 79
Red Chevalier wheat 80
red clover 70, 78–80
Red Globe Turnips 79
Red Harty wheat 79
Red Lammas wheat 80
Red Northumberland wheat 80
Redbourne, Lincs. 238
Redgrave, Suffolk 242
regional specialisation:

  and selective breeding 96
rent and rentals 1, 5, 7, 9, 17, 28, 30, 34, 215, 217–19, 220–5
rental value:
  and income tax 37
Report of the Judges of Book-Keeping 40
Rew, R.H., agricultural writer:
  on carcass weights 183–4, 193
Riby, Leics. 238
  seed drills 92
Ricardo, David, economist 1
rice:
  imports 217
Rickinghall, Suffolk 242
Rider Haggard, H., writer:
  on Edward Strutt 46
ridge and furrow drainage 89
Ridley family, Park End, Northumb. 43
rindepest 4
Rivett's wheat 79–80
Robertson, George, writer:
  on post-French wars agricultural
    depression 222
Robinson, G.W., agricultural writer:
  on lime 82
Rochford, Essex 234
  Doggetts Hall Farm 91
  manure 84–5
  use of lime 82
Rodmersham, Kent 237
Romney Marsh:
  and livestock farming 96
Romney sheep 181–2
root crops 2, 35, 37–8, 213
  and barley 156
  as feed 59, 101–2
  in a Norfolk-type rotation 71–2
  and oats 159
  and phosphates 84, 88
  and seed 77
  and soil conditioners 83
  see also under named crops
root fertilizers 87
rotational grasses 55–6
Rothamsted, experimental agricultural station 33
  wheat yields 126
  see also, Lawes, J.B.
Rousseaux:
  agricultural price index 22

Royal Agricultural Society:
  on farm accounts 40
Royal Commission on Agricultural Depression,
  1894–6 38
  on farmers as businessmen 46–7, 50
  reports of assistant commissioners 50
Royal Society, the Georgicall Inquiries 83
Rudby, Yorks. 244
Rufford, Lancs. 237
Rural History Centre, Reading University:
  on farm records 6
  see also, Collins, E.J.T.
Rushbrooke Park, Suffolk:
  experiments with turnips 35
Russell Estate, Hants.:
  land management 105
Russell, N.:
  on carcass weights 176
Rutland 62, 240
  water meadows 104
  see also Preston
rye 18, 20
  in arable rotations 66–8
  as a bread grain 68 n. 2
  as a catch crop 68–9
  in a Norfolk-type rotation 89
  yields 162–3, 165
ryegrass:
  purchased seeds 77
  yields 164–5
Ryton, Dur. 234

sack:
  customary measure 246
sainfoin 71
  in arable rotations 67
  in ley farming 70
  as livestock feed 101
  as nitrogen fixer 85–6, 108
  in a Norfolk-type rotation 73–5
  and soil conditioners 83
Saltmarshe, Yorks. 57, 244
  purchasing manure and fertilizer 87
  seed drills 92
  underdrainage 92
saltpetre:
  as manure and fertilizer 84
salts:
  as a cure for wire worm 94
  on John West's farm 112

salving 93
Sandford, Devon 233
  crop varieties 79
  hand tools 92
  sheep carcass weights 192
  sheep sales 184
Scarisbrick, Lancs. 237
Schofield, R.:
  see Wrigley, E.A.
Scorer Farm, Lincs. 63
Scotch sheep 98
Scotland:
  livestock trade 229
Scottish heifers:
  at Wentworth 42
Scotts cattle:
  carcass weights 195
scythe and scythesman 92
seam:
  customary measure 246
Searby, Lincs. 238
  crop varieties 80
  underdraining 90
seaweed:
  as manure and fertilizer 83–4
Second Report of the Commissioners on Weights
  and Measures, 1820 247
seed drills 91–2
  seeding rates 172
seed varieties 79–81
seeding rates 12, 56, 166–72, 213–14
  wheat 218–19
seedlip 92
seeds 20, 23, 35, 50, 54–6
  at East Sutton, Kent 108
  as feed 101
  on John West's farm 112
  in a Norfolk-type rotation 73–4
Select Committees:
  of the 1830s 222
  see also Royal Commission on Agricultural
    Depression
selective breeding (of animals) 95–7, 99
Shapwick, Dorset 233
Sharnbrook, Beds. 231
  Colworth Farm 54
shearhogs 190–1
  sales 184
sheep 20, 37, 96–7, 100, 174–5, 192–3
  and barley 156

breeding 57-9, 99, 111
breeds 98
  carcass weights 181-93
  at East Sutton, Kent 108
  and feed 60, 103, 108, 111-12
  at Langham, Norfolk-type 89
  and parasites 107
  ratio of meat to offal 177
  at Upton Farm, Sompting, Sussex 113
  for wool 95-6
  see also named breeds; separate animals
sheep accounts 58-9
sheep folding 85
sheep rot 107
Sheffield, Yorks:
  customary measures 246
Shelford, Notts. 240
  crop varieties 80
Shelley Yorks. 244
  crop varieties 80
  use of lime 82
Shilton, Oxon. 240
Shirley, Surrey 107, 242
short wool sheep, see fleeces
Shorthorn cattle 98
shorthorn cows:
  carcass weights 196
shorthorn oxen:
  carcass weights 196
Shropshire 62, 106, 240-1
  bacon pig carcass weights 202
  bullock carcass weights 200-1
  calf carcass weights 196
  carcass weights 195
  cow and heifer carcass weights 199
  porker carcass weights 205-6
  seeding rates 168
  sheep breeding 99
  sheep carcass weights 192
  wethers carcass weights 189
  see also Brandwood; Bridgewater Estate;
    Bridgnorth; Ellesmere; Houlston;
    Moreton Say; West Drayton;
    Whitchurch; Wrockwardine
Shropshire cattle 98
Shropshire Down sheep:
  carcass weights 183
Sibbertoft, Northants. 48
Sibford Ferris, Oxon. 240
sickles 92

Silksworth, Dur. 234
Sinclair, Sir John:
  on cattle carcass weights 194
slaughter weights 58
  ratio of live to slaughter weight 176-80
  see also carcass weights
Sleaford, Lincs. 113
Sledmere, Yorks.:
  estate accounting system 45
Slicher van Bath, B.H.:
  on yield rates 170-1
small-seeded legumes:
  in ley farming 70
  as nitrogen fixers 69
  seeds 77-8
  see also named varieties
Smeeton, William of Sibbertoft, Northants.,
    client of John Ellis 48
Smith, Charles, agricultural writer:
  on agricultural output 20-1
  on grain 150-1
  on grain consumption 216
  on wheat yields 117, 138
Smithfield market, London 194
Smythe, William of Little Houghton Farm,
    Northants:
  on fallowing 75
Snowshill, Glos. 235
soil acidity 81, 83, 89
soil conditioners 35, 56, 81-3
  see also named varieties
soil fertility 83-6
Somerset 63, 241
  cow and heifer carcass weights 198
  ewe carcass weights 185
  lamb carcass weights 185
  Norfolk-type rotation 73
  pig carcass weights 207
  sheep carcass weights 192
  wethers carcass weights 189
  wheat yields 129, 134-5, 145
  see also Bishop's Lydeard; Crowcombe;
    Dulverton; Dunster Castle; Frome;
    Ilminster; Middlezoy; Nettlecombe;
    South Petherton; Taunton; Trull;
    Uphill; Whitstock
Somerset Levels 38
Somersham, Hunts. 236
  crop varieties 79
  land management 106

Sompting, Sussex 242
  crop varieties 80
  Upton Farm 113
soot:
  as manure and fertilizer 83–4
South Down sheep 96–7
South Milton, Devon 233
South Petherton, Som. 42, 241
  carcass weights 177
  sheep carcass weights 192
South Wales cattle 98
South Warnbrough, Hants. 37–8, 236
South Wonston, Hants. 236
  sheep breeding 99
Southdown sheep 98, 100
  carcass weights 182
  ewe carcass weights 185
sowing (of crops) 35, 54, 228
sowing rates, see seeding rates
sows:
  breeding of 58
  carcass weights 186–7
Spalding, Lincs.:
  wheat exports 225
Sparrowbill White oats 79
spring corn:
  in a Norfolk-type rotation 75
  see also named crops
stable dung 106
Staffordshire 63, 241
  customary measures 245
  Norfolk-type rotation 73
  wheat yields 133
  see also Abbots Bromley; Litchfield; Milwich
stall feeding:
  in Norfolk-type 19
Stamp, Sir Lawrence Dudley:
  on land utilization 142–5
Standlake, Oxon. 240
Stansfield, Suffolk 242
Stanton All Saints, Suffolk 242
  sheep carcass weights 183
statistical surveys, see agricultural statistics
statute acre 245–7
steam drainage:
  in Lincolnshire 225
steam machines 93, 225
steelyards 178–9
steers:
  carcass weights 199–200
Stephens, H.:
  on steelyards 178
Steventon, Berks. 231
stock feeding:
  as a farming system 100
Stockbridge, Hants. 236
stocking rates 105
Stoke Abbot, Dorset 233
Stoke Orchard, Glocs. 235
Stoke Wood Farm, Meonstoke, Hants.:
  re. tenant right 38
Stokenchurch, Bucks. 232
  crop varieties 80
Stonehouse, Glocs. 235
Stopsley, Beds. 231
store cattle:
  from Ireland 194
  see also cattle
Stovin, Cornelius, farmer 113 n, 129
  on crop diseases 94
Stratford-upon-Avon, War.:
  report on in RC 50
straw 38
  as animal bedding 85
  from barley 156
  as fodder crop 214
  as manure and fertilizer 83, 85
  from oats 81
Street Farm, South Warnborough, Hants.:
  re. tenant right 37–8
Street, A.G., agricultural writer:
  on learning to farm 32
Strelley, Notts. 240
strike:
  customary measure 246
Strutt, Edward, estate owner from Essex 46, 50
stubble:
  as livestock feed 101
Style, Ches. 232
Sudbourne, Suffolk 242
Sudbrooke, Lincs. 238
Suffolk 63, 241
  barley yields 156
  carcass weights 195
  customary measures 246
  pig carcass weights 207
  probate inventories 31
  seed drills 91

threshing machines 93
wheat yields 15, 117, 123, 125, 129, 132–5, 147
  see also Badmondisfield; Bulcamp; Bungay; Bury St Edmunds; Buxhall; Coddenham; Darsham; Flempton; Hensham; Ickworth; Kessingland; Mettingham; Playford; Redgrave; Rickinghall; Rushbrooke Park; Stansfield; Stanton All Saints; Sudbourne; Swilland; Wortham
Suffolk cattle 98
Suffolk pigs 98
Suffolk sheep 98
sugar beet:
  re. soil conditioners 83
Sulham, Berks. 231
superphosphates 82, 86, 88
  on John West's farm 112
Superfine Early peas 79
Surrey 63, 242
  carcass weights 195
  sowing rates 168
  threshing machines 93
  see also Pyrford; Shirley
Sussex 34, 242
  calf carcass weights 196–7
  carcass weights 195
  customary measures 246
  Danny Estate 56
  pig carcass weights 207
  wheat yields 128–9, 133–5, 146
  see also Chailey; Goring; Hurstpierpoint; Sompting; West Dean
Swallowfied, Berks. 60, 98, 231
  crop varieties 79
  hand tools 92
  hog carcass weights 203–4
  porker carcass weights 205–6
swedes 73, 89
  in arable rotations 67
  as feed 101–2
  in a Norfolk-type rotation 71–2, 111
  seeds 77
  special varieties 80
Swilland, Suffolk 242
Swindon, Wilts.:
  wheat yields 126
Swing Riots 228–9

Tabor family, Bocking, Essex 56
Talavera wheat 79
tallow 42, 176–7
  ratio of meat to offal 193
Tankard turnips 79–80
Tankersley, Yorks. 244
tares:
  as livestock feed 101
  as nitrogen fixers 108
  in a Norfolk-type rotation 74
Tarrant Monkton, Dorset 234
Tartarian oats 79–80, 159
task work 51–3
Taunton wheat 79–80
Taunton, Som. 241
Tavistock, Devon 233
Tawneys barley 79
tax, *see* income tax
technology 226
  see also named implements; mechanization
temporary grasses 71
  in arable rotations 70
  in a Norfolk-type rotation 72
tenant farmers 5, 9–10, 16, 32, 36–8
  their accounting systems 46–50
  their records 34, 36–42
tenant right 37–8
tending the soil 81–8
tenure 10, 30–1, 228
Thirsk, Joan 13
Thompson, F.M.L.:
  on agricultural inputs 224
  on agricultural revolution 11
  on imported artificial fertilizers 140
Thornbrough, Northumb. 98, 239
  crop varieties 80
  reaper-binders 93
Thornton, Lincs. 238
threaves 190
three-course rotation 66, 180
  see also arable, rotations
threshing 47, 52, 54
threshing machines 93, 228
Thurgarton, Norfolk-type 239
Thurvaston, Derby. 233
Tithe Commutation Act, 1836:
  re. wheat yields 125
tithes and tithe commutation 3, 28, 77
Tonbridge, Kent:
  wheat yields 126

Tonge, Leics. 237
  land management 105
tovet:
  customary measure 246
Towcester, Northants. 239
  crop varieties 80
Townshend, Lord 'Turnip':
  re. innovation 230
transport, *see* marketing
transport costs 81
tree bark:
  as manure and fertilizer 83
trefoil 70–1
  in arable rotations 67
  as feed 60, 101
  as nitrogen fixer 70, 85–6
  in a Norfolk-type rotation 74–6, 89
  see also clover
Trifolium, *see* clover
Trow-Smith, R.:
  on carcass weights 176
Trull, Som. 241
Trump wheat 80
Trumpington, Cambs. 232
tuberculosis in cows 107
Tudeley, Kent 237
  accounting method employed 50
Tull, Jethro, innovator 230
tups 190–1
turf:
  as manure and fertilizer 84
Turkey long pod beans 79
Turner, Jabez, RC assistant commissioner 50
turnip cutters 92, 228
turnip manure 87
turnips 54, 73, 213
  in arable rotations 67, 111
  in East Anglia 19 n. 42
  at East Sutton, Kent 108
  and fallowing 75
  as feed 101–2, 108
  in Norfolk-type 19
  in a Norfolk-type rotation 71–2, 75–6, 89
  and phosphate 84
  as seeds 77–8
  special varieties 80–1
Turnor, Christopher, Lincs., landowner 33
  on keeping accounts 46
  on manure and fertilizer 85
turnpike roads:
  and marketing 229
Twynam, John of Whitchurch, Hants., sheep breeder 57, 99

underdrainage 35, 90, 227–8
  see also drainage
undersowing 56
  in a Norfolk-type rotation 75
United Kingdom:
  grain yields 160–2
  wheat yields 128, 135
Uphill, Som. 241
Upminster, Essex 234–5
  crop varieties 79
  hand tools 92
  land management 106
Upottery, Devon 233
Upton Farm, Sompting, Sussex 113–15
urbanization:
  implications for agriculture 223, 229
  see also population

Vale of Aylesbury, Bucks.:
  open fields 88 n. 47
  water meadows 104
valuations 36, 64
vetches 75
  in arable rotations 67
  as feed 101–2
  as nitrogen fixer 86, 108
  in a Norfolk-type rotation 75–6, 89
  and oats 160
  and wire worm 94
  yields 164–6
veterinary science 106–7
Victoria wheat 80

Wacton, Norfolk-type 239
Wade Martins, S. (and Williamson, T.)
  on a barter system 48 n. 79
  on East Anglian agriculture 19
wage books 51–3
wages 1, 5, 51–3, 60
Wakefield, Yorks. 58
Wales:
  livestock trade 229
Wallop, Hants. 235–6
Walter of Henley:
  on marl 83
Warbstow, Corn., *see* Fentrigan
Ware, Herts.:

INDEX

for malting barley 155 n. 12
Warrington, Lancs.:
　wheat yields 126
Warwickshire 63, 242–3
　calf carcass weights 196
　cow and heifer carcass weights 199
　ewe carcass weights 185
　lamb carcass weights 185
　pig carcass weights 207
　sheep carcass weights 191–2
　threshing machines 93
　water meadows 104
　see also Castle Bromwich; Coventry;
　　Eathorpe; Stratford-upon-Avon;
　　Weethley; Wootton
Washingborough, Lincs. 238
water meadows 101, 103–4
　and water-borne disease 107
Wath upon Dearne, Yorks. 244
weather diaries 57
　see also farm diaries
Webb's Farm Account Book 39
weeds and weeding 69, 93
　in a Norfolk-type rotation 72
　and seed drills 91
　and soil conditioners 83
Weethley, War. 243
weighbridges 42, 177–9
weights (of carcasses), see named animals
weights and measures 165–6, 245–7
Welbury, Herts. 236
　wheat yields 128, 132
Welsh borders:
　re. mechanization 228
Welsh cattle 98
　carcass weights 195
Welsh sheep 98
Wentworth Estate, Yorks. 42, 244
Wessex:
　Norfolk-type rotation 72
Wessex Downs:
　barley 156
West Bridgford, Notts. 240
West Brunton, Northumb. 239
West Country sheep 98
West Dean, Sussex 42, 242
　customary measures 245–6
West Drayton, Shropshire 241
West Mersea, Essex 234
West, John of Dunholme, Lincs.:

　during High Farming 111–13
　family farm accounts 63
　purchasing feeds 103
Westgate-in-Weardale, Dur. 37, 234
Westington, Glocs. 235
Westmorland 63, 243
　pig carcass weights 207
　see also Kirkby Thore
Weston, Sir Richard:
　on the introduction of red clover 70
wethers 184
　carcass weights 186–7, 189–90
whale blubber:
　as manure and fertilizer 87
wheat 47, 55–6, 12
　acreages 18, 141, 151, 215, 218–19,
　　226–7
　in arable rotations 66–8, 75, 111
　compared with barley 152, 154
　customary measures 245–6
　cutting of 92
　and fallowing 75
　fertilizer 87
　imports of 223
　in a Norfolk- type rotation 71–2, 74–6, 89
　output 16, 18, 20, 141, 215–30
　reaping 92
　and seed drills 91
　seed varieties 78–81
　seeding rates 166–72, 213
　at Upton Farm 113
　and wire worm 94
　yields 4, 7, 42, 54, 56, 116–49, 160–3, 165,
　　212–13, 218–19, 223–7
　yields and soil types 142–9
Whitchurch, Hants. 236
　sheep breeding 99
　steam cultivation 93
　John Twynam 57
Whitchurch, Shropshire 241
　Bridgewater Estate 106
　underdraining 105
White Chevalier wheat 80
White clover 78–9
White Dutch oats 79
White Essex wheat 79
White Hedge wheat 79
White loaf turnips 79
White Round turnips 79–80
White Thanet barley 79

Whitstock, Som. 38
Whittington wheat 80
Wickham, Northants. 239
Wickham Bishops, Essex 235
Wickham St Pauls, Essex 235
Wilkes, A.R.:
  on wheat yields 126 n. 13
Willand, Devon 233
Williamson, T.:
  see Wade Martins, S.
Wilson Fox, A., RC assistant commissioner 50
  on farm labour and wages 53
Wilson, Leics. 237
  land management 105
Wilson, R.G.:
  on malting barley 154, 156
Wilson, Thomas:
  on landholders 30
Wiltshire 32, 63, 243
  customary measures 245–7
  cow and heifer carcass weights 199
  dairying 99
  pig carcass weights 207
  seeding rates 168
  sheep breeding 98–9, 193
  sheep carcass weights 182
  water meadows 104
  wheat yields 146
  see also Aldbourne; Baydon; Bishops Cannings; Bratton; Crudwell; Enford; Kingston Deverill; Marlborough (Fair); Swindon; Wingfield; Wishford
Winchester measure 245–6
Wingfield, Wilts. 243
winnowing machines 92–3, 228
winter grains:
  as feed 101
Winterborne St Martins, Dorset 234
Winterslow wheat 79
wire worm 94
Wishford, Wilts. 243
Witham, Essex 235
Wittering, Northants. 239
Wolverley, Worcs. 243
Wood, Thomas, of Didsbury, Lancs. 58
Woodstreet Farm, Wool, Dorset:
  manure 85
wool 7, 18, 95–6, 175

see also fleeces
Wool, Dorset 85, 234
woollen industry 193
Wootton, War. 243
Worcester, Worcs.:
  wheat yields 126
Worcestershire 63, 243
  bacon pig carcass weights 202
  calf carcass weights 196–7
  cow and heifer carcass weights 199
  customary measures 246
  hog carcass weights 203
  lamb carcass weights 185
  pig carcass weights 207
  porker carcass weights 205–6
  probate inventories 31
  sheep carcass weights 192
  wheat yields 145
  see also Castlemorton; Church Lench; Clifton; Ombersley; Wolverley; Worcester
Worlidge, J.:
  on manure and fertilizer 83
Wormhill, Derby. 233
Worsbrough Bridge, Yorks. 244
Wortham, Suffolk 242
Wrigley, E.A.:
  on labour productivity 227
  population estimates 215
Wrockwardine, Shropshire 241
Wybunbury, Ches. 232
Wymondham, Norfolk-type 239
  underdraining 90

Yamey, B.:
  on accounting systems 45
Yaverland, Isle of Wight 236
Yellow barley 79
Yellow turnips 79
yeomen 9, 15, 30–1
yield rates (and ratios) 167–72, 214
yields, see named crops
York, Yorks.:
  supplying manure 87
  wheat yields 126
Yorkshire 243–5
  carcass weights 42, 195
  pig carcass weights 207
  threshing machines 93

water meadows 104
wheat yields 133–5, 146
see also Badsworth; Bradfield; Bradford; Bramley; Burton Constable; Campsell; Cleveland; Heslington; High Hoyland; Hook; Houndhill (in Darfield); Hoyland; Hull; Loftus; Londesborough; Market Weighton; North Frodingham; North Holme; Rudby; Saltmarshe; Sheffield; Shelley; Sledmere; Tankersley; Wakefield; Wath upon Dearne; Wentworth; Worsbrough Bridge; York

Yorkshire cattle 98

Young, Arthur, agricultural writer 2–3, 11
  on agricultural output 20–2
  on carcass weights 175, 181, 195
  on enclosure 88
  on farmers 5, 28
  on innovation 230
  on wheat yields 116–17, 125, 127, 133, 138
  on yield rates 170–1